Cloud of Things

Cloud computing is the on-demand availability of computer system resources, especially data storage and computing power, without direct active management by the user. This book explores algorithms, protocols, and system design principles of key smart technologies including cloud computing and the internet of things.

- Discusses the system design principles in cloud computing along with artificial intelligence, machine learning, and data analytics applications
- Presents blockchain-based solutions, cyber-physical system applications, and deep learning approaches to solving practical problems
- Highlights important concepts including the cloud of things architecture, cloud service management and virtualization, and resource management techniques
- Covers advanced technologies including fog computing, edge computing, and distributed intelligence
- Explores cloud-enabling technology, broadband networks and internet architecture, internet service providers, and connectionless packet switching.

This book is primarily written for graduate students, academic researchers, and professionals in the field of computer science and engineering, electrical engineering, and information technology.

Cloud of Things
Foundations, Applications, and Challenges

Edited by
Jitendra Kumar, G. R. Gangadharan,
Ashutosh Kumar Singh,
and Chung-Nan Lee

CRC Press
Taylor & Francis Group
Boca Raton London New York

CRC Press is an imprint of the
Taylor & Francis Group, an **informa** business

A CHAPMAN & HALL BOOK

Front cover image: Andrey Suslov/Shutterstock

First edition published 2024
by CRC Press
2385 NW Executive Center Drive, Suite 320, Boca Raton FL 33431

and by CRC Press
4 Park Square, Milton Park, Abingdon, Oxon, OX14 4RN

CRC Press is an imprint of Taylor & Francis Group, LLC

ISBN: 978-1-032-48441-9 (hbk)
ISBN: 978-1-032-48824-0 (pbk)
ISBN: 978-1-003-39095-4 (ebk)

DOI: 10.1201/9781003390954

Typeset in Times
by Newgen Publishing UK

Contents

About the Editors

Jitendra Kumar is an Assistant Professor in the Department of Mathematics, Bioinformatics and Computer Applications at Maulana Azad National Institute of Technology Bhopal, Madhya Pradesh, India. Prior to this, he was with the Department of Computer Applications, National Institute of Technology Tiruchirappalli, Tamilnadu, India. Kumar earned his PhD in Machine Learning and Cloud Computing from the National Institute of Technology Kurukshetra, Haryana, India. He has authored and published many research articles in peer-reviewed and indexed journals and conferences of high repute. He is a senior member of the Institute of Electrical and Electronics Engineers (IEEE) and several other IEEE technical committees, including Computational Intelligence Society (CIS), Technical Community on Cloud Computing (TCCLD), Technical Committee on Services Computing (TCSVC), and Special Interest Group on Humanitarian Technology (SIGHT). He is also a member of the ACM and MIR Labs. He is an active review board member of various journals and prestigious conferences like *IEEE Transactions on Parallel and Distributed Systems, IEEE Transactions on Computers, IEEE Systems Journal, IEEE Access, International Joint Conference on Neural Networks (IJCNN), FUZZ-IEEE, Congress on Evolutionary Computation (CEC)*, etc. His research interests include Cloud Computing, Machine Learning, Computational Intelligence, Time Series Forecasting, and Optimization.

G. R. Gangadharan is working as Associate Professor in the Department of Computer Applications, National Institute of Technology, Tiruchirappalli, Tamilnadu, India. His research interests are mainly focused on the interface between technological and business perspectives. His research interests include Artificial Intelligence, Machine Learning, Big Data Analytics, and Cloud Computing. He received his PhD degree in Information and Communication Technology (2008) from the University of Trento, Trento, Italy and European University Association and an MS in Information Technology (2004) from Scuola Superiore Sant'Anna, Pisa, Italy. He is a senior member of the IEEE and the ACM.

Ashutosh Kumar Singh is an esteemed researcher and academician in Electrical and Computer Engineering. Ashutosh Kumar Singh is the Director of the Indian Institute of Information Technology Bhopal, India. He has more than 20 years of research, teaching, and administrative experience in various university systems in India, the UK, Australia, and Malaysia. Dr Singh obtained his PhD degree in Electronics Engineering from the Indian Institute of Technology-BHU, Varanasi, India; Post doc from the Department of Computer Science, University of Bristol, United Kingdom; and Charted Engineer from the United Kingdom. He is a recipient of the Japan Society for the Promotion of Science (JSPS) fellowship for a visit to the University of Tokyo and other universities in Japan. Singh was a visiting scientist in Frankfurt Institute for Advanced Studies, Goethe University, Germany.

His research interests include Verification, Synthesis, Design and Testing of Digital Circuits, Predictive Data Analytics, Data Security in the Cloud, and Web Technology. He has more than 380 publications, which include peer-reviewed journals, books, conferences, book chapters, and news magazines in these areas. He has co-authored 11 books, including *Machine Learning for Cloud Management*, *Web Spam Detection Application Using Neural Network*, *Digital Systems Fundamentals*, and *Computer System Organization & Architecture*. Singh has worked as a principal investigator or investigator for six sponsored research projects and was a key member of a project from Engineering and Physical Sciences Research Council (EPSRC) (United Kingdom) entitled "Logic Verification and Synthesis in New Framework". Singh has visited several countries, including Australia, the United Kingdom, South Korea, China, Thailand, Indonesia, Japan, Germany, and the USA, for collaborative research and invited talks, and presented his research work. He is also serving as an editorial board member, guest editor, and associate editor in various journals of high repute.

Chung-Nan Lee received his BS and MS degrees, from National Cheng Kung University, Tainan, Taiwan, in 1980 and 1982, respectively, and his PhD degree from the University of Washington, Seattle, in 1992, all in Electrical Engineering. Since 1992, he has been with National Sun Yat-Sen University, Kaohsiung, Taiwan, where he was an Associate Professor in the Department of Computer Science and Engineering from 1992 to 1999; he was the Chairman of the Department of Computer Science and Engineering from August 1999 to July 2001; Director of Wireless Broadband Communication Protocol Cross-Campus Research Center from 2013 to 2017; President of Taiwan Association of Cloud Computing from 2015 to 2017; currently, he is a Distinguished Professor and Director of Cloud Computing Research Center; Member and Fellow of Asia Pacific Signal and Information Processing Association (APSIPA) from 2017. In 2016 he received the outstanding engineering professor award from the Chinese Institute of Engineers, Taiwan. His current research interests include Multimedia over Wireless Networks, Cloud Computing, and Evolutionary Computing.

Contributors

Vamshi Adouth
National Institute of Technology,
Tiruchirappalli
India

Michael Arock
National Institute of Technology,
Tiruchirappalli
India

Chandra Shekar Besta
National Institute of Technology,
Calicut
India

Ghanshyam S. Bopche
National Institute of Technology,
Tiruchirappalli
India

Elif Bozkaya-Aras
National Defence University Turkish
Naval Academy, Istanbul
Turkey

Berk Canberk
Edinburgh Napier University,
Edinburgh, Scotland,
United Kingdom

Gayatri K. Chaturvedi
Veermata Jijabai Technological
Institute, Mumbai
India

Selvam Durairaj
National Institute of Technology,
Tiruchirappalli
India

Ramprasad Ohnu Ganeshbabu
Cloudera Inc., Aurora, Illinois
USA

Anitya Kumar Gupta
Shoolini University of Biotechnology
& Management, Solan
India

Fredrick Ishengoma
The University of Dodoma, Dodoma
Tanzania

Sataditya Jana
Shoolini University of Biotechnology
& Management, Solan
India

Aman Jolly
KIET Group of Institutions,
Ghaziabad
India

Ananya Kaim
Thapar Institute of Engineering &
Technology, Patiala
India

Danish Ali Khan
National Institute of Technology,
Jamshedpur
India

Jitendra Kumar
Maulana Azad National Institute of
Technology,
Bhopal
India

Mohit Kumar
Dr. B. R. Ambedkar National Institute
of Technology, Jalandhar
India

Rakesh Kumar
GLA University, Mathura
India

Lokraj
Shoolini University of Biotechnology
& Management, Solan
India

Sudakshina Mandal
Techno India University, Kolkata
India

Vinaytosh Mishra
Gulf Medical University, Ajman
United Arab Emirates

Kuldeep Mohanty
Kalinga Institute of Industrial
Technology University, Bhubaneshwar
India

Kunwar Pal
Dr. B. R. Ambedkar National Institute
of Technology, Jalandhar
India

Vikas Pandey
Babu Banarasi Das University,
Lucknow
India

Yashwant Singh Patel
Thapar Institute of Engineering &
Technology, Patiala
India

Nilin Prabhaker
National Institute of Technology,
Tiruchirappalli
India

Suhailam Pullanikkattil
National Institute of Technology,
Calicut
India

Harsh Rai
National Institute of Technology,
Tiruchirappalli
India

Vinay Raj
National Institute of Technology,
Tiruchirappalli
India

Eswari Rajagopal
Department of Computer
Applications, National Institute of
Technology, Tiruchirappalli,
Tamilnadu,
India

Mothiram Rajasekaran
Cloudera Inc., Florida
USA

Rakesh D Raut
IIM Mumbai, Mumbai
India

Md Kamar Raza
Dr. B. R. Ambedkar National Institute
of Technology, Jalandhar
India

Chitra Sabapathy
Cloudera Inc., Arizona
USA

Kshira Sagar Sahoo
Umea University, Umea
Sweden

Sarita
Kirorimal College, University of
Delhi, New Delhi
India

Nickolas Savarimuthu
National Institute of Technology,
Tiruchirappalli
India

Deepika Saxena
University of Aizu, Aizuwakamatsu
Japan

Stefan Schmid
Technical University of Berlin, Berlin
Germany

Chandra Sekhar
National Institute of Technology,
Kurukshetra
India

Suseela Sellamuthu
Vellore Institute of Technology,
Chennai Campus
India

Mahesh Shirole
Veermata Jijabai Technological
Institute, Mumbai
India

Ulka Shirole
A.C. Patil College of Engineering,
Navi Mumbai
India

Ashutosh Kumar Singh
National Institute of Technology,
Kurukshetra
India

Indrasen Singh
Vellore institute of Technology,
Vellore
India

Shailu Singh
Hansraj College, University of Delhi,
New Delhi
India

Surjit Singh
Thapar Institute of Engineering &
Technology, Patiala
India

Vivek Kumar Singh
Sharda University, Greater Noida
India

Saurabh Singhal
GLA University, Mathura
India

Rajeswari Sridhar
National Institute of Technology,
Tiruchirappalli
India

Om Prakash Suman
Dr. B. R. Ambedkar National Institute
of Technology, Jalandhar
India

Uma Sankararao Varri
Neerukonda, Mangalagiri Mandal,
Guntur District, Andhra Pradesh,
India

Nitesh Kumar Vashisht
Shoolini University of Biotechnology
& Management, Solan
India

Ashutosh Yadav
Hansraj College, University of Delhi,
New Delhi
India

Deepshikha Yadav
Hansraj College, University of Delhi,
New Delhi
India

Raju Yerolla
National Institute of Technology,
Calicut
India

1 Cloud of Things
Architecture and Industrial Applications

*Suhailam P, Raju Yerolla, and
Chandra Shekar Besta*

1.1 INTRODUCTION

The term "Cloud of Things" (CoT) describes the fusion of cloud computing and Internet of Things (IoT) hardware. This architecture enables data to be saved, processed, and analyzed in the cloud by connecting IoT devices (such as sensors, actuators, and other smart devices) to the cloud via the Internet. It enables remote access to IoT devices and real-time data streams in addition to more effective and scalable data processing and analysis. Other features offered by the cloud infrastructure, like security, data storage, analytics, and machine learning, can be utilized to enhance the usefulness and performance of IoT devices. Healthcare, agriculture, manufacturing, and smart cities are just a few of the sectors and businesses that use CoT applications [1]. Common applications of CoT include the following:

1. IoT devices like smart thermostats, lighting controls, and security cameras can be connected to the cloud for centralized management and control in order to construct intelligent homes and buildings.
2. CoT enables real-time monitoring, data analysis, and control by connecting IoT devices such as sensors, actuators, and controllers to the cloud in the automation of industrial processes.
3. By linking IoT devices like weather sensors, soil moisture sensors, and drones to the cloud, which offers real-time data analysis and decision support, CoT can be utilized to boost agricultural production.
4. CoT can be used to remotely monitor patient health, giving healthcare professionals access to real-time monitoring of vital signs, medication adherence, and other health data.
5. CoT can be used to build IoT-enabled cities with real-time monitoring and management capabilities, such as those with traffic sensors, intelligent lighting, and waste management systems.

CoT applications offer a wide range of advantages, including improved productivity, scalability, and real-time data analysis, which allows businesses to make better decisions and enhance their operations. The importance of cloud computing and IoT applications is rising as more businesses try to use these technologies. The significance of CoT applications [2] are as follows it must:

DOI: 10.1201/9781003390954-1

1. **Scalability:** CoT makes it simple for businesses to scale their IoT deployments by utilizing cloud resources. Cloud computing offers the required infrastructure to store and handle the massive volumes of data that IoT devices create.
2. **Savings:** Businesses can save money by using cloud resources instead of developing and maintaining their own Information Technology IT infrastructure. Moreover, CoT offers a pay-per-use approach that enables businesses to only pay for the services they actually utilize.
3. **Real-Time Data Analysis:** Data analysis in real-time is made possible through CoT, which enables enterprises to make decisions and solve problems more quickly. This is especially significant in sectors like manufacturing and healthcare where quick judgements can make all the difference.
4. **Enhanced productivity:** CoT gives businesses the ability to automate procedures, cutting down on manual labor and enhancing productivity. Increased productivity and cost savings may result from this.
5. **Improved security:** By utilizing cloud resources like data encryption, access control, and threat detection, CoT helps enterprises to strengthen their security posture. This is crucial in sectors like finance and healthcare, where data security and privacy are essential.

The growth of IoT devices has resulted in an unprecedented amount of data being generated. While these IoT devices collect useful data, it can be difficult to efficiently process it and derive insights that can be put to use. Additionally, it becomes difficult to manage and secure the enormous amounts of data produced by IoT devices, necessitating a stable and scalable infrastructure. The difficulties that organizations encounters are further exacerbated by the requirement for real-time data analysis, seamless connection with existing systems, and data privacy protection. Scalable data processing, cost savings, real-time insights, increased productivity, and tighter security are all made possible by CoT by utilizing cloud resources. These significant contributions open the door for businesses to fully utilize IoT and cloud computing, which will ultimately spur innovation, increased productivity, and better results across industries. This chapter emphasizes the major contributions of CoT applications while addressing the aforementioned challenges.

1.2 CLOUD OF THINGS ARCHITECTURE

CoT architecture is made up of a number of parts that work together to let IoT devices connect to the cloud for data archiving, processing, and analysis. Figure 1.1 shows the general architecture of CoT. There are three layers: application layer, network layer, and perception layer. IoT devices are tangible objects like sensors, actuators, and intelligent gadgets that gather data and connect to the cloud. IoT devices and the cloud are connected through a gateway, which serves as a middleman. It offers a secure communication channel for data transmission and makes it possible for IoT devices to connect to the cloud. The cloud infrastructure offers the tools like processing, storage, and analytics capabilities that make it possible to store, examine, and act on CoT data. Applications and services are the programs and services

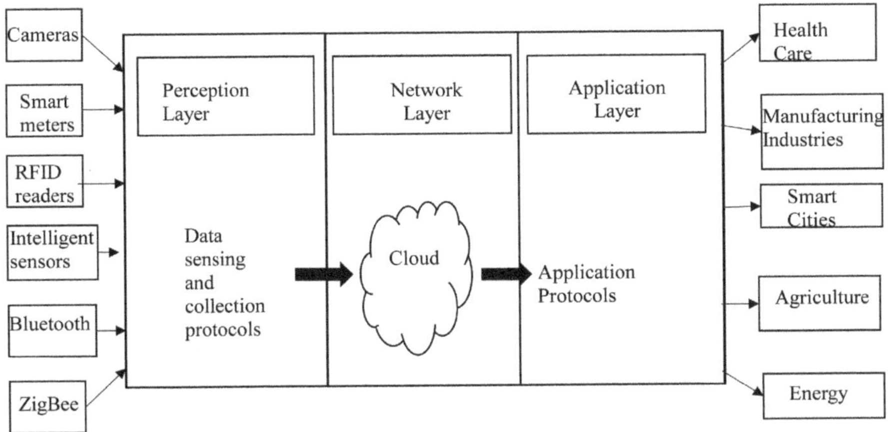

FIGURE 1.1　　Cloud of things architecture.

that make use of the cloud infrastructure to offer value-added features like predictive maintenance, machine learning, and data analytics. Essential elements of the CoT architecture include security and privacy. To protect the security, integrity, and availability of CoT data, measures like data encryption, access control, and threat detection are used [3].

1.3　CLOUD OF THINGS APPLICATIONS IN THE INDUSTRY

CoT is a flexible technology that may be used in a variety of industries to boost productivity, cut costs, and enhance efficiency. The following are some of the sectors that have used CoT applications:

Healthcare: CoT applications are used in the healthcare sector to track patients' health problems and give providers of care real-time feedback. Wearable technology can gather information such as heart rate, blood pressure, and activity levels, for instance, through smartwatches and fitness trackers. Healthcare professionals have the capacity to access this data and analyze it in order to offer personalized care and treatment suggestions for improvement.

Manufacturing: Implementing CoT (IoT) applications within the manufacturing sector aims to enhance operational processes, reduce idle periods, and improve the overall quality of the final product. For instance, sensors can monitor the machinery's operational state and forecast maintenance needs, consequently diminishing periods of inactivity and enhancing production efficiency. To enhance the quality of products and reduce the occurrence of defects, CoT can be implemented to automate the quality assurance procedures.

Transportation: CoT applications in transportation increase safety, efficiency, and cost. Telematics solutions like CoT may help fleet managers increase safety, efficiency, and fuel efficiency. These devices monitor vehicle performance, fuel usage, and driver behavior. Connected Vehicle Technology (CoT) can also give up-to-date traffic information, enhancing route choices and minimizing traffic congestion.

Agriculture: CoT increases crop output, decreases water use, and improves agricultural practices. Sensors can measure soil temperature, humidity, and wetness, helping farmers optimize irrigation and fertilizer use. CoT can track crop growth and estimate agricultural production, improving harvesting efficiency and minimizing waste.

Energy: The application of CoT (condition-based monitoring and maintenance) is commonly observed in the power industry, where the main goal is to optimize energy utilization, minimize expenses, and improve overall operational efficiency. Facility managers possess the capability to employ the IoT paradigm to monitor and control energy usage within buildings efficiently. By employing this technology, facility managers can optimize the operation of lighting and cooling and heating systems, resulting in reduced energy usage as well as costs. Additionally, CoT can be employed to evaluate the efficacy and waste mitigation of energy generation derived from renewable sources such as solar and wind.

CoT technology finds applications across different industries such as manufacturing, transportation, energy, agriculture, and healthcare. CoT is a significant technology for companies and organizations since it can enhance productivity, decrease expenses, and improve efficiency in these areas.

1.4 HEALTHCARE APPLICATIONS

Remote patient monitoring is possible because of CoT, which enables medical professionals to keep an eye on their patients' health and give them feedback in real time. Smartwatches and fitness trackers are examples of wearable technology that can gather data on heart rate, blood pressure, and activity levels and communicate to healthcare professionals for analysis. Personalized care and treatment strategies can be created using this data. The general architecture of CoT in health care system is shown in Figure 1.2.

CoT can be utilized to offer telemedicine services, enabling patients to communicate with medical professionals at a distance. Patients who live in rural or isolated places and may not have access to medical services may find this to be especially helpful. CoT can also be utilized to offer in-the-moment medical advice, cutting down on wait times and enhancing patient outcomes. Remote healthcare monitoring and management information services based on the cloud and the IoT can successfully offer early identification and treatment of chronic diseases that have a substantial influence on people's health. In other words, wearable or implantable IoT body sensors collect the necessary data from a person, which can be processed and analyzed in the cloud.

Utilizing CoT in the healthcare industry can improve healthcare services and open up new prospects for the medical IT infrastructure. By streamlining the process of collecting patients' essential data and transferring them to a medical center on the Cloud for storage and processing needs, CoT also enhances the quality of the real healthcare services. In other words, the application of CoT in Body Sensor Network

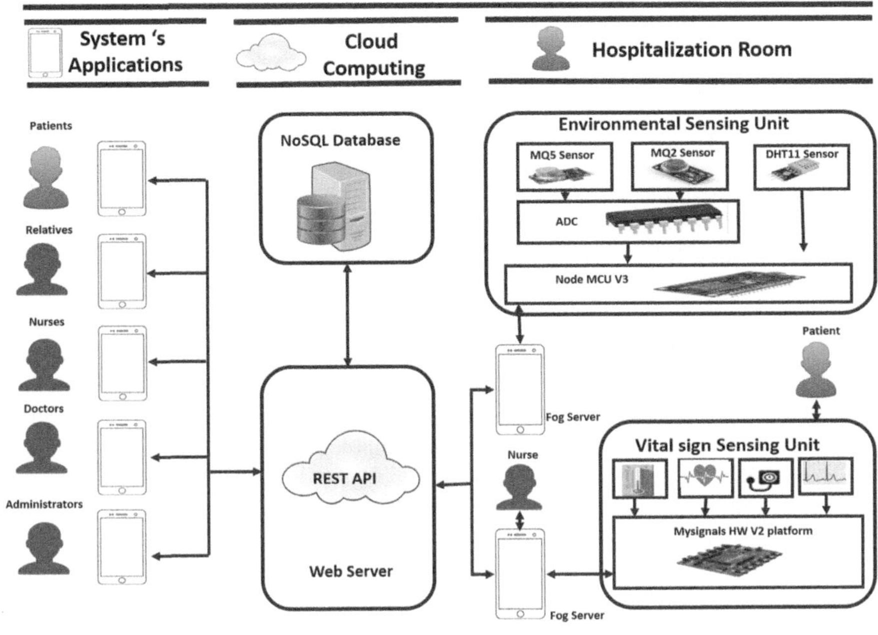

FIGURE 1.2 CoT in healthcare system.

(BSN)-based healthcare aids in the scalable processing, processing, and analysis of acquired data.

Medical asset tracking: CoT can be used to monitor medical supplies and equipment, lowering the risk of theft or loss. This is particularly crucial in healthcare settings because the disappearance of essential tools or supplies may compromise patient care. A platform for tracking and monitoring patients at home Ambient Assisted Living (AAL) is called Sensor Platform for Healthcare in a Residential Environment (SPHERE) Project. Utilizing useful data sets that were derived from the sensor data collection, it enhances healthcare detection and management. Smartphone-based gateways can successfully provide transparent communication between IoT devices and the Internet, as well as IoT device administration, in addition to the recognition and detection of patients' state changes. As an illustration, work suggested a high-level design of an opportunistic gateway architecture focused on smartphones to facilitate flexible and transparent interoperability.

Health data analytics: CoT enables healthcare providers to find trends and patterns in patient health by gathering and analyzing health data from various sources. The creation of patient-specific care plans and treatment choices is possible using this data.

Patient involvement: By providing patients the tools and resources needed to manage their health conditions, CoT can be utilized to get patients more involved in their own care. Patients can track their drug schedules, get reminders, and keep an eye on their symptoms using smartphone apps, for instance.

1.4.1 EXAMPLES OF ACTUAL CLOUD OF THINGS APPLICATIONS IN THE HEALTHCARE INDUSTRY

A CoT-based technology called **Philips eICU** enables medical professionals to remotely monitor seriously ill patients in intensive care units (ICUs) from a central location. The technology employs sensors and other equipment to keep track of patients' vital signs and send data to a central monitoring station, where medical professionals may examine it and give on-site care teams immediate feedback [4].

Proteus Digital Health: Proteus Digital Health is a CoT-based system that tracks medication adherence and gives patients and healthcare professionals real-time feedback. The method enables patients to track their medication schedules and get reminders by providing them with a sensor-enabled pill, a wearable sensor patch, and a smartphone app [5] .

VitalPatch: VitalPatch is a wearable, CoT-based gadget that continuously monitors patients' vital signs and communicates information to medical professionals. The chest-worn gadget has sensors for measuring temperature, respiration rate, heart rate, and other vital signs [6, 7].

Healthcare professionals can access patient data and other clinical information using the mobile app **AirStrip ONE**, which is based on the CoT. This app gives healthcare professionals immediate access to patient data so that they can decide how best to treat patients [8].

The ReSound LiNX Quattro is a hearing aid that utilizes sensors and machine learning algorithms based on the concept of CoT to autonomously adapt to the user's surrounding conditions. The device is capable of perceiving alterations in the environment, such as variations in wind or ambient noise, by means of its sensors. Subsequently, it adjusts the settings of the hearing aid in response [11].

The IoT has been used in healthcare to monitor patients remotely, track medication adherence and vital signs, provide mobile app-based accessibility to patient information, and even more to monitor vital indicators. The utilization of these applications holds promise in enhancing patient outcomes, mitigating expenses, and optimizing facility operations, thereby establishing CoT as a valuable instrument for both patients and healthcare practitioners.

1.5 APPLICATIONS IN THE MANUFACTURING SECTOR

CoT can be utilized to keep an eye on production tools together with procedures in real-time which makes it possible for business to discover ineffectiveness and also simplify their procedures (see Figure 1.3). CoT can help producers in equipment improvement and also its capabilities, thereby lowering downtime and improving efficiency by using information analytics and artificial intelligence formulas. CoT can be used to check the problem of manufacturing tools coupled with foretell when upkeep is needed. By this companies can reduce tools, downtime, reduce upkeep costs, and lengthen the life of their equipment.

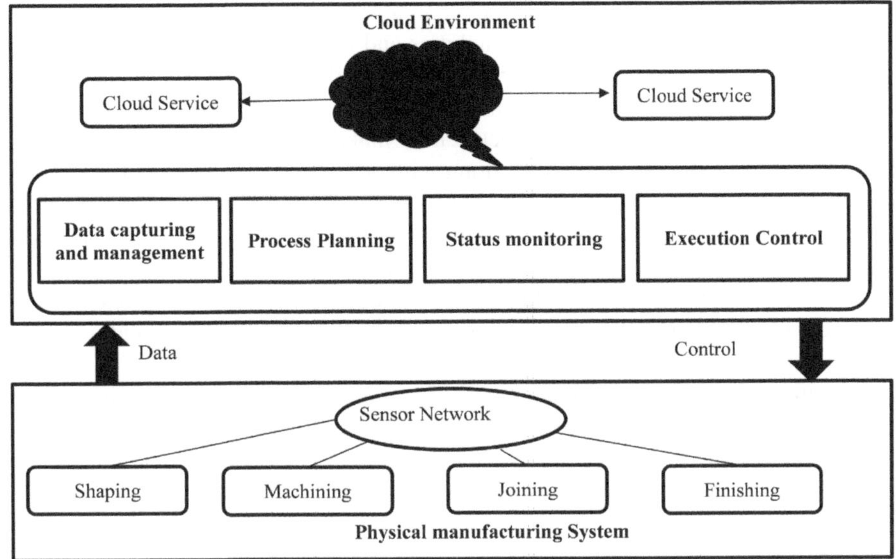

FIGURE 1.3 CoT in manufacturing.

Along with these the concept CoT can be utilized for the function of checking the activity of products throughout the supply chain. This assists in the stipulation of present details on supply amounts, delivering progression as well as approximated times for distribution. This can improve client satisfaction, lower costs, and promote reliable company supply chain monitoring. The application of CoT allows the constant tracking of item top quality at different phases of the production procedure, which permits the early discovery and recognition of problems before the items are delivered to customer. Furthermore, it can be used for keeping track of employee safety and security in producing centers determining possible dangers, plus supplying timely comments to employees. This can help with services in increasing staff member security, reducing the possibility of mishaps and abiding by security guidelines.

CoT provides significant advantages to the production field, such as improved efficiency optimization of the supply chain, greater degrees of control, and improved employee safety and security. The execution of CoT has the prospective to improve functional effectiveness, minimize prices, as well as enhance economic efficiency for producers with the exercise of artificial intelligence, information analytics, and also real-time tracking.

1.5.1 APPLICATIONS OF THE CLOUD OF THINGS IN THE MANUFACTURING SECTOR IN THE REAL WORLD

Common Electric (GE) Digital: GE Digital is a software application system developed on the idea of the Industrial Internet of Things (IIoT), which permits makers to track the performance of procedure of their equipment as well as to optimize their procedures. The platform forecasts when maintenance is necessary and spots chances for process improvement using data analytics and machine learning algorithms [9].

The Bosch Industry 4.0 is a solution that is based on the concept of the IoT and aims to facilitate the tracking and enhancement of production processes for manufacturers. The system allows manufacturers to make informed decisions to improve their operations by utilizing sensors along with additional devices to collect data on production efficiency, quality control, and supply chain activities. Siemens Mind-Sphere is a technology based on the concept of CoT, which allows the connection and ongoing monitoring of manufacturing machinery for manufacturers [10]. The platform has machine learning algorithms and data analytics capabilities that may be used to find process inefficiencies, improve equipment efficiency, and anticipate when a repair is necessary [11].

Manufacturers can connect to and track their manufacturing equipment in real time using the Amazon Web Services IoT platform, which is based on the CoT. The platform has machine learning algorithms and data analytics capabilities that may be utilized to speed up production, decrease downtime, and boost output [12]. PTC ThingWorx is a platform built on the CoT that enables manufacturers to track and improve their processes in real-time. The platform has augmented reality capabilities, machine learning algorithms, and data analytics tools that may be utilized to increase worker productivity, decrease downtime, and enhance product quality [13].

The industrial sector has several practical uses for the CoT, including supply chain optimization, quality control, worker safety, and equipment monitoring. CoT is an important technology for manufacturers trying to streamline their operations and boost their bottom line because of these applications' ability to raise productivity, decrease costs, and improve production efficiency.

1.6 APPLICATIONS IN TRANSPORTATION INDUSTRY

CoT may be used to track the whereabouts and condition of vehicles and cargo in real time, giving logistics and transportation businesses useful data. It is possible to use this data to enhance delivery times, save transportation costs, and optimize routes. CoT can be used to monitor the condition of cars and foretell when maintenance is necessary. By doing this, transportation companies may decrease maintenance expenses, prevent unscheduled downtime, and lengthen the life of their vehicles.

CoT can be used to track the security of cargo and vehicles, spotting potential dangers and security risks in real time. This can assist transportation firms in enhancing the security of their operations, lowering the danger of accidents and theft, and adhering to security and safety standards. By giving real-time updates on the location and status of delivery, CoT can be utilized to increase customer happiness. This can assist transportation providers in enhancing client satisfaction and lowering complaints. CoT can be used to track how transportation operations affect the environment and spot possibilities to cut back on fuel use and emissions. This can aid transportation firms in lowering their carbon emissions and strengthening their sustainability procedures [14].

The transportation sector benefits greatly from the CoT, including real-time monitoring, preventative maintenance, safety and security, customer happiness, and environmental sustainability. CoT can assist transportation firms in streamlining their operations, cutting expenses, and boosting their bottom line by utilizing the power of data analytics, machine learning, and real-time monitoring.

1.6.1 APPLICATIONS OF THE CLOUD OF THINGS IN THE TRANSPORTATION SECTOR

Uber: Uber tracks the whereabouts and condition of its cars and drivers in real time using CoT. The utilization of machine learning and data analytics in this technology facilitates the optimization of routes, reduction of waiting times, and improvement of the overall consumer experience [15].

DHL: DHL employs CoT technology to track and monitor the precise geographical location as well as the state of its freight in real time. The platform has various sensors and other devices that collect data on factors such as temperature, humidity, and additional environmental variables. This ensures cargo transportation is conducted under ideal conditions, facilitating its successful delivery to the intended destination [16].

FedEx: FedEx utilizes a system called Continuous Optimization Technology which is based on CoT to consistently assess and evaluate the operational efficacy of their fleet of delivery vehicles. The platform uses fuel consumption, engine efficiency, and other sensors to improve vehicle performance and save maintenance costs [17].

Waze: Waze utilizes Crowd-Sourced Traffic in which data processing and analysis are cloud-based , typically involves the integration of IoT devices and sensors into a cloud-based ecosystem in order to provide its users with real-time updates on traffic conditions. To enhance drivers' navigating efficiency, the platform incorporates data from various sensors and devices to provide real-time information on traffic patterns, incidents, and possible dangers [18].

The transportation sector presents various practical applications for the CoT, encompassing real-time monitoring, preemptive maintenance, route optimization, and customer experience enhancement. The implementation of CoT has the potential to improve the operational efficiency, cost reduction, and financial performance of transportation companies through the utilization of machine learning, data analytics, and real-time monitoring.

1.7 APPLICATIONS IN AGRICULTURE

With the use of CoT, farmers may monitor soil characteristics, weather patterns, and other environmental aspects in real time and gain insightful information. Using this information will increase overall crop health, optimize crop yields, and use less water.

Monitoring of livestock: CoT can be used to keep an eye on the well-being and behavior of cattle, giving prospective health problems a head start. Farmers may be able to do this to lower death rates, boost productivity, and enhance animal welfare. CoT can be used to monitor the condition of farm equipment and predict when maintenance is necessary. Farmers may be able to avoid unplanned downtime, lower maintenance expenses, and extend the life of their equipment by doing this.

Supply chain optimization: CoT may be used to track the whereabouts and condition of livestock and crops in real time, giving logistics and transportation businesses important data. This data can be used to enhance delivery times, save transportation costs, and optimize routes.

Environmental sustainability: CoT can be used to track how agriculture operations affect the environment and spot possibilities to use fewer pesticides and water. This can aid farms in improving their sustainability practices and lowering their carbon impact.

The agriculture sector benefits greatly from the CoT, including supply chain efficiency, environmental sustainability, precision agriculture, animal monitoring, and predictive maintenance. CoT can assist farmers in streamlining their operations, cutting expenses, and increasing their bottom line by utilizing the power of data analytics, machine learning, and real-time monitoring.

1.7.1 Use Cases for the Cloud of Things in the Agricultural Sector

John Deere uses CoT to continuously track the condition and efficiency of its farm machinery. The platform has sensors and other gadgets that collect data on things like fuel usage, engine performance, and other things, enhancing equipment performance and lowering maintenance costs [19]. **Cargill** employs CoT to continuously track the whereabouts and health of its livestock. The platform has sensors and other gadgets that collect information on the condition and behavior of cattle, enhancing both efficiency and animal welfare [20].

Trimble leverages CoT to provide farmers with solutions for precision agriculture. The platform has sensors and other devices that offer information on crop health and growth, assisting in crop production optimization and waste reduction. Thermal imaging can be used in smart irrigation to monitor crop water stress, gas-exchange rate, evapotranspiration rate, stomatal conductance, and closing of stomata. It provides a temperature value for all pixels within the sensor's field of view and can discriminate between different components such as sunlit versus covered plant portions and wet against dry soil surfaces. This information can be used to improve irrigation scheduling and decrease crop water stress (see Figure 1.4).

A CoT network can also be deployed in smart agriculture to make energy use more efficient and less costly. The CoT network can collect data analytics from sensors measuring heat, moisture, chemicals, water stress, pump status, level of water resources, etc. This allows farmers to utilize water, fertilizer, and pesticides in more precise quantities and positions and with better time scheduling to increase yields. It can also provide more effective energy uses for pumps, lighting, boosters, and other purposes and remotely control the status, working conditions, and performance of equipment [21].

In conclusion, the CoT has a wide range of practical applications in the agricultural sector, including supply chain optimization, environmental sustainability, animal monitoring, precision agriculture, and predictive maintenance. CoT can assist

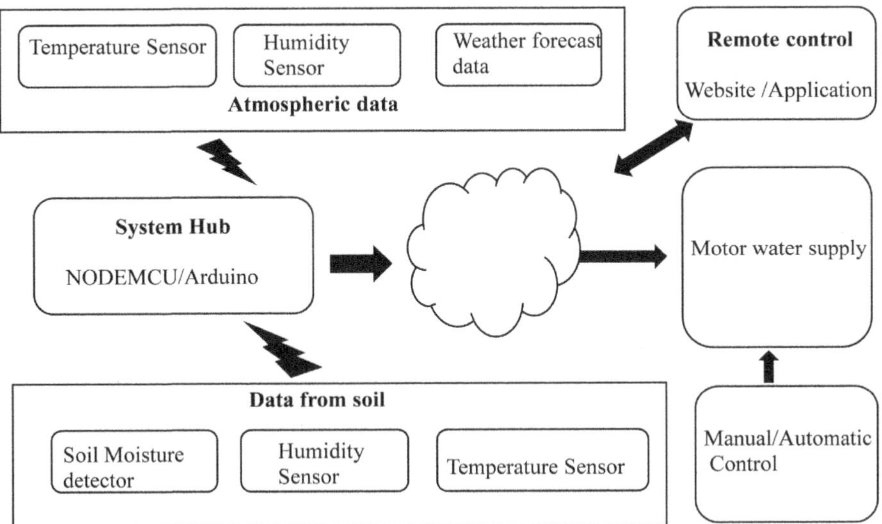

FIGURE 1.4 CoT-based smart irrigation system.

farmers in streamlining their operations, cutting expenses, and increasing their bottom line by utilizing the power of data analytics, machine learning, and real-time monitoring.

1.8 APPLICATIONS IN THE ENERGY SECTOR

CoT may be utilized to monitor and control the power grid in real time, thereby spotting opportunities to boost productivity and cut costs through smart grid management. Utility companies may be able to reduce outages, increase dependability, and streamline operations as a result of implementing such technologies [22].

Integration of renewable energy sources: CoT can be used to continuously monitor and control renewable energy sources like solar and wind power. This can assist utilities in more efficiently integrating alternative sources into the grid, lowering their dependency on fossil fuels and increasing sustainability.

Energy effectiveness: CoT can be used to track and control how much energy is consumed in buildings, finding chances to cut back and boost effectiveness. This can lower energy costs for businesses and individuals while also lessening their environmental impact. CoT can be used for predictive maintenance to track the condition of energy infrastructure, including power plants and transmission lines, and to foretell when maintenance is necessary. This can assist utilities in reducing maintenance expenses, avoiding unscheduled downtime, and enhancing the dependability of their operations.

Demand response: CoT can be used to oversee programs that encourage consumers to use less energy during periods of high demand. By doing so, utilities may be able to lessen their reliance on pricey peaker plants and prevent blackouts.

The energy sector benefits greatly from the CoT, including demand response, smart grid management, integration of renewable energy sources, and energy efficiency. CoT can assist utilities and consumers manage their energy usage, lower costs, and enhance sustainable practices by utilizing the power of data analytics, machine learning, and real-time monitoring.

1.8.1 EXAMPLES OF REAL-WORLD CLOUD OF THINGS APPLICATIONS IN THE ENERGY SECTOR

GE Digital Power Plant: GE's Digital Power Plant makes use of CoT to continuously monitor and improve the operation of power plants. This involves minimizing fuel use and cutting emissions, as well as applying advanced analytics to detect equipment faults and prescribe maintenance activities [23].

Enel X: Enel X manages demand response programs for clients in the business and consumer sectors using CoT. Enel X is able to assist utilities prevent blackouts and lessen the need for pricey peaker plants by providing incentives to energy users to limit their usage during periods of high demand [24].

Tesla Powerwall: Using CoT to optimize energy use and save costs, the Tesla Powerwall is a battery storage system for homes. The Powerwall can assist customers in lowering their energy bills and enhancing their sustainability practices by storing extra solar energy throughout the day and using it during times of high demand [25].

The performance of energy infrastructure, such as power plants and transmission lines, is optimized through CoT at Schneider Electric, a global energy management firm. Schneider Electric can forecast equipment failures, prescribe maintenance steps, optimize energy use, and cut costs by utilizing advanced analytics and real-time monitoring.

1.9 SMART CITY APPLICATIONS

CoT is a promising technology that might alter healthcare, manufacturing, transport, energy, agriculture, and smart cities. CoT allows IoT devices and cloud computing infrastructure to work together. Real-time data processing and analysis improve operational efficiency, cost-effectiveness, and sustainability. The CoT, or IoT paradigm, improves operational efficiency, resource allocation, safety and security, sustainability, and citizen experiences. Examples of CoT use in diverse industries and real-world scenarios demonstrate its benefits. CoT is helping cities and businesses become smarter and more efficient. CoT also makes the global society more sustainable and integrated. Illustrative instances encompass intelligent traffic control systems implemented in Barcelona and energy management systems employed in Amsterdam.

CoT is a fast-growing technology that can alter healthcare, manufacturing, transportation, agriculture, energy, and smart cities. Integrating CoT lets IoT devices and cloud computing infrastructure work together. The IoT paradigm, also known as the CoT, improves operational efficiency, resource allocation, safety and security, sustainability, and citizen experiences through real-time data processing and analysis.

1.9.1 EXAMPLES OF CLOUD-BASED APPLICATIONS IN SMART CITIES IN THE REAL WORLD

The municipality of Barcelona has implemented a comprehensive City Operating System CoT called "CityOS" to streamline the collection and analysis of data from various sources, including sensors, social networking sites, and open databases. This versatile platform supports innovative city applications like waste management, energy conservation, and traffic regulation [26]. In a similar vein, the municipality of Amsterdam utilizes CoT technology to develop an advanced infrastructure known as a "smart grid." This system allows real-time monitoring and control of energy consumption by employing intelligent meters, sensors, and a cloud-based data management and analysis system. Smart meters and sensors provide valuable information on energy usage [27].

Singapore has implemented a system known as "Smart Nation" to enhance the city's infrastructure and services. CoT utilizes advanced technologies like IoT sensors, data analytics, and artificial intelligence. Its applications span various domains such as public safety, healthcare, and transportation [28]. Similarly, San Diego has introduced the "Smart Streetlights" system, which optimizes street lighting efficiency through sensor-driven data analytics while minimizing energy usage. With over 3,000 interconnected streetlights that can be remotely controlled and monitored, it also collects valuable information on traffic patterns and air quality [29]. Another notable example is Songdo in South Korea—an intentionally designed smart community that operates harmoniously with its meticulously planned urban layout.

The municipality employs a CoT framework for various purposes, encompassing waste management, optimization of energy use, and control of traffic flow. The system is equipped with sensors and data analytics capabilities that enable the real-time monitoring and enhancement of city services [26].

1.10 CONCLUSION

CoT is a powerful and rapidly emerging technology that might transform healthcare, manufacturing, transportation, agriculture, energy, and smart cities. CoT integrates IoT devices with cloud computing infrastructure for real-time data processing and analysis for more efficient, cost-effective, and sustainable operations. CoT improves efficiency, resource management, security, sustainability, and citizen experiences. These benefits are seen in real-world CoT utilization across sectors and purposes. CoT is making cities and industries brighter, more efficient, and more sustainable and connected. Barcelona has sophisticated traffic control, while Amsterdam has energy management.

Examples of CoT usage in real-world settings across a range of sectors and uses are demonstrating these advantages. CoT is transforming cities and industries, making them more intelligent and efficient, and assisting in the creation of a more sustainable and connected world. Examples include smart traffic control systems in Barcelona and energy management systems in Amsterdam. While there are many advantages and contributions provided by CoT applications, it is necessary to be aware of their limitations.

CoT application implementation frequently calls for a stable and scalable cloud infrastructure. The efficiency of CoT applications may be jeopardized in locations with weak connectivity or sporadic network availability. CoT uses security features from cloud infrastructure, but maintaining data security and privacy is still a top priority. CoT applications frequently contain a wide variety of IoT devices from many manufacturers, each with their own interfaces and protocols. The widespread adoption and full potential of CoT may be hampered by organizational inertia and resistance to change. Applications based on CoT may give rise to ethical questions, especially in industries like healthcare where the gathering and processing of private patient data are involved. To foster trust among stakeholders, it is crucial to address ethical issues linked to data usage, permission, and transparency.

In addition to continuing research and development initiatives in fields like connection, security, interoperability, and ethical frameworks, addressing these limits necessitates a thorough awareness of the unique difficulties faced by organizations. Although CoT applications have many benefits, it is crucial to be aware of their limits and work to overcome them in order to enable the effective deployment and uptake of CoT across a variety of industries.

1.10.1 DISCUSSION

CoT applications have a bright and exciting future. CoT is anticipated to advance and integrate even further as technology develops, opening up new opportunities and applications in a variety of sectors and use cases. Increased integration with artificial intelligence and machine learning technologies will enable more complex and automated data analysis and decision-making. This is only one of the significant trends and advances anticipated for CoT in the future. The development of increasingly decentralized CoT systems, facilitated by edge computing, enables data processing and analysis at the device level, eliminating the need for centralized cloud infrastructure. Increasing the use of blockchain technology is helpful to improve CoT system security, privacy, and data sharing. As CoT becomes more widely available and affordable, it will be embraced by smaller and medium-sized businesses, leading to the creation of new CoT applications and use cases in fields including smart homes, wearable technology, and autonomous vehicles. Therefore, CoT has a promising future with the ability to significantly progress several global sectors and applications. The continuing advancement and use of CoT will contribute to the development of more intelligent, effective, and sustainable systems, enhancing the quality of life for both individuals and communities.

1.11 RECOMMENDATIONS

Although CoT technology has advanced significantly in recent years, additional study and development are still required in a number of areas to fully realize the potential of CoT applications. CoT applications work with sensitive data, thus security and privacy are top priorities. To create more sophisticated and reliable security and privacy methods for CoT systems, more research is required. The requirement for standardized interfaces and protocols that enable seamless interoperability across various devices and systems is growing as the number of IoT devices and cloud platforms rises. The energy needed to run CoT systems is substantial. Future studies should concentrate on creating energy-saving tools and procedures that reduce CoT systems' negative environmental effects. Lack of standardization in CoT technologies might make it challenging for various systems to operate well together. Future studies should concentrate on creating industry norms and best practices that guarantee the compatibility and interoperability of various CoT systems. CoT apps are created keeping the needs of both individuals and communities in mind. To ensure that CoT technologies are created and used in ways that are advantageous to individuals and society as a whole, future research should consider the social, ethical, and cultural implications of these technologies. By concentrating on these research areas, we can ensure that CoT develops and continues to provide major advantages to businesses and society in the years to come.

REFERENCES

1. Alhaidari, F., Rahman, A., Zagrouba, R. Cloud of Things: Architecture, Applications and Challenges. *J Ambient Intell Human Comput*, 2020, 14(5), 5957–5975. https://doi.org/10.1007/s12652-020-02448-3
2. Ning, Z., Kong, X., Xia, F., Hou, W., Wang, X. Green and Sustainable Cloud of Things: Enabling Collaborative Edge Computing. *IEEE Communications Magazine*, 2019, 57, 72–78. https://doi.org/10.1109/MCOM.2018.1700895
3. Nguyen, D. C., Pathirana, P. N., Ding, M., Seneviratne, A. Integration of Blockchain and Cloud of Things: Architecture, Applications and Challenges. *IEEE Communications Surveys & Tutorials*, 2020, 22, 2521–2549. https://doi.org/10.1109/COMST.2020.3020092
4. Thangaraj, M., Ponmalar, P. P., Anuradha, S. Internet of Things (IOT) Enabled Smart Autonomous Hospital Management System - A Real World Health Care Use Case with the Technology Drivers. In *2015 IEEE International Conference on Computational Intelligence and Computing Research (ICCIC)*, 2015, pp 1–8. https://doi.org/10.1109/ICCIC.2015.7435678
5. Puntoni, S., Kleinsmith, N. Proteus Digital Health: Healthcare for Everyone, Everywhere. 2022.
6. Zia, T., Liu, P., Han, W. Application-Specific Digital Forensics Investigative Model in IoT. In *Proceedings of the 12th International Conference on Availability, Reliability and Security (ARES '17)*. Association for Computing Machinery: New York, NY, USA, 2017, pp 1–7. https://doi.org/10.1145/3098954.3104052
7. Aljahdali, A., Aldissi, H., Banafee, S., Sobahi, S., Nagro, W. IoT Forensic Models Analysis. *RRIA*, 2021, 31, 21–34. https://doi.org/10.33436/v31i2y202102
8. Wong, S. Y., Soh, M. Y., Wong, J. M. Internet of Medical Things: Brief Overview and the Future. In *2021 IEEE 19th Student Conference on Research and Development (SCOReD)*, 2021, pp 427–432. https://doi.org/10.1109/SCOReD53546.2021.9652784
9. Chen, Y. Integrated and Intelligent Manufacturing: Perspectives and Enablers. *Engineering*, 2017, 3, 588–595. https://doi.org/10.1016/J.ENG.2017.04.009
10. Nergiz, E., Barutcu, H. C. The Impact of Industry 4.0 Applications on Production Processes: The Case of Bosch Industry and Trade Corporation. *Econder*, 2020, 4, 47–71. https://doi.org/10.35342/econder.666369

11. Petrik, D., Herzwurm, G. IIoT Ecosystem Development through Boundary Resources: A Siemens MindSphere Case Study. In *Proceedings of the 2nd ACM SIGSOFT International Workshop on Software-Intensive Business: Start-ups, Platforms, and Ecosystems (IWSiB 2019)*. Association for Computing Machinery: New York, NY, USA, 2019, pp 1–6. https://doi.org/10.1145/3340481.3342730

12. Muhammed, A. S., Ucuz, D. Comparison of the IoT Platform Vendors, Microsoft Azure, Amazon Web Services, and Google Cloud, from Users' Perspectives. In *2020 8th International Symposium on Digital Forensics and Security (ISDFS)*, 2020, pp 1–4. https://doi.org/10.1109/ISDFS49300.2020.9116254

13. Cao, D., Xue, D., Ma, Z., Mei, H. XiUOS: An Open-Source Ubiquitous Operating System for Industrial Internet of Things. *Sci. China Inf. Sci.*, 2022, 65, 117101. https://doi.org/10.1007/s11432-021-3294-y

14. Stergiou, C., Psannis, K. E., Kim, B.-G., Gupta, B. Secure Integration of IoT and Cloud Computing. *Future Generation Computer Systems*, 2018, 78, 964–975. https://doi.org/10.1016/j.future.2016.11.031

15. Xu, X., Dou, W., Zhang, X., Hu, C., Chen, J. A Traffic Hotline Discovery Method over Cloud of Things Using Big Taxi GPS Data. *Software: Practice and Experience*, 2017, 47(3), 361–377. https://doi.org/10.1002/spe.2412

16. Al-Turjman, F. Mobile Couriers' Selection for the Smart-Grid in Smart-Cities' Pervasive Sensing. *Future Generation Computer Systems*, 2018, 82, 327–341. https://doi.org/10.1016/j.future.2017.09.033

17. Weinman, J. *Digital Disciplines: Attaining Market Leadership via the Cloud, Big Data, Social, Mobile, and the Internet of Things*, John Wiley & Sons, 2015.

18. Jha, S., Jha, N., Prashar, D., Ahmad, S., Alouffi, B., Alharbi, A. Integrated IoT-Based Secure and Efficient Key Management Framework Using Hashgraphs for Autonomous Vehicles to Ensure Road Safety. *Sensors*, 2022, 22, 2529. https://doi.org/10.3390/s22072529

19. Alam, T. Cloud-Based IoT Applications and Their Roles in Smart Cities. *Smart Cities*, 2021, 4, 1196–1219. https://doi.org/10.3390/smartcities4030064

20. Schrage, M. Digital-(Rst Enterprise Success Demands Clarity- (Rst Supply Chain Design. *MIT Sloan Management Review*, 2020.

21. Roopaei, M., Rad, P., Choo, K.-K. R. Cloud of Things in Smart Agriculture: Intelligent Irrigation Monitoring by Thermal Imaging. *IEEE Cloud Computing*, 2017, 4 (1), 10–15. https://doi.org/10.1109/MCC.2017.5

22. Hossein Motlagh, N., Mohammadrezaei, M., Hunt, J., Zakeri, B. IoT and the Energy Sector. *Energies*, 2020, 13, 494. https://doi.org/10.3390/en13020494

23. Steiber, A., Alänge, S., Ghosh, S., Goncalves, D. Digital Transformation of Industrial Firms: An Innovation Diffusion Perspective. *European Journal of Innovation Management*, 2020, 24, 799–819. https://doi.org/10.1108/EJIM-01-2020-0018

24. When Gradual Change Beats Radical Transformation - ProQuest www.proquest.com/openview/529b4f499a48ecee8a51724b82d1a4e8/1?cbl=26142&pq-origsite=gscholar&parentSessionId=C9N%2FnDRKIUDamfRgMqj30owuFO%2By1kVGDSw7wg2i%2Fh0%3D (accessed Apr 25, 2023).

25. Joseph, A., Balachandra, P. Energy Internet, the Future Electricity System: Overview, Concept, Model Structure, and Mechanism. *Energies*, 2020, 13, 4242. https://doi.org/10.3390/en13164242

26. Carrera, B., Peyrard, S., Kim, K. Meta-Regression Framework for Energy Consumption Prediction in a Smart City: A Case Study of Songdo in South Korea. *Sustainable Cities and Society*, 2021, 72, 103025. https://doi.org/10.1016/j.scs.2021.103025

27. Tao, F., Cheng, Y., Xu, L., Zhang, L., Li, B. CCIoT-CMfg: Cloud Computing and Internet of Things-Based Cloud Manufacturing Service System. *IEEE Transactions on Industrial Informatics*. 2014, 10, 1435–1442. https://doi.org/10.1109/TII.2014.2306383
28. Ng, R. Cloud Computing in Singapore: Key Drivers and Recommendations for a Smart Nation. *Politics and Governance*, 2018, 6(4), 39–47.
29. Tripathi, N. G., Malti Goel, M. Citizen-Centric Smart Planning: Case Study San Diego, California. *Smart Master Planning for Cities: Case Studies on Domain Innovations*. Springer Nature Singapore: Singapore, 2022, 271–300.

2 Development of Digital Twin System for Cloud of Things

Elif Bozkaya-Aras, Berk Canberk, and Stefan Schmid

2.1 INTRODUCTION

New technologies emerging with Industry 4.0 have led to an increase in capabilities in communication networks and a change in lifestyle. In particular, Internet of Things (IoT) services have reshaped the way people interact with the world. IoT establishes a network of intelligent devices, referred to as "things," encompassing sensors, actuators, machine, software, and other technologies, enabling them to connect and share data with other devices and systems via the Internet. These devices can range from basic household appliances to industrial tools used in manufacturing. Thus, IoT can provide smarter, faster, cheaper, and more sustainable operation together with effective data sharing and management.

According to the Ericsson Mobility Report, it is expected that all growth in mobile data traffic will come from 5G, and total mobile traffic will reach approximately 453 EB per month by the end of 2028 and video traffic is expected to account for 80% by 2028 [2]. In addition, upcoming 6G technology is envisioned to further integrate the emerging technological advances in wireless communication through the digitization and connectivity of everything, offering superior performance than 5G. Despite all these developments, meeting the service quality requirements and processing of such big data are main issues that need to be addressed. How we will process the data obtained from so many IoT devices is another important challenge. In addition, this digital transformation causes the emergence of new technologies and protocols, and redesigns of the applications depending on all these developments.

Cloud is an obvious technology for diverse potential IoT applications and big data. Cloud computing has developed to serve, acquire, manage, process, compute, and store big data. It has become the leading communication infrastructure for supporting new and flexible functions, providing smart services and increasing service quality due to the increase in the requirements (such as low latency, increase in data size, etc.) in IoT applications. Thus, Cloud of Things (CoT), meaning integrating the IoT with cloud computing, can manage both growth rates of big data and stringent Quality of Service (QoS) requirements.

Today, CoT-based systems play an important role in a wide range of applications, such as personal healthcare, smart city, smart home systems, and autonomous cars. Such application scenarios have been developed for a single purpose and are

DOI: 10.1201/9781003390954-2

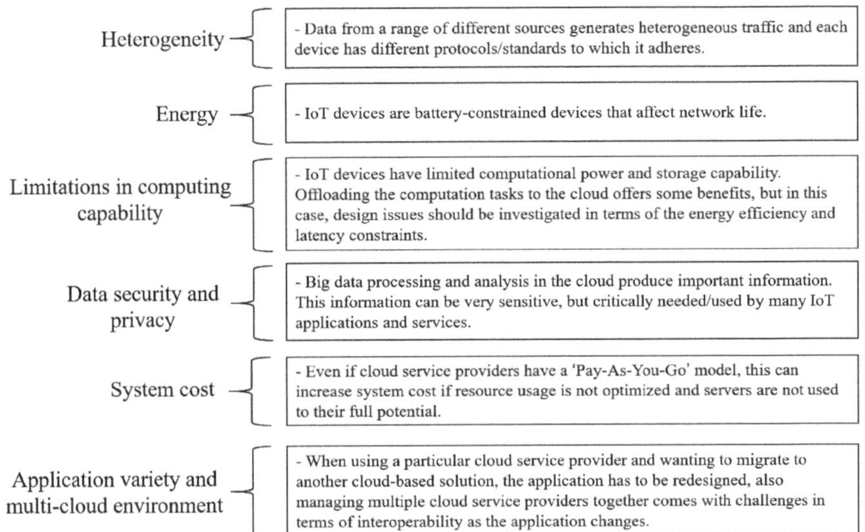

FIGURE 2.1   Key challenges faced by CoT systems.

often designed on a sensor/actuator-based structure. However, CoT is not only about sensors/actuators-based systems, but it also includes applications, servers, etc. Thus, it is the large-scale integration of potentially all technologies [40]. Therefore, we first summarize the main challenges of CoT systems as shown in Figure 2.1. The potential challenges can be listed as heterogeneity, energy management, limitations in computation capabilities of devices, data security and privacy issues, system cost, and application variety and multi-cloud environment.

In practice, heterogeneity can be considered from two perspectives. The first perspective is that each device is of different specifications e.g. CPU, RAM etc., and they are subject to different standards/protocols. The second perspective is data heterogeneity. Heterogeneous data means high variability of data types and formats and dealing with big data analytics. This can be managed with efficient energy management mechanisms, but the IoT devices are both battery-constrained devices and have the limitations in terms of computation and storage. Given the limitations in computation capability of IoT devices, edge computing is a promising approach where edge nodes can perform data processing close to the end devices. However, data security and privacy are still challenging as many CoT applications, such as applications related to healthcare, analyze sensitive and critical data. The current developments may result in increased system cost, and this requires the emergence of new design and operational strategies. Finally, managing multiple cloud environments is challenging in terms of interoperability of different CoT applications.

The current challenges can be addressed by developing new technologies. In this regard, Digital Twin (DT) technology is transforming the industry into a new form by replicating every entity in the physical world in the digital world and providing feedback from the digital world. Thus, the goal of this chapter is to design an architecture integrated with cloud computing and IoT. In this context, we propose a DT-empowered system architecture that can collect, analyze, process, and manage massive IoT data.

2.1.1 Overview of Digital Twin Concept

DT represents an advancement of the continuous research and practical implementations in Industry 4.0. DT technology is a real-time virtual representation of a physical system or process. The digital models enable the detection of problems, testing of new configurations, and analysis to take place on a virtual copy. Therefore, it can also be considered as a bridge between the real world and virtual world. Within the scope of the DT concept, physical machines are complemented by virtual twins, and a digital model of the physical system used in the transactions to be made is created and intelligent decision-making mechanisms are used over this model [23]. By creating virtual twins of models in the physical world using DT technology, performance evaluation can be executed without human intervention with the support of machine learning and artificial intelligence (AI) approaches, and network management can be performed with a zero-touch approach with intelligent decision-making mechanisms. Modeling of such a system requires analyzing the behavior of IoT devices, following interaction between them, predicting their future actions, and taking an appropriate reaction accordingly. From this perspective, DT and cloud computing could help improve planning, design, operation, and management processes.

The first definition of DT was proposed by NASA [39]. It was suggested that DT could be integrated with data mining and text mining technologies using sensor data from the vehicle, maintenance history, and other acquired data. The aim was to combine all these data to continuously predict the system's health, remaining operational lifespan, and the likelihood of mission success. Thus, the performance is increased by generating recommendations over DT.

We will present an example to clarify the importance of the DT concept and its difference from the traditional simulation environment. For example, users can observe the traffic density from source to destination and receive recommendations for a specific route through map applications. This can be achieved by running the shortest path algorithms through a simulation environment with real-time monitoring. As shown in Figure 2.2(a), traffic density seems to be quite intense at the times marked in the circle. This analysis can be obtained through a traditional simulation environment. When the same problem is analyzed via DT concept, DT analyzes the vehicle density by using both historical and real-time data in this region, and focuses on reducing the traffic congestion with advanced analytics and gives recommendation to make dynamic changes as shown in Figure 2.2(b), for example, in the flashing times of traffic lights. It improves system performance by providing feedback as well as route planning.

As a result, due to the integration of cloud computing and DT, the cloud services can provide intelligent decision-making strategies and real-time monitoring. In this chapter, a DT-empowered cloud platform is proposed with its challenges and advantages.

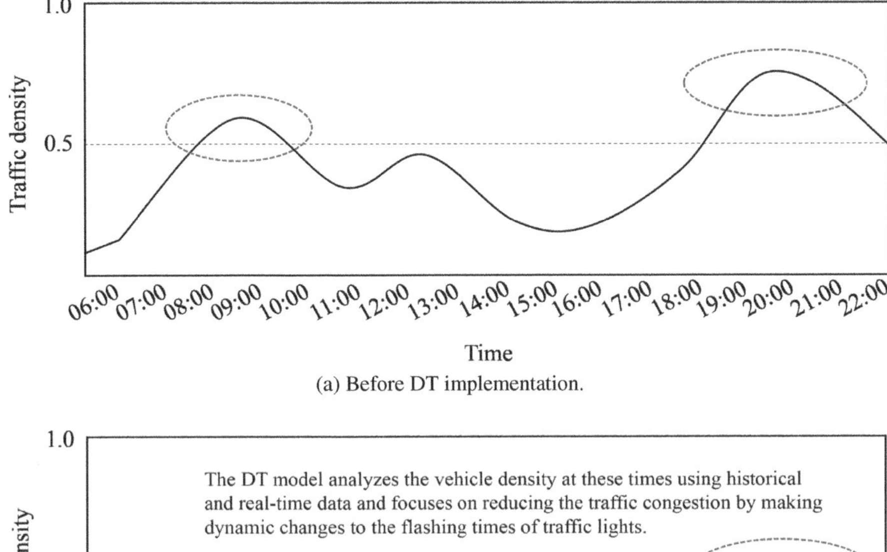

(a) Before DT implementation.

(b) After DT implementation.

FIGURE 2.2   An example of DT implementation.

2.1.2 CONTRIBUTIONS

In this subsection, we highlight the key contributions of this chapter:

- This chapter presents a DT-empowered CoT architecture for cloud services. The benefits of such an architecture are also highlighted.
- This chapter also discusses the state-of-the-art analysis of CoT applications from DT perspective and with new emerging challenges.
- The combination of DT and AI-powered models is assessed and discussed in different CoT application areas, and finally some important future directions in this domain are emphasized.

The remainder of the chapter is organized as follows: Section 2.2 gives a comprehensive overview of DT-empowered CoT architecture, which consists of three components: data layer, virtualization layer, and service layer. Then, we discuss the benefits of this architecture with the combination of DT and AI methods. Section 2.3

presents the examples of CoT applications. Finally, Section 2.4 summarizes the contribution and discuses future research directions.

2.2 DT-EMPOWERED COT ARCHITECTURE DESCRIPTION

In this section, we will describe a DT-empowered CoT architecture, which supports a system capable of collecting, analyzing, processing, and managing massive IoT data. There is no standardized approach for the DT architecture since the requirements of each application have different contents [12]. Figure 2.3 gives an overview of the proposed CoT architecture, which consists of three main components: data layer, virtualization layer, and service layer.

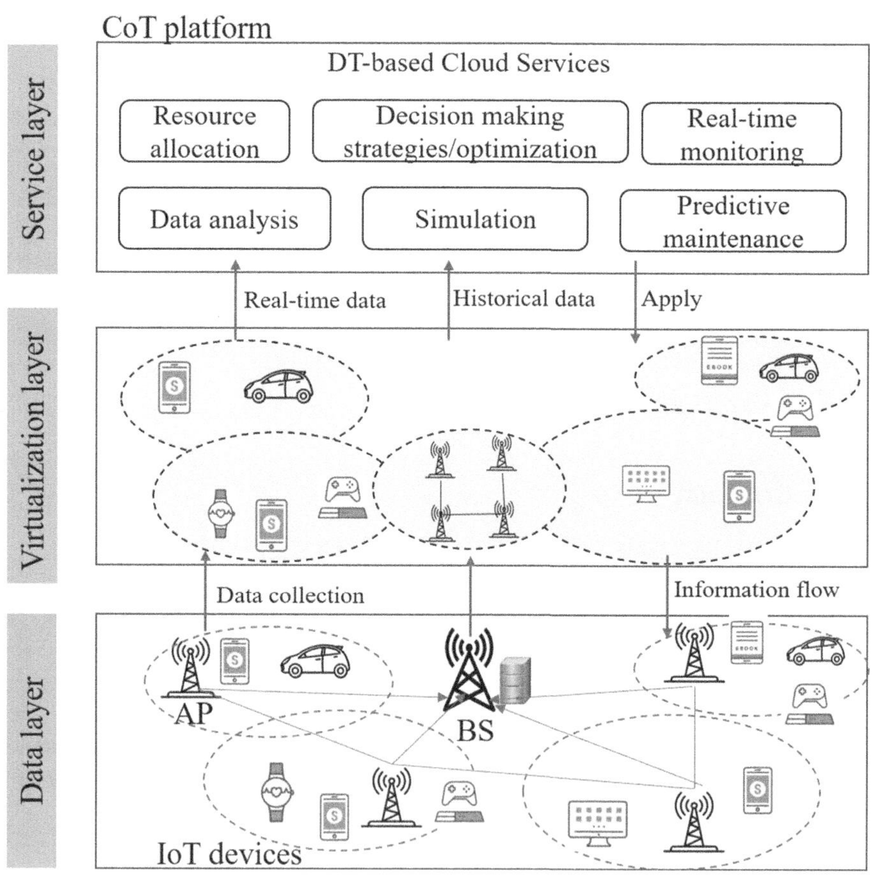

FIGURE 2.3   The proposed DT-empowered CoT architecture.

2.2.1 DATA LAYER

Data layer consists of multiple IoT devices including sensors, actuators, cameras, smart phones, and more. These devices can sense, measure, and convert the physical properties of the environment into the electrical signals [10]. Data layer may include several kinds of data such as video, audio, image, sensor data, etc. Thus, this layer is responsible for collecting heterogeneous data and handling the imprecise or uncertain data and requires extremely well-organized structure to handle massive number of data packets. The collected data is transferred to the upper layer through the gateway.

The potentially massive number of devices forms the data layer in the cloud. These devices may have some limitations related to energy, heterogeneity, and computational capability despite their wide range of applications and uses. Therefore, we will first discuss the architectural issues that can affect the performance of the data layer.

- *Heterogeneous IoT systems:* IoT systems consist of both resource/computation-constrained devices and more powerful devices. In addition, these devices can use different communication protocols and technologies, including Bluetooth, WiFi, and cellular (such as 5G RAN). The choice of a technology depends on various factors such as the range and speed of data transmission required and the power and cost constraints of the devices. IoT devices may also have different data formats. Thus, one of the main reasons to implement the data layer in the cloud is the heterogeneity-based challenges. This can be classified into three domains: (1) heterogeneity of IoT devices, (2) heterogeneity in data formats, and (3) heterogeneity in communication [34].
- *Limitations in computational capability of IoT devices:* IoT devices have limited processing power, memory, and storage capabilities. On the other hand, IoT-based applications are mostly massive in number and computation, and latency sensitive. Due to the strict QoS requirements, computation offloading is a promising solution to migrate the computation tasks partially or completely to the edge of the network.
- *Energy:* IoT systems consist of mostly battery-powered devices to communicate and interact with the environment. These devices may be simple sensors (e.g. humanity and temperature sensors), complex sensors to provide more information (e.g. ultrasound and vision sensors), actuators, or more intelligent devices. Thus, IoT applications may range from an environmental monitoring application to the smart cities and data can be transmitted real time or on demand. As a result, several algorithms/methods have been designed to save the energy efficiency and improve network lifetime and sustainability. For example, scheduling the data transfer is one of the methods to save the energy consumption. This approach schedules the transmission of data at particular intervals. However, the transmitted data can be classified as critical or non critical data depending on the IoT applications; therefore, critical data is prioritized. Energy harvesting in sensors is also a promising technology to prevent the physical battery change and ensure continuous operation [33]–[32].

- *QoS requirements:* IoT technology relies on the communication between intelligent devices with various application domains, such as smart cities, smart transportation, smart grid, and more. IoT devices can share data with the communication equipment and collect/store the data from the environment so that data analysis can be performed. The IoT applications have different QoS requirements. Therefore, in order to guarantee QoS, it is necessary to implement the suitable mechanisms at each layer of the CoT architecture.

2.2.2 Virtualization Layer

Virtualization layer is the core of the DT implementation. It lies between data layer and service layer. Each physical entity such as IoT devices, servers, storage, and network in the CoT environment is all virtualized as a digital twin in the cloud. As shown in Figure 2.3, digital twins are the virtual representations of the physical entities.

Based on the illustration provided in Figure 2.3, IoT devices are necessary for the realization of DT entities [30]. It is also important to emphasize that cloud-based connections are necessary for the communication between DT and data and service layers. Both real-time and historical data collected from data layer are transferred to the upper layers. While real-time data continuously provides information to track the network states, historical data is stored to describe the characteristics of IoT devices so that these data can be used for predictions about the maintenance and optimizations to improve system performance. The advantage of creating such a virtualization layer is that IoT devices use a variety of protocols and standards to connect to the network [35]. This complicated structure should not affect the service layer functionalities. The virtualization layer provides an abstraction independent of the data layer, which increases the efficiency of the services in the service layer.

In order to create a sophisticated digital representation of IoT devices in virtualization layer, there are different tools, such as Microsoft Azure Digital Twins and Oracle IoT Digital Twin. They use JavaScript Object Notation (JSON)-based model which contains attribute values and uses a semantic model. Microsoft Azure Digital Twins creates the twin graphs based on digital models of physical world, which can be buildings, bridges, factories, networks, and more [31]. These digital twins are used for better operations and reduced cost. Digital models are determined by Digital Twins Definition Language, which is a variant of JSON, and it describes the content of the IoT devices, called as interface according to their properties, telemetry events, commands, components, and relationships. An interface is described as follows [31].

- *Property:* Property describes the state of any digital twin. This field can be read-only or read and write. For example, an identification number of any device may be read-only property, but humidity on an environment may be read and write property. This field can be also read at any time.
- *Telemetry:* Telemetry describes the measured data with its data type and unit.
- *Command:* Command describes the functions that can be performed by any digital twin. At given times, the digital twin starts and completes a function.
- *Relationship:* Relationship describes the links among digital twins and provides a relationship graph among them.

- *Component:* Component describes the interfaces to be composed of other interfaces.

2.2.3 SERVICE LAYER

Service layer is defined as the cloud computing services in IoT-based applications. IoT devices migrate data to the service layer though the virtualization layer. Cloud server has powerful computing capabilities. Thus, service layer builds an analytical or statistical model and provides cloud services such as resource allocation, anomaly detection, decision optimization, data analysis, modeling and simulation, monitoring, prediction, etc., by using advanced data analytics, AI, machine learning, and other related technologies as shown in Figure 2.3. The service layer is responsible for providing specific services for each domain. For example, when healthcare services are considered, real-time health monitoring, human activity recognition, remote diagnosis, emergency warning, and more can be included in the services. Users can interact with the services through mobile phones, wearable devices, personal computers, etc., via Bluetooth, WiFi, ZigBee, 4G, and wired network.

Using DT implementation in the service layer provides the following benefits:

- With DT implementation in the service layer, the design, operation, maintenance, and predictions on the system performance can be simplified, allowing real-time monitoring [44].
- By using both real-time and historical data, system predictions can be applied with advanced analytics, and this enables to improve system performance throughout its lifecycle, providing service cost reduction.
- DT can be implemented for different types of services by creating different DT components, each designed for multiple services, and simplifying tasks [44].
- By analyzing user characteristics, in multiple requests for the same service, instead of receiving the data each time, it is retrieved once, stored in the cloud, and thus the user demand is satisfied. This reduces both network workload and the need for multiple similar data offloading.

2.2.4 BENEFITS OF DT-EMPOWERED CoT ARCHITECTURE

The benefits of the proposed DT-empowered CoT architecture are discussed as follows:

- *Reusability:* Data generated by IoT devices/end users is stored, processed, and analyzed when offloaded to the cloud. When same computation tasks are requested, it can be performed without the need for data transfer again by analyzing the user characteristics.
- *Manageable:* IoT services can be run faster on the cloud platform with more manageability and less maintenance [48].
- *Extendable:* DT-empowered CoT architecture can provide low-cost general-purpose IoT system to manage and control different services on the cloud.

- *Powerful:* DT-empowered CoT architecture allocates the resources to meet fluctuating demands by predicting the system state in the future and providing high computing power [48].
- *Adaptability:* Cloud services may be operating on different standards/protocols. The variety of standards/protocols can limit the effectiveness of cloud services. In a DT-empowered CoT architecture, there is no need to design a new model as the types of data change or the use case changes due to heterogeneity.

2.2.5 COMBINATION OF DT AND AI-POWERED MODELS WITH CLOUD SERVICES

AI is one of the main technologies for DT and CoT environment to learn, analyze, model, and predict the system behavior. In addition, machine learning, one of the subset of AI, is a powerful method for solving specific tasks by learning from real-time and historical data and making predictions. With the use of cloud platforms, AI has started to provide convenience to developers in problem-solving. At the same time, cloud providers now offer AI services within their public clouds as well. Google Cloud and Amazon are some examples of public cloud AI services. Both train AI models cost-effectively and iterate faster with high performance cloud Central Processing Units (CPU) and Tensor Processing Units (TPUs). In this section, we will not discuss the AI methods in detail, but will examine some CoT application scenarios where AI/machine learning methods are used. In Table 2.1, we summarize DT-assisted CoT applications with challenges and advantages in terms of transportation, industrial IoT (IIoT), healthcare, city monitoring and environmental monitoring, and 5G and beyond networks.

The transportation industry has achieved significant breakthroughs in recent years. Especially in the last decade, vehicular networks integrate innovative technologies, such as computer vision, machine learning, and IoT, for real-time interactive applications and traffic safety and efficiency. However, limited vehicular computing, high mobility, and diversity of application requests raise big challenges, and AI-powered cloud computing provides resources for intelligent transportation services. For example, Ren et al. [37] define a software-defined vehicular networks architecture to manage the network resources. A controller in the cloud is responsible for obtaining all the network information and managing the vehicular computing nodes and applying an intelligent service offloading model to minimize the task delay.

In IIoT applications, the main challenges lie in resource constraints (i.e., power, short battery lifetime, and maintenance difficulties), reliability and latency requirements, and stringent communication availability. In industrial platforms, IoT devices continuously monitor environmental factors and enable information sharing with remote servers with any event triggering [41]. However, the industrial outcome can require complex computational process and continuous battery replacement may not be possible. Therefore, energy-efficient optimization and battery lifetime extension are key components in industry IoT applications, increasing system productivity by supporting continuous real-time monitoring and intelligent decision-making. For example, Fahim et al. [14] predict the wind speed and then the power generation

TABLE 2.1
Summary of DT-assisted CoT applications

CoT Applications	Challenges	DT Advantages
Transportation	1. Distribution of massive vehicles 2. Traffic density 3. Unpredictable vehicle topology 4. Reliability	1. Better mobility management 2. Enabling self-driving vehicle 3. Higher quality of experience (QoE) 4. Increased traffic safety and efficiency
Industrial IoT	1. High investment cost 2. Continuous monitoring 3. Security risks and storage 4. Reliability	1. Real-time monitoring and analysis 2. Intelligent decision-making 3. Improved productivity 4. Energy efficiency
Healthcare	1. Leakage of sensitive data 2. High healthcare investments 3. Complex data 4. Human factor	1. Better accuracy ratio 2. Advanced analytics 3. Remote monitoring and diagnosis 4. Better collaboration between healthcare providers and patients
City Structure Monitoring and Environmental Monitoring	1. Poor intelligent decision-making 2. Spatio temporal fluctuations 3. Big data 4. Heterogeneous data	1. Predictive maintenance 2. Lower maintenance 3. Energy efficiency 4. 3D modeling
5G and Beyond Networks	1. Increasing number of IoT devices 2. Increasing traffic requests 3. Complicated and unpredictable network environment 4. Mobility management	1. Low latency 2. High reliability 3. Energy efficiency

based on a wind turbines data set to improve system performance with a deep learning model.

AI-powered models have an important role in health monitoring and diagnosis, making recommendations for both patients and healthcare providers. For example, Zhang et al. [50] focus on predicting lung cancer diagnosis by designing healthcare DT platform. In traditional methods, such a diagnosis is based on clinical symptoms, physical signs, and unknown causes of patients. Machine learning is likely to help reveal relationships between data-driven medical examinations and individualized treatment so that doctors can track the patient records, monitor the patient body, and detect the problems remotely through DT. It is important to emphasize that the patient-related data can be stored, processed, and used to provide feedback in the cloud with high computing capabilities. Latest advancements in AI, big data, and IoT also develop more secure solutions and eliminate the privacy and security concerns.

The focus of intelligent infrastructures is sensing that can monitor itself and act intelligently on its own. IoT devices can monitor public infrastructures, such as bridges, roads, buildings, and more. This provides efficient resource usage based on the collected data and analysis with AI methods. Real-time monitoring eliminates the necessity for periodic inspections, resulting in cost reduction. Additionally, measuring energy consumption in households enables precise load forecasting. Furthermore, the deployment of sensors on roads for traffic monitoring gathers essential data needed for the implementation of intelligent city infrastructure. The energy system

TABLE 2.2
AI-powered methods in CoT applications

Paper	Application Area	Case Study	Used Method
[11]	Transportation	Traffic flow detection	Deep Learning
[51]	Transportation	Computation task offloading and edge resource allocation	Multiagent Deep Reinforcement Learning
[37]	Transportation	Intelligent service offloading decision	Deep Reinforcement Learning
[15]	Industrial IoT	Anomaly detection	Deep Learning
[14]	Industrial IoT	Cloud-based predictive model in wind turbines to estimate the speed of wind and the generated power	Deep Learning
[17]	Industrial IoT	Resource allocation for D2D	Federated Reinforcement Learning
[42]	Healthcare	Emotion recognition system	Deep Learning
[25]	Healthcare	Cloud healthcare system: Elderly healthcare services	Big Data-Based Cloud Platform
[28]	Healthcare	Intelligent data duplication and classification in the cloud-enabled healthcare environment	Deep Transfer Learning
[12]	City Structure Monitoring and Environmental Monitoring	Cloud-based DT for structural health monitoring: Damage detection of digital bridge and real bridge structures	Deep Learning
[49]	City Structure Monitoring and Environmental Monitoring	Air quality prediction	Deep Neural Networks
[19]	City Structure Monitoring and Environmental Monitoring	Remote monitoring of parking spaces	Deep Learning
[38]	City Structure Monitoring and Environmental Monitoring	Water pollution detection	Support Vector Machine, k-nearest Neighbor, Single Layer Neural network and Deep Neural network
[27]	5G and Beyond Networks	DT placement and DT migration strategy	Deep Reinforcement Learning
[43]	5G and Beyond Networks	Mobile offloading decision for Digital Twin Edge Network in 6G	Actor-Critic-Based Deep Reinforcement Learning
[26]	5G and Beyond Networks	Real-time data processing and computation	Federated Learning

plays a vital role in building a smart city. The primary steps in energy management during urban planning involve the planning, design, and operational processes of energy supply systems, encompassing electricity, cold, heat, and gas [18].

5G and beyond networks aim to provide full-scale wireless connectivity with the concept of digitizing everything. AI and cloud computing are key innovative research trends for network management. AI methods imitate the human brain to process data, create patterns for decision-making process, and propose innovative solutions in specific domains to improve system performance. Cloud computing, on the other hand, provides fast computing services, in particular for delay-sensitive tasks in 5G and beyond networks.

Some examples of used AI methods in CoT applications are given in Table 2.2. Example application scenarios will be also discussed in Section 2.3.

2.3 COT APPLICATIONS

In recent years, the development of communication technologies is redefining the lifestyle. As the number of devices connected to the Internet increased, the need and requirements also increased and led to new application areas. Innovations in CoT have formed the most fundamental applications of Industry 4.0. In this section, we will review some examples of CoT applications.

2.3.1 TRANSPORTATION APPLICATIONS

The emergence of intelligent transportation system improves traffic safety and efficiency due to various transportation applications, such as traffic information system, emergency management [5], medical waste transportation, and supply chain policy determination [21, 9]. Connected vehicle technology allows vehicles to communicate with each other through vehicle-to everything communication, thereby enabling vehicles to obtain more information about the traffic, other vehicles, drivers, and many potential conflicts that can be avoided through transportation applications [7] and cloud computing is used for these collected massive amount of data in different transportation problems.

In recent years, various transportation application areas have been proposed using cloud computing capabilities. Wang et al. [48] propose a mobility DT framework which consists of three planes: (1) human, vehicles, and traffic infrastructure in the physical space, named as the lower plane, (2) human DT for user management, vehicle DT for cloud-based driver-assistance systems and traffic DT for traffic monitoring and speed limit in the digital space named as the upper plane, and (3) communication plane between these two planes. The benefits of such a model can be as follows: (1) high computational tasks can be offloaded to the cloud for more powerful computations, (2) the model can be easily implemented to various vehicles, and (3) any update can be quickly applied.

Liao et al. [24] present a DT framework based on vehicle-to-cloud communication to implement cooperative ramp merging design. Data is offloaded to the cloud server via 4G/LTE cellular network. Cloud server creates the DTs of vehicles and drivers, processes the data, and then sends the feedback to the vehicles. A real-world implementation was conducted with three vehicles in a ramp merging scenario.

The proposed model was compared with no feedback approach and the results are discussed in terms of energy consumption and communication delay.

Gu et al. [16] focus on computation-intensive and service-sensitive Internet of Vehicles applications. Both the collaboration between edge and cloud computing and the collaboration among small cell eNodeBs are investigated. The authors construct a DT model to define task offloading and resource allocation strategy. Zhang et al. [51] investigate the challenges of different resource demands generated by smart vehicles and propose a multiagent learning scheme for task offloading so that it could minimize the offloading latency. Any vehicle can offload its computing task to another vehicle or road side units and the authors manage the resource scheduling.

2.3.2 Industrial IoT Applications

In recent years, cloud computing and DT technology have been used in industrial applications for the processes of product development and manufacturing, testing, and deploying complex industrial systems. IIoT is one of the applications of IoT in the industry and performs monitoring, collecting, and analyzing processes. IIoT devices may range from simple environmental sensors to the industrial robots. While the term "industrial" brings to mind warehouses, shipyards, and factories, IIoT technologies also have potential for a wide variety of industries, including agriculture, healthcare, and financial services.

Large numbers of heterogeneous IoT devices generate massive amounts of data. Due to the different capabilities of IoT devices, maintaining QoS requirements on these data and incoming requests and efficient management for resource allocation are challenging issues in IoT networks. Privacy is also one of the important concerns in IIoT. The authors of [17] propose a DT edge network in 6G IIoT scenario, where edge computing is responsible for computational tasks; DT is a bridge between the physical and virtual worlds and device-to-device (D2D) communication manages resource-limited IoT devices. It is aimed to provide real-time monitoring and constitute resource allocation strategies by minimizing the communication overhead. Similarly, Van Huynh et al. [46] propose a DT wireless edge networks for IIoT to reduce the latency in computation offloading. DT models the capacity of edge servers and defines resource allocation strategies via ultra-reliable and low-latency communications (URLLC) links.

Dang et al. [12] propose a DT framework based on cloud computing for health monitoring. Cloud-based DT architecture consists of three components: physical object, virtual object, and connected data interface. The framework continuously monitors the physical objects and performs proactive maintenance; therefore, maintenance strategy is defined, reliability is increased, and remaining service time is extended. As a case study, damage detection of digital bridge and real bridge structures is tested and cloud computing is used for fast computation.

Ji et al. [20] propose an intelligent management system for thermal power plant water resource using DT, IoT, and cloud computing technologies. In order to enhance water use efficiency and minimize waste emissions, a comprehensive investigation of water consumption and drainage at a plant and power plant is conducted and a DT-based framework is proposed.

2.3.3 HEALTHCARE APPLICATIONS

Cloud computing in the healthcare industry provides a connected, accessible, manageable, and collaborative environment for better monitoring and analysis of healthcare-related data. With the integration of IoT and cloud computing, patients can track their physiological data by wearing some IoT devices and send reports to the healthcare centers. These data can be used for the diagnosis and treatment of diseases with the help of data analytics and machine learning algorithms. More importantly, this can minimize the unnecessary visits to healthcare centers, saving time and cost.

Zheng et al. [52] propose the utilization of cloud computing to store healthcare data and provide query services, alongside DT technology for creating digital representations of patients. This approach enables continuous monitoring of the healthcare status of patients. Cloud computing and DT-based healthcare monitoring system also discusses data privacy due to the fact that private health records consist of personal and highly sensitive data. In addition, real-time monitoring requires the interaction between patients and healthcare centers and the capabilities of the centers should be adequate for such a interactive service. Here emerges the importance of cloud computing, IoT, and DT.

Subramanian et al. [42] investigate emotion recognition in healthcare, where the DT of a person in real time can monitor and understand the physical entity's capabilities and improve the quality of life for personalized healthcare. Such an emotion recognition system can be useful for patient health monitoring and early diagnosis of diseases and defining the effective treatment. However, there are some technical challenges to achieve this, such as limited data set, high cost, and defining the important features. These challenges are overcome by the implementation of DT concept.

Elderly people are more prone to diseases that reduce their quality of life, such as chronic illnesses, physical disabilities, and mental illnesses. Preventing, monitoring, and controlling the health problems of the elderly people require a multifaceted approach. Combining cloud computing and DT will provide an efficient way for accurate and fast services. Liu et al. [25] propose a cloud healthcare system based on DT to monitor, diagnose, and predict the heath status of elderly people. In this framework, wearable IoT devices play a significant role toward the goal of personal health management.

Laamarti et al. [22] use sensors for human's five senses. These sensors are controlled by the DT to collect the information about environment. The data is stored in cloud, which serves as memory of the DT. Data can be processed using AI algorithms to detect patterns and then provide the feedback to the real object.

In recent years, research in the Robotic Internet of Things (IoRT) has gained momentum and it is believed that IoRT will digitize healthcare. Das et al. [13] discuss the integration of DT and IoRT in surgical sector. Collecting data by DT and then evaluating system dynamics through a machine, process, or living organism can revolutionize healthcare in reducing risks and finding the best possible solution, the most important element in the surgical sector. In this way, using the human body's DT and connecting to the Internet can create new and modern health applications.

2.3.4 City Structure Monitoring and Environmental Monitoring

A smart and sustainable city is often described as an example application, integrating ubiquitous connectivity, data management approaches, and information technologies for next-generation wireless networks. Creating a smart and sustainable city is a complex, long-term, and ever-evolving process. The main purpose of the IoT-based smart and sustainable cities is to use public resources efficiently through big data, thereby providing a wide range of smart applications, including smart manufacturing, smart home, autonomous vehicles, augmented/virtual reality, smart healthcare, and so on [36]. To achieve these, it is necessary to make smart decisions for both sustainability and climate change through big data and citizen engagement.

We live in an unprecedented and exciting era of connectivity, where billions of sensors, devices, and machines work together and rapidly share information. Considering a large number of new computation-intensive applications, it is unfortunately inevitable that there will be an increase in traffic demands and the complexity of activities related to network management. However, the computational capacity of IoT devices is not sufficient for these operations and high traffic volume reduces the QoS. On the other hand, a key requirement for IoT-based smart and sustainable cities is to have a rapid fulfillment of requests and ultra-reliable low-latency communication. Smart city applications are addressed to city authorities, infrastructure and utilities managers, as well as various city stakeholders interested in adopting smart city solutions to enhance the city's competitiveness, cohesion, and sustainability.

DT is an emerging trend, where a digital replica of a physical object can be created. It has been also started to use in smart manufacturing to monitor, analyze, simulate, and visualize the system. The digital twins can continuously monitor the physical machines, using both historical and real-time data, and as a result new configurations can be designed and tested; more importantly, predictive maintenance can be applied to the system to improve lifetime.

Marai et al. [29] discuss the implementation of the digital twin concept specifically for roads in smart cities. They suggest deploying a DT Box equipped with a camera and a set of IoT devices. The DT box monitors the road asset, sends real-time data to the cloud, including live stream, measurements of the environment, such as temperature and humidity. The data is used for real-time monitoring as well as object detection and recognition, such as vehicle and human. The paper highlights that the proposed DT Box is a step toward autonomous vehicles and smart mobility.

Fahim et al. [14] investigate wind turbines which is one of the sources of renewable energy. In order to understand and analyze the wind farms, cloud-based DT framework is proposed and this approach makes pay-as-you-go cloud services possible. In the proposed model, univariate time series data of wind is processed to estimate its speed. Then, it predicts the power generation. While the cloud platform provides real-time services with lower latency, DT approach reduces the cost of deployment and achieves better resource utilization.

2.3.5 5G AND BEYOND NETWORKS

Industry 4.0 is a rapidly growing field and has gained significant research attention in various fields. From these fields, the 5G and beyond wireless network is expected to strengthen cloud computing and DT in the concept of digitizing and connecting everything. The increase in both the number of IoT devices and mobile data traffic has an important role in the advent of 5G networks. With multiple devices, the number of connections for each user is also rapidly increasing. This triggers the development of new services and applications, revealing the potential of 5G and beyond networks. Radiocommunication Sector of International Telecommunication Union (ITU-R) has determined the following scenarios for 2020 and beyond [1]:

- *Enhanced Mobile Broadband (eMBB):* eMBB focuses on addressing the surge in data rates, high user density, and providing substantial traffic capacity for hotspots in addition to ensuring seamless coverage and supporting medium to high mobility.
- *Ultra Reliable and Low Latency Communications (URLLC):* URLLC has stringent requirements such as latency, reliability, throughput, etc., based on mission critical and time-sensitive applications.
- *Massive Machine Type Communications (mMTC):* mMTC is characterized by a massive number of connected devices and requires low-cost solutions with long battery life to transmit low volumes of time-insensitive data.

To meet the stringent requirements of the new applications and services, Multi-access Edge Computing (MEC) is determined by the European Telecommunications Standards Institute as a fundamental communication infrastructure for the transition to 5G and beyond [47]. MEC reduces the network operation and service latency by allowing data to be processed on nodes close to the user. Thus, MEC can be considered as a cloud server at the network edge deployed to the nodes such as Access Points, Base Stations, etc. MEC servers generally have high computation capabilities, making them suitable for analyzing and processing large amounts of data. This not only reduces task completion time for resource-constrained IoT devices but also provides energy efficiency [6].

DT is also attracting attention toward 5G and beyond networks. It has the potential to replicate the physical assets in the virtual world and obtain the recommendation from the virtual world. It has the potential to replicate the assets in the physical world in the digital world and obtain the recommendation from the digital world [3]. With the integration of MEC and DT technology, accuracy can be increased in the decision-making process by making analysis on digital models based on both historical and real-time data.

Sun et al. [43] design an offloading scheme by utilizing MEC technology and show the relationship between computation latency and energy consumption of the user devices. DT-assisted model estimates the workload of edge servers and defines a series of offloading decisions. Similarly, Lu et al. [26] propose a blockchain-based federated learning framework in the DT networks. This approach aims to enhance the system's reliability and security while also supporting data privacy between end users and edge servers. Van Huynh et al. [45] present MEC-based URLLC DT architecture for metaverse applications. Edge caching, task offloading, and resources allocation problems are modeled to reduce the latency and guarantee stringent requirements of reliability. In the studies of [4] and [8], mobile users intelligently offload the computation tasks to the mobile edge servers/cloud server and identify the selection of edge servers/cloud server with the assistance of DT.

2.4 CONCLUSION AND FUTURE DIRECTIONS

In this chapter, we investigated the implementation of DT technology in CoT system architecture. We first described how to benefit from the integration of DT technology in CoT and discussed some typical challenges related to cloud platform (i.e., heterogeneity, energy, limitations in computing capabilities, data security and privacy, system cost, etc.) that draw attention both from industry and academia. Next, we proposed a DT-empowered CoT architecture and described a layer-based framework, called data layer, virtualization layer, and service layer. We also detailed some examples from CoT applications with DT perspective as well as the combination of DT and AI-powered models for cloud services.

DT implementation is promising in various IoT applications because of its features such as advanced data analytics, system predictions, decision-making strategies, simulation, and so on. A DT-empowered CoT is more effective, manageable, and powerful with reduced system cost, less energy consumption, and improved system performance. All these features, along with cloud computing capabilities, can still face potential research challenges. However, the collaboration of DT technology and CoT platform may suffer from some limitations, especially in the context of the requirements of upcoming 6G networks. Due to the limitations on battery lifetime of IoT devices, increased complexity of computations and meeting QoS requirements become more challenging and may be more complicated for large-scale networks.

In addition, different requirements in each application domain such as energy efficiency, latency, fault-tolerance, and potential AI methods for these requirements can be taken into account. Replacing human involvement in controlling with the zero-touch management and autonomous systems has some challenges. In this context, self-learning-based methods for zero-touch management are still in its early stage.

One potential future work direction may tackle DT standardization. Currently, the International Organization for Standardization (ISO), Digital Twins Consortium (DTC), and the Industrial Digital Twins Association (IDTA) are still developing DT concept for unifying digital twins into systems. ISO-23247-1 (Automation systems and integration) — Digital twin framework for manufacturing) provides the general principles of a DT framework for manufacturing.

The use of digital models can also cause a significant security issue. Now, it is necessary to ensure the security of not only physical assets but also digital twins and connection between them. Because DTs are used to interact with cloud services on behalf of physical assets, they are more vulnerable to the external environment compared to the physical assets themselves. Consequently, ensuring the security of DTs becomes even more crucial [44].

ACKNOWLEDGMENT

This research was supported by The Scientific and Technological Research Council of Turkey (TUBITAK) 1515 Frontier R&D Laboratories Support Program for BTS Advanced AI Hub: BTS Autonomous Networks and Data Innovation Lab. Project 5239903. This research was supported by the German Federal Ministry of Education and Research (BMBF), project 6G-RIC (grant no. 16KISK020K), 2021–2025.

REFERENCES

1. (09/2015). IMT vision-framework and overall objectives of the future development of IMT for 2020 and beyond. *Recommendation ITU-R M.2083-0.*

2. (November 2022). Ericsson mobility report.

3. Attaran, M. and Celik, B. G. (2023). Digital twin: Benefits, use cases, challenges, and opportunities. *Decision Analytics Journal*, 6:100165.

4. Bozkaya, E. (2023). Digital twin-assisted and mobility-aware service migration in mobile edge computing. *Computer Networks*, 231:109798.

5. Bozkaya, E., and Canberk, B. (2015). Robust and continuous connectivity maintenance for vehicular dynamic spectrum access networks. *Ad Hoc Networks*, 25:72–83.

6. Bozkaya, E., Canberk, B., and Schmid, S. (2023a). Digital twin-empowered resource allocation for 6G-enabled massive IoT. In *Workshop on the Evolution of Digital Twin Paradigm in Wireless Communications, IEEE International Conference on Communications (ICC 2023)*, pages 1–6.

7. Bozkaya, E., Erel, M., and Canberk, B. (2014). Connectivity provisioning using cognitive channel selection in vehicular networks. In Mitton, N., Gallais, A., Kantarci, M. E., and Papavassiliou, S., editors, *Ad Hoc Networks*, pages 169–179, Springer International Publishing.

8. Bozkaya, E., Erel-Özçevik, M., Bilen, T., and Özçevik, Y. (2023b). Proof of evaluation-based energy and delay aware computation offloading for digital twin edge network. *Ad Hoc Networks*, 149:103254.

9. Bozkaya, E., Eriskin, L., and Karatas, M. (2022). Data analytics during pandemics: a transportation and location planning perspective. *Annals of Operations Research*.

10. Bozkaya, E., Karatas, M., and Eriskin, L. (2023). *Heterogeneous wireless sensor networks: Deployment strategies and coverage models*. In Ahuja, K., Nayyar, A., and Sharma, K., editors, Comprehensive Guide to Heterogeneous Networks, pages 1–32, Academic Press.

11. Chen, C., Liu, B., Wan, S., Qiao, P., and Pei, Q. (2021). An edge traffic flow detection scheme based on deep learning in an intelligent transportation system. *IEEE Transactions on Intelligent Transportation Systems*, 22(3):1840–1852.

12. Dang, H. V., Tatipamula, M., and Nguyen, H. X. (2022). Cloud-based digital twinning for structural health monitoring using deep learning. *IEEE Transactions on Industrial Informatics*, 18(6):3820–3830.

13. Das, C., Mumu, A. A., Ali, M. F., Sarker, S. K., Muyeen, S. M., Das, S. K., Das, P., Hasan, M. M., Tasneem, Z., Islam, M. M., Islam, M. R., Badal, F. R., Ahamed, M. H., and Abhi, S. H. (2022). Toward IoRT collaborative digital twin technology enabled future surgical sector: Technical innovations, opportunities and challenges. *IEEE Access*, 10:129079–129104.

14. Fahim, M., Sharma, V., Cao, T.-V., Canberk, B., and Duong, T. Q. (2022). Machine learning-based digital twin for predictive modeling in wind turbines. *IEEE Access*, 10:14184–14194.

15. Ferrari, P., Rinaldi, S., Sisinni, E., Colombo, F., Ghelfi, F., Maffei, D., and Malara, M. (2019). Performance evaluation of full-cloud and edge-cloud architectures for industrial IoT anomaly detection based on deep learning. In *2019 II Workshop on Metrology for Industry 4.0 and IoT*, pages 420–425.

16. Gu, L., Cui, M., Xu, L., and Xu, X. (2023). Collaborative offloading method for digital twin empowered cloud edge computing on internet of vehicles. *Tsinghua Science and Technology*, 28(3):433–451.

17. Guo, Q., Tang, F., and Kato, N. (2023). Federated reinforcement learning-based resource allocation for D2D-aided digital twin edge networks in 6G industrial IoT. *IEEE Transactions on Industrial Informatics*, 19(5):7228–7236.

18. Huang, W., Zhang, Y., and Zeng, W. (2022). Development and application of digital twin technology for integrated regional energy systems in smart cities. *Sustainable Computing: Informatics and Systems*, 36:100781.

19. Iqbal, R., Maniak, T., and Karyotis, C. (2019). Intelligent remote monitoring of parking spaces using licensed and unlicensed wireless technologies. *IEEE Network*, 33(4):23–29.

20. Ji, H., Li, J., Zhang, S., and Wu, Q. (2021). Research on water resources intelligent management of thermal power plant based on digital twins. In *2021 IEEE 6th International Conference on Cloud Computing and Big Data Analytics (ICCCBDA)*, pages 557–562.

21. Karatas, M., Eriskin, L., and Bozkaya, E. (2022). Transportation and location planning during epidemics/pandemics: Emerging problems and solution approaches. *IEEE Transactions on Intelligent Transportation Systems*, 23(12):25139–25156.

22. Laamarti, F., Badawi, H. F., Ding, Y., Arafsha, F., Hafidh, B., and Saddik, A. E. (2020). An ISO/IEEE 11073 standardized digital twin framework for health and well-being in smart cities. *IEEE Access*, 8:105950–105961.

23. Lehner, D., Pfeiffer, J., Tinsel, E.-F., Strljic, M. M., Sint, S., Vierhauser, M., Wortmann, A., and Wimmer, M. (2022). Digital twin platforms: Requirements, capabilities, and future prospects. *IEEE Software*, 39(2):53–61.

24. Liao, X., Wang, Z., Zhao, X., Han, K., Tiwari, P., Barth, M. J., and Wu, G. (2022). Cooperative ramp merging design and field implementation: A digital twin approach based on vehicle-to-cloud communication. *IEEE Transactions on Intelligent Transportation Systems*, 23(5):4490–4500.

25. Liu, Y., Zhang, L., Yang, Y., Zhou, L., Ren, L., Wang, F., Liu, R., Pang, Z., and Deen, M. J. (2019). A novel cloud-based framework for the elderly healthcare services using digital twin. *IEEE Access*, 7:49088–49101.

26. Lu, Y., Huang, X., Zhang, K., Maharjan, S., and Zhang, Y. (2021a). Low-latency federated learning and blockchain for edge association in digital twin empowered 6G networks. *IEEE Transactions on Industrial Informatics*, 17(7):5098–5107.

27. Lu, Y., Maharjan, S., and Zhang, Y. (2021b). Adaptive edge association for wireless digital twin networks in 6G. *IEEE Internet of Things Journal*, 8(22):16219–16230.

28. Mageshkumar, N., and Lakshmanan, L. (2023). Intelligent data deduplication with deep transfer learning enabled classification model for cloud-based healthcare system. *Expert Systems with Applications*, 215:119257.

29. Marai, O. E., Taleb, T., and Song, J. (2021). Roads infrastructure digital twin: A step toward smarter cities realization. *IEEE Network*, 35(2):136–143.

30. Masaracchia, A., Sharma, V., Canberk, B., Dobre, O. A., and Duong, T. Q. (2022). Digital twin for 6G: Taxonomy, research challenges, and the road ahead. *IEEE Open Journal of the Communications Society*, 3:2137–2150.

31. Microsoft (2022). Azure digital twins documentation. 1–782. https://learn.microsoft. com/en-us/azure/digital-twins/.

32. Mohanti, S., Bozkaya, E., Naderi, M. Y., Canberk, B., Secinti, G., and Chowdhury, K. R. (2021). Wifed mobile: WiFi friendly energy delivery with mobile distributed beamforming. *IEEE/ACM Transactions on Networking*, 29(3):1362–1375.

33. Mohanti, S., Bozkaya, E., Yousof Naderi, M., Canberk, B., and Chowdhury, K. (2018). Wifed: WiFi friendly energy delivery with distributed beamforming. In *IEEE INFOCOM 2018 - IEEE Conference on Computer Communications*, pages 926–934.

34. Noaman, M., Khan, M. S., Abrar, M. F., Ali, S., Alvi, A., and Saleem, M. A. (2022). Challenges in integration of heterogeneous internet of things. Hindawi Scientific Programming, 2022, Article ID 8626882, 1–14, https://doi.org/10.1155/2022/8626882

35. Oracle Corporation (2023). Developing Applications with Oracle Internet of Things Cloud Service, 23.3.1, https://docs.oracle.com/en/cloud/paas/iot-cloud/iotgs/developing-applications-oracle-internet-things-cloud-service.pdf, pages 1–148.

36. Qian, L. P., Wu, Y., Ji, B., Huang, L., and Tsang, D. H. K. (2019). Hybridiot: Integration of hierarchical multiple access and computation offloading for IoT-based smart cities. *IEEE Network*, 33(2):6–13.

37. Ren, Y., Yu, X., Chen, X., Guo, S., and Xue-Song, Q. (2020). Vehicular network edge intelligent management : A deep deterministic policy gradient approach for service offloading decision. In *2020 International Wireless Communications and Mobile Computing (IWCMC)*, pages 905–910.

38. Shafi, U., Mumtaz, R., Anwar, H., Qamar, A. M., and Khurshid, H. (2018). Surface water pollution detection using internet of things. In *2018 15th International Conference on Smart Cities: Improving Quality of Life Using ICT & IoT (HONET-ICT)*, pages 92–96.

39. Shafto, M., Conroy, M., Doyle, R., Glaessgen, E., Kemp, C., LeMoigne, J., and Wang, L. (2010). Draft modeling, simulation, information technology & processing roadmap, technology area 11. National Aeronautics and Space Administration (NASA).

40. Singh, J., Pasquier, T., Bacon, J., Ko, H., and Eyers, D. (2016). Twenty security considerations for cloud-supported internet of things. *IEEE Internet of Things Journal*, 3(3):269–284.

41. Sodhro, A. H., Pirbhulal, S., and de Albuquerque, V. H. C. (2019). Artificial intelligence-driven mechanism for edge computing-based industrial applications. *IEEE Transactions on Industrial Informatics*, 15(7):4235–4243.

42. Subramanian, B., Kim, J., Maray, M., and Paul, A. (2022). Digital twin model: A real-time emotion recognition system for personalized healthcare. *IEEE Access*, 10:81155–81165.

43. Sun, W., Zhang, H., Wang, R., and Zhang, Y. (2020). Reducing offloading latency for digital twin edge networks in 6G. *IEEE Transactions on Vehicular Technology*, 69(10):12240–12251.

44. Vaezi, M., Noroozi, K., Todd, T. D., Zhao, D., Karakostas, G., Wu, H., and Shen, X. (2022). Digital twins from a networking perspective. *IEEE Internet of Things Journal*, 9(23):23525–23544.

45. Van Huynh, D., Khosravirad, S. R., Masaracchia, A., Dobre, O. A., and Duong, T. Q. (2022a). Edge intelligence-based ultra-reliable and low-latency communications for digital twin-enabled metaverse. *IEEE Wireless Communications Letters*, 11(8):1733–1737.

46. Van Huynh, D., Nguyen, V.-D., Khosravirad, S. R., Sharma, V., Dobre, O. A., Shin, H., and Duong, T. Q. (2022b). Urllc edge networks with joint optimal user association, task offloading and resource allocation: A digital twin approach. *IEEE Transactions on Communications*, 70(11):7669–7682.

47. Wang, Y. and Zhao, J. (2022). Mobile edge computing, metaverse, 6G wireless communications, artificial intelligence, and blockchain: Survey and their convergence.

48. Wang, Z., Gupta, R., Han, K., Wang, H., Ganlath, A., Ammar, N., and Tiwari, P. (2022). Mobility digital twin: Concept, architecture, case study, and future challenges. *IEEE Internet of Things Journal*, 9(18):17452–17467.

49. Yi, X., Duan, Z., Li, R., Zhang, J., Li, T., and Zheng, Y. (2022). Predicting fine-grained air quality based on deep neural networks. *IEEE Transactions on Big Data*, 8(5):1326–1339.

50. Zhang, J., Li, L., Lin, G., Fang, D., Tai, Y., and Huang, J. (2020). Cyber resilience in healthcare digital twin on lung cancer. *IEEE Access*, 8:201900–201913.

51. Zhang, K., Cao, J., and Zhang, Y. (2022). Adaptive digital twin and multiagent deep reinforcement learning for vehicular edge computing and networks. *IEEE Transactions on Industrial Informatics*, 18(2):1405–1413.

52. Zheng, Y., Lu, R., Guan, Y., Zhang, S., and Shao, J. (2021). Towards private similarity query based healthcare monitoring over digital twin cloud platform. In *2021 IEEE/ACM 29th International Symposium on Quality of Service (IWQOS)*, pages 1–10.

3 A Critical Analysis of Enhanced Virtual Machine Selection and Planning Using Statistical Approaches in Cloud Data Centers

Selvam Durairaj and Rajeswari Sridhar

3.1 INTRODUCTION

A virtual machine (VM) is a computer that uses software rather than a physical computer to run programs and organize apps [1], [2]. It runs on a host server with a modified basic operating system. The levels of intelligence in the grading levels, which are lower for hardware and higher for software, realize the hardware and software in the computing system. All the physical constituents have actual belongings in the hardware levels, and the boundaries are distinct in numerous parts that are substantially associated ([3], [4], [5]). The software levels of constituents are reasonable, with fewer limitations based on the physical, individually tailored VM realized by adding a software layer to an actual machine to support the wanted VM's construction [6], [7], [8], [9], explain VMs are commonly classified into two classes namely process VM and categories system VM. The system progression is from a high level of programming to an executable code arrangement using object code. However, in the VM cloud computing environment, the situation has changed from a high level of programming to a virtual memory image arrangement known as host education format using a virtual translator or compiler. In the clouds, usually, ([73]), VMs are delineated as homogeneous VMs and heterogeneous VMs. The VM has exclusive cloud computing data center characteristics, concepts, and technologies. The process VM is given in Figure 3.1. It agrees to run a single progression as an application on a host machine. A good illustration of a process virtual machine is the Java VM (JVM). Figure 3.2 shows a system VM. It is known as a "VM monitor" (VMM) in cloud computing, and the operating system runs on the guest machine. By using virtualization software in the cloud data center (CDC), the primary physical server is separated into several virtual servers, based on server virtualization ([10], [11]).

The mapping ([12], [13]) amongst VM to physical machines (PM) or servers, and the VM allocating specific missions to VMs are what they're called "VM placement"

DOI: 10.1201/9781003390954-3

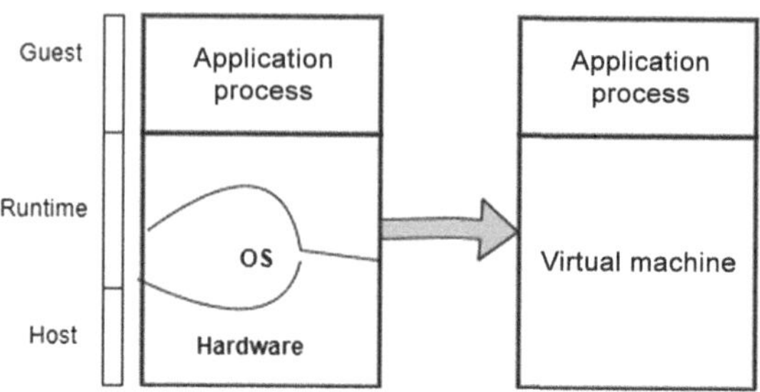

FIGURE 3.1 Process virtual machine.

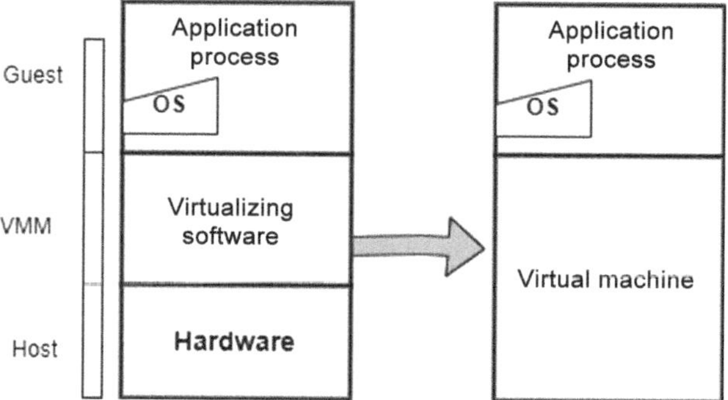

FIGURE 3.2 System virtual machine.

if a VM's mission is located on a specific PM from a virtual server or application. In the cloud environment, computing [14], [15], VM selection, and placement is essential. VMs translate computing data centers' characteristics, concepts, and technologies. This presents a comparative algorithm classification using a statistical approach of various algorithm-based VM selection and placement solutions in CDC. Furthermore, the VMSAP virtual system objectives are examined, and various VMSAP keyword factors are to VMSAP schemes.

Motivation of the field in VMSAP in CDCs continuously evolving. By conducting a survey, researchers and practitioners can identify the latest statistical approaches, algorithms, and methodologies. This survey acts as a knowledge synthesis tool, providing an overview of the state-of-the-art techniques and facilitating the adoption of innovative strategies. A survey on enhanced VM selection and placement using statistical approaches can help identify research gaps and areas for improvement. By examining the existing literature, the survey can highlight the limitations and challenges faced by previous studies.

The scope of work outlines the general structure and content for a survey on enhanced VM selection and planning using a statistical approach CDCs. The following are the key objectives of this survey:

- Identify previous studies' key concepts, methodologies, and statistical approaches.

- Explain the selection criteria for identifying relevant research papers and articles.
- Investigate various planning and optimization strategies for VM placement and allocation in CDC.
- Analyze and summarize the findings from the literature, highlighting the gaps and limitations in current research.
- Exploring different techniques and algorithms for VM selection in the CDC.

The remaining segment part of this chapter is organized as follows: Section 3.2 delineates the literature review of various survey papers depending on VMSAP. Section 3.3 presents the background of VMSAP. Section 3.4 describes the VMSAP schemes based on objectives. Section 3.5 illustrates VMSAP algorithm-based classification. Section 3.6 demonstrates the results and discussion. Finally, Section 3.7 concludes the manuscript.

3.2 SURVEY

Among the various research works on VM selection and placement, some of the most recent works are reviewed here,[16] presented the VM migration survey was presented. Cloud computing was a hopeful pattern in evaluating the oldest data technology. So, to decrease costs in administration and creativity, computing models accept elastic cloud services in a pay-per-use mode. Virtualization is used to diminish source utilization. For power organizations, server consolidation and VM migration. The VM method was exhaustive to avoid saturating network bandwidth and needed intellectual methods. [17] provided the VM migration and placement in cloud environments. VM placement and VM migration took a lot of works to attain their complex and differing aims. The presented survey evaluates numerous proposals, cloud computing contextual issues, problematic originations, and the reviewed works' benefits and failings. It failures presents the challenges of new explanations and delivers numerous open problems. For providing computing resources, cloud computing was a model that faced more tasks in managing virtualized resources.

Xu et al. [18] presented the load-balancing approaches for VM placement in cloud computing. The development of cloud computing, which depends on virtualization expertise, carries enormous chances for virtual host sources in terms of reduced cost and actual payload. But CDCs typically include various product servers hosting numerous VMs under various conditions. Considering the updating and optimization restrictions indicated in this study, the objective function determines if heterogeneous VMs possess the most noticeable difference in completion time. We created a method for updating the particle placements with load balancing [19].

Large amounts of data can be managed, decentralized, and centralized through load balancing. A load-balanced architecture for resource scheduling distribution in cloud-based healthcare systems is presented in this study. Use reinforced learning techniques like State-action-reward-state-action (SARSA), Q-learning, and Genetic Algorithm (GA) for resource scheduling [20].

Bermejo et al. [21] presented a systematic review on VM consolidation, here examining overhead manipulating factors of the VM. Based on these influences, the researchers presented a classification that categorizes the most significant vital investigation as working in virtualization, including VM merging overhead.

Le [22] suggested live VM migration methods. Here, the authors discusses the shortcomings and assets of advanced live migration methods and the influence of workload features on the appropriateness of altered migration apparatuses. Live migration was difficult to instantaneously meet the aims of reducing downtime and total migration time, and the CPU and network resources used were the critical challenges. Live VM migration is the progression of a functional VM from one physical host to another, with the least disruption to continuing facilities. Amri et al. [23] presented Interference-Aware VM Placement. Here, a thorough analysis of VM placement techniques from various VM-related research articles is offered. Cloud workers accept server consolidation and virtualization technologies to maintain energy efficiency and optimize source usage in CDCs. Server consolidation involves employing as many VMs on as few physical servers as possible, exploiting idle servers, and reducing energy intake. Also, they reviewed the best related interference-aware VM placements and gave a comparative study among them.

Khan et al. [24] presented a review of VM consolidation algorithms. The authors have categorized and analyzed VMC algorithms from various perspectives. They have resolved with upcoming instructions that it has fellow researchers' method to provide additional underwriting in this zone. Here, VM consolidation was the most prevalent research issue. In Zhang et al. [25] presented a review on VM migration. VM migration was the essential equipment for cloud management responsibilities. It releases a VM from the essential hardware. Cloud suppliers and users were features that carried sufficient profits. The authors provided an outline of VM migration and discussed both its benefits and challenges. Also, they explained the VM migration to user mobility well. Finally, they listed the open problems that must wait for solutions or additional optimizations on the live VM migration.

3.2.1 VM Placement

A group of virtual machines (VMs) that are specifically designated to run on a defined set of physical machines (PMs) [26], [27], [28]. It is known as VM placement to boost resource usage while lowering the total energy consumed using PMs on the knowledge center for mapping progression. The crucial action in any data center is the placement of the VM over the actual equipment. VM placement is the most crucial issue in every data center as shown in Figure 3.4. The clients yield their resource necessities regarding the essential resources (e.g., memory, network, and CPU) to the cloud system, and the cloud system wants to choose the resource allocation. Figure 3.3 shows the diagram of VM selection and placement. The classification of VM placement structures is given in Figure 3.5.

The VM placement systems could be classified into two types, namely static VM placement and dynamic VM placement.

1. **Static VM placement**: By the end of the VM's life, the mapping of the VMs is static. It must be recalculated for a limited period.

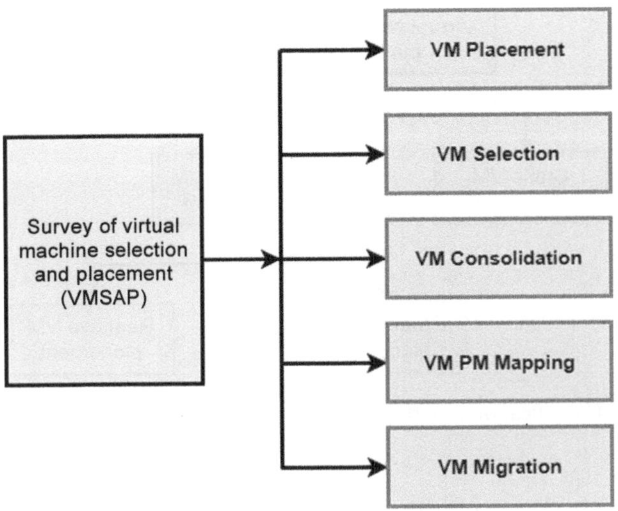

FIGURE 3.3   Survey of virtual machine selection and placement (VMSAP).

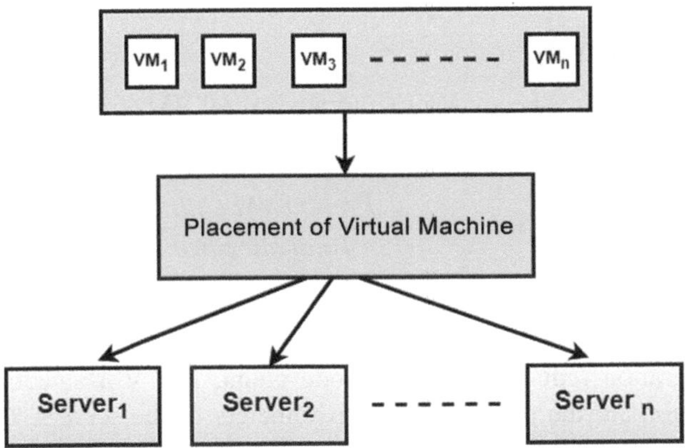

FIGURE 3.4   Virtual machine placement.

2. **Dynamic VM placement**: A few modifications in the system load allow for the preliminary placement to be modified. Adaptive VM placement and progressive VM placement are two subcategories of dynamic VM placement.

   (a) **Reactive VM placement**: Reactive VM placement creates a variation for early placement afterward; the system influences a few undesirable. The variations can be due to maintenance, efficiency, load or power problems, or specific SLA (Service-Level Agreement) SLA destructions.

   (b) **Proactive VM placement**: Proactive VM placement refers to anticipating future resource demands and deploying VMs accordingly before

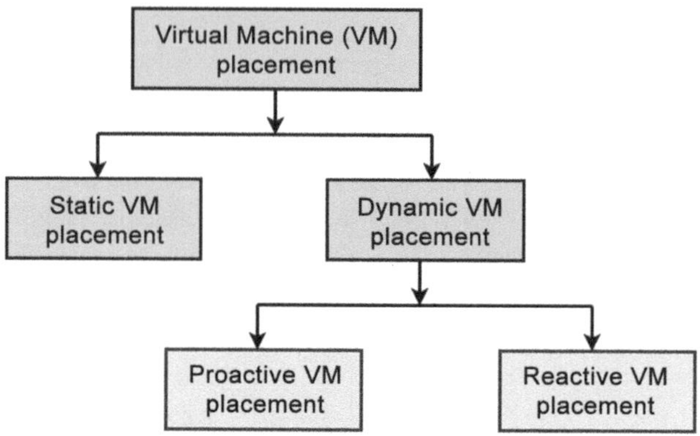

FIGURE 3.5   Classification of virtual machine placement.

those demands arise. This approach aims to optimize resource utilization, enhance performance, and preemptively address potential issues such as overloading or underutilizing resources.

3.2.2   VM SELECTION

In every data center, it is necessary to migrate several VMs from the overloaded PM to reduce their overloading hazard. Equation 3.1 gives the VM selection in the CDC.

$$VM\ Selection = \frac{Present\ VM\ Utilization}{Total\ allocated\ size} \tag{3.1}$$

where present VM utilization indicates the CPU usage of the VM, and total allocated size represents the RAM allocated to the VM. Optimal selection ([29]) of VMs for migration reduces power-on-node count. For VM selection, over- and under-utilization are the two significant parameters of the servers. If the server is underutilized, then choose every VM in the PM. If the server is overutilized, then choose the VM for migration.

3.2.3   VM CONSOLIDATION

VM consolidation is essential to enhance energy efficiency in cloud computing systems. [30] discussed that VM consolidation is broadly used for the following tasks to decrease energy consumption, to increase resource application, to decrease the cost of cloud providers, to assure maximal performance, and to avoid service level agreement (SLA) violations. Further ([31]), to improve resource utilization and power efficiency inside CDCs, VM consolidation involves live migration of VMs among actual servers [20]. Some tasks are directed to improve the VM consolidation progression to reduce resource wastage. Live migration delivers the capability to transfer VMs (VMs) without ending them. This method [32] is called VM consolidation, which involves efficiently decreasing the power feeding of a CDC by shutting down vacant servers. A VM consolidation technique that is both energy-efficient and quality of service (QoS)-aware is suggested to solve the problems. A combined prediction model based on the grey model and ARIMA(autoregressive integrated moving average) is used to detect host status. We have found the best hosts, as well as a policy known as AUMT that chooses VMs with low average CPU utilization and short migration times [33].

3.2.4 VM-PM Mapping

VM-PM mapping is the main process in CDC consumption. The mapping of ([?]) VM-PM is denoted as j^{th} VM mapped to i^{th} PM with $b_i = \{b_1, b_2, b_3...b_m\}$ request being owed to server S_j and this is a random process in the cloud. Therefore, the mapping procedure given by $w(g)$ indicates the count of VM requests acknowledged by the CDC, which is calculated

$$g(j,i) = \begin{cases} 1 & b_i \text{ is allocated } s_j \\ 0 & \text{otherwise} \end{cases} \tag{3.2}$$

The VM-PM mapping procedures run on the local source directory with changed VM and PM parameters: memory size, CPU number, and bandwidth. There are three steps in VM-PM mapping procedures. The first step is monitoring memory, CPU, and bandwidth. The second step is determining whether the PM is overloaded or underloaded and which VMs are mapped, along with resource usage calculations. The last step is listing the VM mapped to PM. The present CPU usage is CPU_i of PM_i, and it could be imitative by collective CPU usage of all VMs, and it is denoted by,

$$CPU_i = \sum_{j=1}^{m} CPU_j \tag{3.3}$$

where m denotes the VMs running on PM, the CPU usage of jth VM. The overall bandwidth is given by

$$BW_i = \sum_{j=1}^{m} BW_j \tag{3.4}$$

where, BW_j denotes the communication bandwidth of the jth VM. High bandwidth capacity among all VMs and the overall memory is calculated by,

$$MR_i = \sum_{j=1}^{m} MR_j \tag{3.5}$$

where, MR_j denotes the normalized memory of jth VM.

3.2.5 VM MIGRATION

VM migration involves transferring a VM from one physical host to another. VM migration involves moving the VM along with its contents, which may include a live process or executed content, or a combination of all. At the same time, VM placement involves the decision-making process of selecting an appropriate host for a VM based on various factors. VM migration and placement are essential in managing virtualized environments and ensuring efficient resource utilization. VM migration is a dynamic technique used in virtualized environments to optimize resource utilization, load balancing, fault tolerance, and energy efficiency. When a VM migration occurs, the state and execution context of the VM are transferred from the source host to the destination host.

The process of VM migration typically involves the following steps:

- **Live Migration:** Live migration allows VM migration while VMs run without disrupting their services or causing downtime.
- **Pre migration Checks:** Before initiating the migration, various pre migration checks are performed to ensure that the destination host meets the requirements to accommodate the VM. These checks include verifying available resources, network connectivity, and compatibility with the target host.
- **Migration Execution:** During the migration execution, the memory pages of the VM are transferred from the source host to the destination host.
- **Post-migration Checks:** After the migration completes, post-migration checks are performed to verify the successful migration of the VM.

The critical aim of VM migration is to decrease the number of PM and VM migration. It also involves improving CSP (Cloud Service Provider) profit, decreasing the number of transferred pages, preventing SLAV, reducing traffic on the DNC network, improving QoS, and minimizing energy consumption in the DC network ([34]). The VM migration metrics are given,

- VM downtime
- Application degrade time
- Preparation time
- Total migration time
- Number of transferred pages
- Resume time
- Migration overhead
- Page dirty rate
- Link speed

3.3 VMSAP SCHEMES BASED ON OBJECTIVES

After discussing the steps in mapping the VM to PM, this section details the VM placement and selection techniques. The VM placement and selection processes are

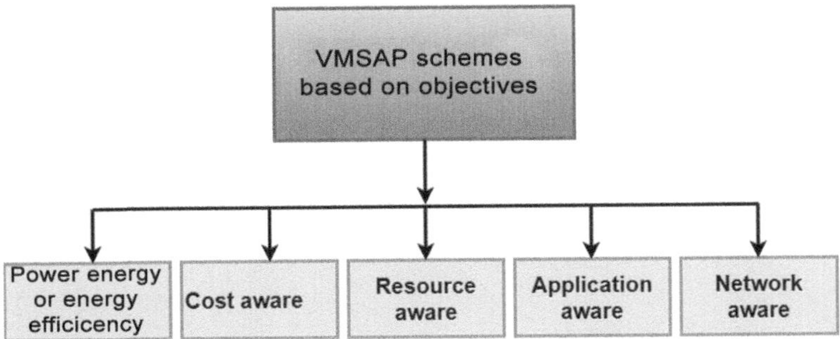

FIGURE 3.6   VMSAP schemes based on objectives.

complex operations regulating a suitable PM called VM-PM mapping. Choosing an appropriate host is essential to increasing the efficiency of Power, resource application, and QoS assistance in the cloud. This manuscript discusses the VMSAP system method based on five organizations using their objectives, as shown in Figure 3.6.

3.3.1 POWER ENERGY

Nowadays, energy efficiency is an essential task in data centers. Based on their objectives, the energy efficiency or power energy could be categorized based on their objectives as CDC power usage, server power usage, VM, PM states usage, Resource CPU utilization, and network as shown in Table 3.1. Several models are designed to decrease power consumption from various characteristics. A VM is the best utility that can improve the power efficiency in the data center [35], and an energy efficient architecture based on VMs is designed. The authors have designed virtual power to the improvement of power organization strategies. Furthermore, they have improved the virtual power management and the power state to the actual variations of the fundamental virtualized resources. An energy-efficient arrangement for higher-performance computing in the VM was proposed here. They provide an energy-saving server consolidation technology as a solution. They use less power overall as well. In ([36]), a CPU reallocation method is designed using a VM live migration technique for energy efficiency in real-time services. The CPU reallocation method chooses the migrated VM for lower CPU usage and power con-sumption. They designed the best method to reduce power usage. In ([37]), an energy-efficient VM placement optimization algorithm in CDC is proposed. This method lessens the total power consumption and selects the most power-efficient ones. Furthermore, the proposed algorithm could reduce power consumption by up to **15%**. This algorithm is available in most VMs on G5 servers, reducing power consumption in the data center.

In [38], dynamic voltage and frequency scaling (DVFS) strategy is presented to decrease power consumption. Authors provide an energy-aware dynamic placement approach that deliberates the frequency formation conferring to the distribution of VMs. The purpose of DVFS organizational strategy is to enhance energy efficiency

TABLE 3.1
Power Energy

Reference	cloud data center power usage	VM and PM states usage	Server usage	Power	Resource CPU Utilization	Network Element
[3]					√	
[9]		√				
[60]	√					
[73]		√				
[12]	√		√			√
[13]					√	
[22]	√		√			
[23]					√	
[24]			√		√	
[25]			√			
[74]					√	
[26]			√		Y	
[32]					√	
[30]		√	√			
[37]		√				
[38]					√	
[39]					√	
[40]		√			√	
[41]		√			Y	
[42]		√	√			
[57]	√	√			√	√
[39]		√				
[40]		√	√		√	√
[75]	√	√				

and mitigate degradation, leveraging investments in power consumption through a combined reduction in VM migrations and power-on procedures. They have attained more power savings and developed energy efficiency with this method. Furthermore, it increases the lifetime of the machine. In [39], numerous novel algorithms exist for VM in CDC. The primary goal is to advance the utilization of resources and decrease energy consumption. They have conducted more simulations for the validation of the proposed algorithms. The outcomes show that this algorithm meaningfully decreases the consumption of energy. The power consumption is reduced by up to **28**%. Here, an energy-efficient knee point-driven evolutionary algorithm (KnEA) is introduced. Energy-efficient KnEA is a good performance approach for solving more objective issues. The best VM placement could be a decrease in energy consumption. This algorithm is suitable for reducing power consumption in a VM.

3.3.2 Cost Aware

Table.3.2 categorizes cost awareness based on its objectives, such as VM, PM, cooling, and CDC. The power of [40] includes a cost-aware VM placement model in a cloud environment. A broad simulation is shown using the cloudsim simulator tool to

TABLE 3.2
Resource Utilization

Reference	CPU	Memory	I/O Device	RAM	Disk I/O	Network	PM
[1]		√					
[2]	√		√		√		
[11]		√					√
[74]					√		
[17]		√	√				√
[18]		√	√			√	
[28]				√			
[34]		√					
[36]		√					
[35]							√
[63]				√		√	√

TABLE 3.3
Cost aware

Reference	VM cost	PM cost	Cooling cost	Cloud Datacenter cost
[4]				√
[11]		√		
[30]			√	
[35]	√			
[48]				√
[51]	√			√
[58]				√

estimate the presented design. The results show that the proposed model performed well in minimizing costs. A greedy approach for cost-aware VM allocation in data centers is described in [41]. It is used to reduce fossil fuel energy costs. The simulation outcomes showed that the presented design is adequate to the variations in fossil fuel costs to attain a climbable presentation. In [42], the VM provisioning of data centers reduces the operational and total power costs. In [43], it is suggested that implementing a cost-aware virtual machine (VM) distribution strategy presents challenges specifically for off-grid green data centers. They proposed a greedy algorithm for reducing the cost. This algorithm can attain accessible performance while reducing fossil fuel energy costs and maintaining QoS requirements.

3.3.3 Resource Utilization

Resource utilization could be categorized based on the intentions as memory, disk input and output, CPU, RAM, input or output device, network, and PM in (Table 3.3).

It is an essential performance metric in data centers. In [44], a correlation-aware VM placement arrangement is presented. Here, the design of neural networks and factors are utilized to predict the resource application movement data conferred by historical resource application information. The simulation outcomes show that the proposed method performed well in resource utilization. In [45], a combined VM provisioning strategy for better resource utilization is presented. It can attain **45%** improvements in complete resource utilization associated with VM-by-VM provisioning. In [46], a swam algorithm was introduced to appraise the resource utilization in a VM. Here, the algorithm performs well for reducing resource utilization in a VM. The fitness function is well-determined in the introduced method as the quantity of resource utilization. In [47], an energy-efficient resource position with application factor-based VM selection was proposed. Here, they achieved better performance in resource utilization for all VMs. In [48], a resource development technique was approved to combine resource necessity estimation and the VM resource utilization procedure. For placing a VM, they proposed a genetic algorithm. In this proposed method, fewer resources are utilized in the VM. In [49], a variable resource is a requirement, and a heuristics-based server is described to reduce resource utilization through consolidation approaches. These two methods decrease resource breaking-up procedures based on resource utilization.

3.3.4 Application Aware

Application awareness can be classified based on their objectives as Data center, cloud computing environment, server configuration, PM, and VM-based, and based on the quality of service in Table.3.4. Cloud computing offers data centers, and it has many resources and various types of applications. In the medical application of resources, reallocating the resources using additional VM depends on their processes, memory, and bandwidth speed. In [50], Storage virtualization permits multiple operators to share one specific host in the cloud. Operating system virtualization is used successively to change applications on a single host. In [51], simulating optimal device VM uses the medical application of kidney disease in optimal VM selection using APSO (analogous Particle Swarm Optimization) and NN (Neural Network) based on healthcare service.

TABLE 3.4
Application aware

Reference	Data center based	Cloud Computing based	Server based	PM and VM based	QoS based
[6]	√	√		√	
[26]		√	√		√
[25]		√			√
[43]	√	√		√	

TABLE 3.5
Network Traffic

Reference	VM traffic	PM traffic	Remote cloud traffic	Minimizing data transfer time	Internal and External Gateway VM and PM
[15]	✓			✓	
[21]	✓			✓	
[44]	✓	✓			
[45]	✓	✓			✓
[46]					✓
[43]	✓				
[49]				✓	
[50]			✓	✓	
[52]	✓	✓			
[53]			✓	✓	
[54]		✓			
[55]	✓				
[62]	✓	✓			

Notes: Tick used for available and black is used for not available

3.3.5 Network Traffic

Based on their objectives, network traffic is classified based on their objectives as VM traffic, PM traffic, and remote cloud traffic. Traffic with power-aware VM placement decreases power consumption along with advanced communication traffic using the traffic cost of a VM as shown in Table 3.5. The critical aim of VM placement is to reducing network traffic and system delay. In [52], a VM deployment approach is presented depending on network traffic with energy consumption. The presented approach incorporates a huge traffic VM with two attached physical nodes to decrease the data transmission energy costs. The outcomes demonstrate that the presented approach can achieve optimum network traffic. In [53], dispensed with the rebalancing of workloads through PMs for VM migration is described. The authors' main aim is to decrease the number of migrations and network traffic. Dependent on the capacity of traffic replaced amongst the VMs, candidate VMs is chosen for migration. This technique decreases the network traffic by **81**%. In [54], VM placement based on the Multi-Optimization with Traffic-Aware algorithm (RPMOTA) is proposed. The final results show that the proposed algorithm can attain minor traffic by dynamic adjustment of the VM. [55] presented a priority, and traffic-aware method for powerfully resolving the VM placement difficulty. This method can reduce power consumption, network consumption, and resource wastage. The study results establish that the proposed system decreases the total network consumption to **29**%. [56] the authors have proposed a traffic-aware VM placement. A heuristic algorithm is proposed for minimizing the network traffic in the VM. [57] proposed an effective and efficient method for solving traffic-aware. They proposed a virtual machine

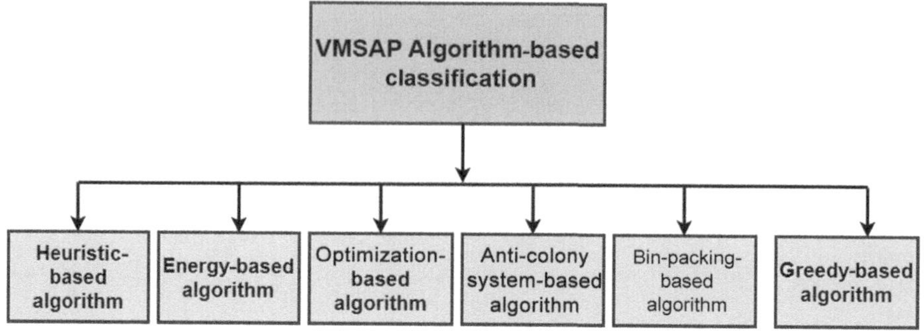

FIGURE 3.7 VMSAP algorithm based classification.

placement particle swarm optimization (VMPPSO) algorithm for reducing network traffic.

3.4 VMSAP ALGORITHM-BASED CLASSIFICATION

The VMSAP algorithm can be classified into six types on their objectives and characteristics: heuristic-based algorithms, energy-based algorithms, optimization-based algorithms, ant-colony system-based algorithms, bin-packing-based algorithms, and greed-based algorithms. VM consolidation is one of the essential objectives for power savings. It uses static and dynamic VMP and provides services based on quality methodologies from one cloud provider to another. The static VMP algorithm mentions a VMP consolidation but does not reflect VMs to the server for new destination PMs. The dynamic VMP algorithm determines the VM-to-server assignment in the VM consolidation process. Figure 3.7 shows the diagram of the VMSAP algorithm-based classification.

3.4.1 HEURISTIC-BASED ALGORITHM

In [58], a nature-inspired meta-heuristic approach for VM placement mechanisms is suggested. VM placement plays a significant part in server consolidation, and mapping is the main task between VMs and PMs in CDC. The authors survey the current VM placement mechanism using nature-inspired meta-heuristic approaches. By decreasing the number of active servers, server consolidation is the primary technique that effectively uses physical resources. In reference paper [59], the multiple objective communication-aware optimizations for VM placement in CDCs are established. Here, the authors express a different multiple objective VM placement difficulty. The established method provides a hybrid multi-objective genetic-based optimization solution to reduce the multi-objective difficulty in VMs. The proposed hybrid meta-heuristic algorithm overtakes existing ant colony optimization-based proficient heuristic-based FFD approaches from VMs to PMs. Including the random-based method in resource wastage, overall power consumption, overall data were transfer rate on the network, and the count of active servers in utilization was achieved.

In [60], an application-aware, energy-efficient VM placement in cloud networks utilizing a meta-heuristic method is introduced. The primary aim is to reduce the utilization of accessible resources while decreasing energy consumption. They have used a genetic algorithm to regulate optimal VM placement. The purpose of this method is to diminish complete resource loss by reducing power consumption and diminishing the count of active PMs. The simulation is made in a cloudsim simulator, and the proposed method is better in terms of implementation time, application, resource wastage, and energy consumption.

3.4.2 ENERGY-BASED ALGORITHMS

In [13], an energy-efficient method for VM placement is presented. The presented method addresses the difficulties of power consumption and resource depletion optimization. Here, MinPR is introduced, i.e., it minimizes the overall power consumption by decreasing the count of active PMs. They have introduced a novel resource application factor mode that accomplishes VM placement in PMs. The simulation outcomes showed that the presented method performed well compared to other methods. In [39] an energy-aware approach for VM placement is suggested. A VM is essential in data centers for developing energy efficiency. Here, GATA is suggested for the VM placement issue. The major purpose of this method is to attain energy efficiency by exploiting load balancing among numerous resources. Finally, the results showed that the suggested method is better than other methods.

3.4.3 OPTIMIZATION-BASED ALGORITHM

In [61], a hybrid multi-verse optimization algorithm for VM placement is presented. It presents a novel method for grouping different hybrid multi-objective whale optimization algorithms. The primary purpose of the presented method is to diminish power consumption. The secondary aim of this method is to reduce resource depletion and maintain possessions by utilizing the optimum placement of VMs over PMs in CDC. The outcomes from the presented method are likened to those from the existing methods. In [62], various algorithms for VM placement are described. Here, they describe the random fit algorithm depending on resource wastage, the best-fit algorithm depending on resource wastage, and an evolutionary algorithm for differential evolution. The outcomes display the differential evolution algorithm with three different transformation methods. By using a random number generator, the data sets are created.

In [63], the particle swarm optimization algorithm based on the multi-objective VM placement algorithm is introduced. This algorithm can diminish packing efficiency while reducing energy consumption. The simulation outcomes show that the introduced algorithm yields better results when compared to other methods. The introduced algorithm needs a lower count of PMs, lower energy consumption, and better-balanced utilization of PMs. In [64], a whale optimization algorithm (WAO) for bandwidth-efficient VM placement in cloud computing conditions is suggested.

Here an innovative bandwidth distribution strategy is suggested, including hybridization and a better variation of WOA named enhanced Levy-based WOA. To test the authority of the presented method, the Cloudsim toolkit is applied. They compared the suggested method with four meta-heuristic algorithms: WOA, PSO, GA, and ITHS. The suggested method applies to medical diagnosis and healthcare delivery establishments.

3.4.3.1 Ant-Colony System-Based Algorithm

In [65], the ant colony method for energy-efficient dynamic VM placement on data centers is introduced. Here, the energy-efficient resolution to the optimization is complex, an ant colony method surrounded by innovative heuristics. To prove the efficiency of the introduced method, they have directed simulation experiments on small-scale, medium-scale, and large-scale data centers. So, the introduced method is the best tool for VM placement in virtualized data. In [66], the ant colony method for VM placement is presented. They presented a modified ant colony optimization method based on chaotic gray-wolf-optimized knowledge for VM selection and placement. Also, they have appraised the presented method using two common network topologies associated with scattered cloudlets via actual traffic workload. The results show that the presented model delivers additional consumption of physical resourcefulness on cloudlets through optimal VM placement and attains low latency and advanced, obtainable bandwidth.

3.4.4 Binpacking-Based Algorithm

In [67], a novel heuristic method in the VM placement is presented. It has established bulk bin-Packing-based VM placement with multiple-capacity bulk VM placement for VM placement and selection. The simulation outcomes on the real-time Amazon EC2 dataset, including synthetic datasets, attain about a **1.6%** decrease in PMs. In [68], a best-fit algorithm for job placement is proposed. It uses VMs that construct processing power, and the category job depends on MIPS. It uses to locate a VM to reduce the number of PMs and exploit resource utilization with decreased power ingestion in the data center. In [69], the authors presented EAGLE and the outline of online first-fit algorithms to answer the bin packing difficulty of multi dimensional space dividers classically with VM placement of energy efficiency on partition symmetrical resource application. The online first fit algorithm is locally optimal if PM has an excellent resource for the VM host and VM place for PM.

3.4.5 Greedy-Based Algorithm

In [70], the establishment of the random greedy random technique in VM placement in cloud servers is discussed. Greedy randomized VM placement stimulates the VMs in the additional power-efficient PM to enhance CDC energy consumption and source consumption. Artificial and real-world manufacturing situations estimate the proficiency of the proposed model with numerous performance matrices. The research outcomes authorize that greedy randomized VM placement together enhances power usage and the complete wastage of source consumption. In [71], the minimum power greedy VM placement is presented. Minor, power-greedy VM placements are the best answer for data center distribution and energy depletion. The greedy algorithm has two methods. The first is Min Power, which challenges reducing power depletion by circumventing and powering unnecessary servers. The second is in Min communication, whose goals are stuffing one user's VM as much as possible on a similar server or below the server stand to reduce network energy costs.

3.5 DISCUSSION

VMSAP methods and VMP migration are a wide-ranging research area with numerous optimizations and objectives, like single or static optimization and multi- or dynamic optimization, on the research of VM placement and VM security in a CDC. It requires CDC cloud users and providers to work together owing to the many parameters. This section suggests the analysis of several criteria, such as timeline, static and dynamic VMSAP approaches, analysis of VMSAP algorithm-based classification, analysis of the VMSAP scheme based on objectives, and analysis of the VMSAP algorithm based on the VMSAP key factor.

3.5.1 TIMELINE ANALYSIS

This section discusses the timeline analysis of the research papers taken for the survey. For this survey, 89 research papers on VM selection and placement considered for the review were published from 2017 to 2021. Figure 3.8 shows the timeline analysis of the reviewed papers. As shown in Figure 3.8, 2 research papers were taken in 2017, 18 research papers were taken in 2018, 28 research papers were taken in 2019, 20 research papers were taken in 2020, 18 research papers were taken in 2021, 3 papers were taken from 2022 and 2 papers were taken from 2023. The popularity of VM placement can be justified by its time-related analysis, which includes resource optimization, performance enhancement, load balancing, fault tolerance, high availability, and dynamic adaptation. These factors collectively contribute to

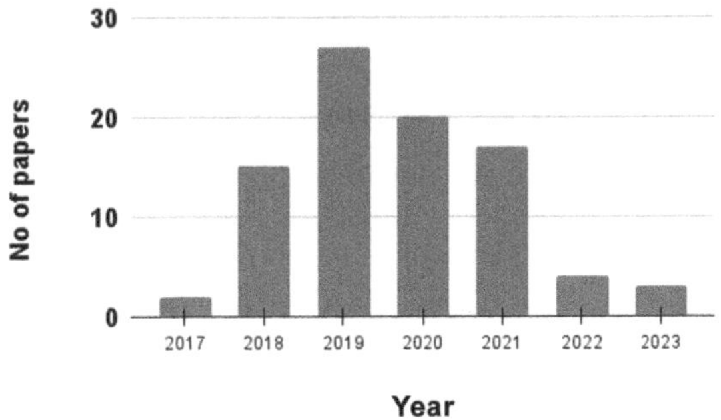

FIGURE 3.8 Timeline analysis of reviewed papers.

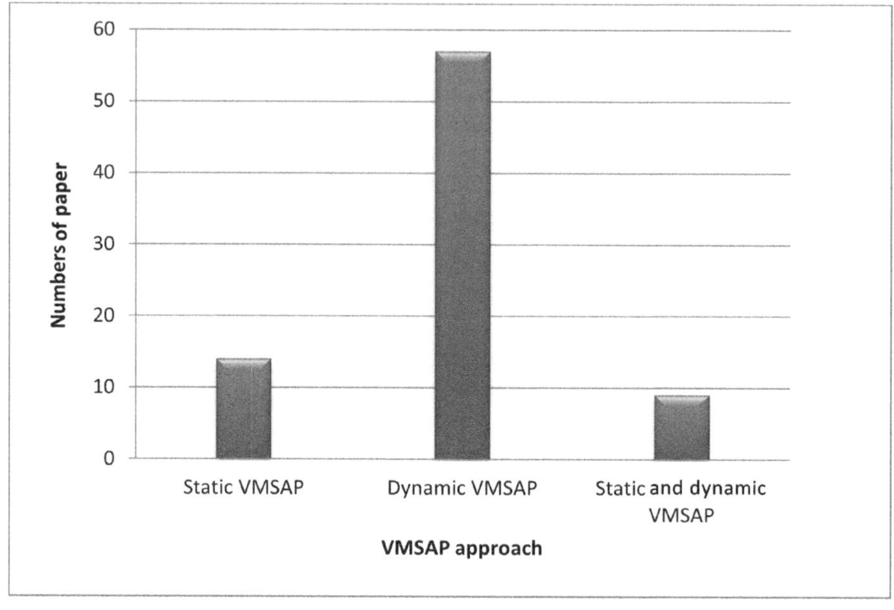

FIGURE 3.9   Static and dynamic VMSAP approach.

TABLE 3.6
VMSAP approach static and dynamic

S.No	VMSAP APPROACH	Percentage
1.	Static VMSAP approach	12%
2.	Dynamic VMSAP approach	78.67%
3.	Static and Dynamic VMSAP approach	9.33%

the efficient utilization of resources, improved performance, and flexibility in managing virtualized environments, making VM placement a widely adopted and valuable practice.

3.5.2 ANALYSIS OF STATIC AND DYNAMIC VMSAP APPROACH

This section discusses the static and dynamic approaches to VMSAP. The static VM selection and placement method is a single server with multiple PMs. The dynamic VM selection and placement method has multiple servers with multiple PMs. This chapter reviews 12 static VMSAP research papers, 78 dynamic VMSAP research papers, and 49 static and dynamic VMSAP research papers for the survey in Table 3.6 and Figure 3.9 show the analysis of the static and dynamic VMSAP approaches.

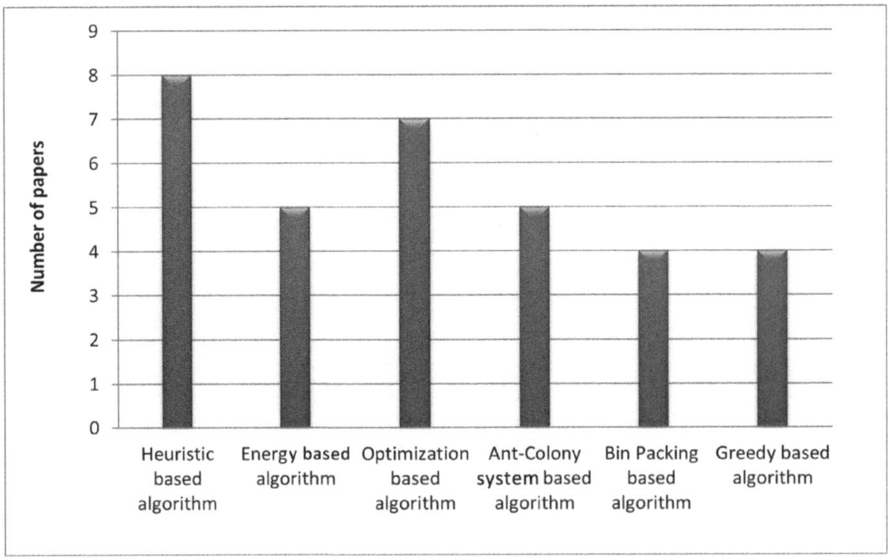

FIGURE 3.10 VMSAP scheme based on objectives.

3.5.3 Analysis of VMSAP Scheme Based on Objectives

In this section, the VMSAP scheme based on objectives is analyzed. There are five objectives: the energy efficiency scheme, the cost-aware scheme, the resource utilization scheme, the application-aware scheme, and the network traffic scheme in VMSAP. Of the papers reviewed, **44%** papers are reviewed on the energy efficiency scheme, **6%** research papers are reviewed on the cost-aware scheme; **28%** research papers are reviewed on the resource utilization scheme; **10%** research papers are reviewed on the application-aware scheme and **12%** research papers are reviewed on the network traffic scheme, as shown in Figure 3.10.

3.5.4 Analysis of VMSAP Algorithm-Based Classification

In this section, VMSAP algorithm-based classification is analyzed. This chapter uses six algorithms for the survey: a heuristic-based algorithm, an energy-based algorithm, an optimization-based algorithm, an ant-colony-based algorithm, a bin-packaging-based algorithm, and a greed-based algorithm. In Figure 3.11, eight research papers are reviewed in the heuristic-based algorithm, five research papers are reviewed in the energy-based algorithm, seven research papers are reviewed in the optimization-based algorithm, five research papers are reviewed in the ant-colony-based algorithm, four research papers are reviewed in the bin-packing-based algorithm, and four research papers are reviewed in the greed-based algorithm.

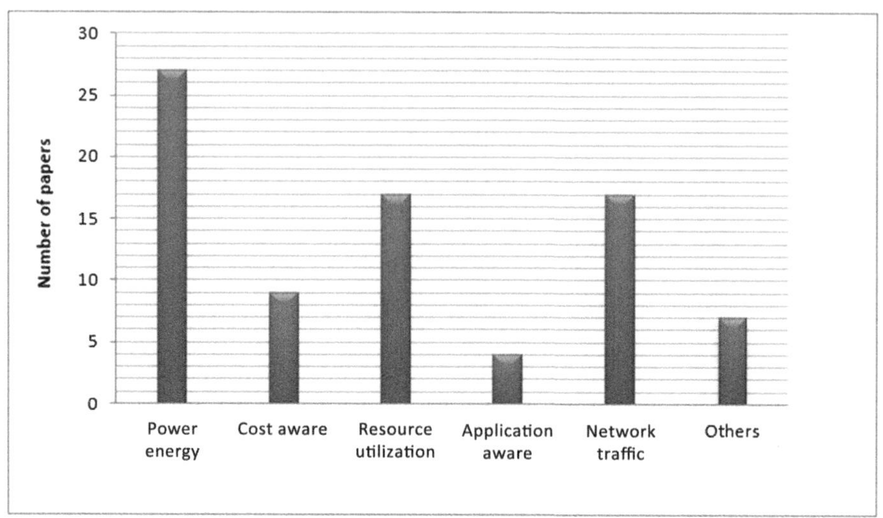

FIGURE 3.11   VMSAP algorithm-based classification.

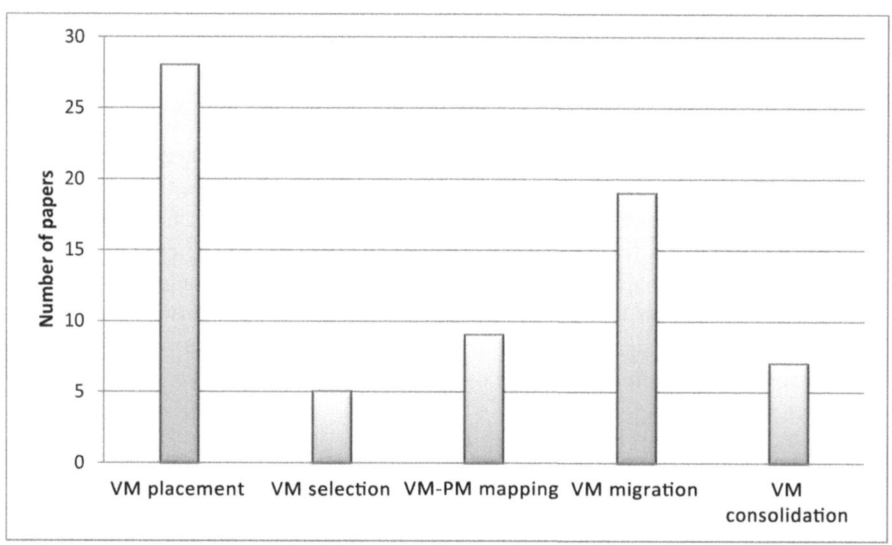

FIGURE 3.12   VMSAP algorithm-based on VMSAP key factor.

3.5.5 ANALYSIS OF VMSAP ALGORITHM BASED ON VMSAP KEY FACTOR

This section analyzes the VMSAP algorithm based on VMSAP key factors: VM placement, VM selection, VM-PM mapping, VM migration, and VM consolidation. Figure 3.12 depicts several papers used for the survey based on VMSAP key factors. As shown in Figure 3.12, 31 research papers are reviewed in VM placement; 6 research papers are reviewed in VM selection; 11 research papers are reviewed in VM-PM mapping; 23 research papers are reviewed in VM migration; and nine research papers are reviewed in VM consolidation.

3.5.6 Analysis and Limitation of VMSPM Objectives

Resource-aware, network-aware, application-aware, and energy-aware factors in VM selection and placement can significantly improve the efficiency and performance of CDCs [72], [76], [77], [78]. Incorporating these factors makes it possible to enhance resource utilization, improve application performance, optimize network usage, and increase energy efficiency.

- **Resource-Aware:** Resource-aware VM selection and placement aim to allocate VMs based on their resource requirements and availability. This approach helps optimize resource utilization and minimize resource contention. By analyzing CPU, memory, storage, and other resource demands, VMs can be placed on hosts that can satisfy their requirements. Resource awareness can result in improved performance and better resource allocation efficiency.
  Limitations: Limited availability of real-time resource utilization data, dynamic workload patterns, and scalability challenges in large-scale environments can limit the effectiveness of resource-aware decisions. Predicting future resource demands accurately is challenging, and incomplete or outdated data can lead to suboptimal VM placements.
- **Network-Aware:** Network-aware VM selection and placement Consider network characteristics, such as bandwidth, latency, and topology, to optimize communication performance between VMs. By placing VMs closer to each other or allocating them to hosts with better network connectivity, network latency can be minimized, and communication bottlenecks can be avoided.
  Limitations: Gathering real-time and comprehensive network data is challenging, and the dynamic nature of network conditions adds complexity. Network topology changes, varying network loads, and limited visibility into network performance can hinder the accuracy of network-aware placement decisions. Furthermore, scalability issues arise when dealing with large-scale deployments with complex network architectures.
- **Application-Aware:** Application-aware VM selection and placement focus on matching the characteristics and requirements of applications with suitable hosting environments. By considering factors like application dependencies, performance demands, security requirements, and QoS constraints, VMs can be placed on hosts that provide the environment for optimal application execution.
  Limitations: Capturing all application-specific requirements accurately can be difficult, and the diversity of applications makes it challenging to create comprehensive models. Different applications may have conflicting requirements, and the trade-offs between application performance, resource utilization, and other factors must be carefully considered. Additionally, predicting future application demands and adapting placement decisions are complex.

- **Energy-Aware:** Energy-aware VM selection and placement aim to minimize energy consumption in CDC. By identifying energy-efficient hosts or consolidating VMs on a smaller number of hosts, it is possible to reduce power usage and operational costs. Energy-aware decisions consider power usage effectiveness (PUE), server power profiles, and workload characteristics.
  Limitations: Energy-aware placement decisions need accurate energy consumption models, which can be challenging to develop due to variations in hardware, power management mechanisms, and environmental factors. The trade-offs between energy efficiency and other objectives, such as performance or fault tolerance, must be carefully evaluated. Balancing energy efficiency with workload demands and QoS requirements is a complex task.

3.5.7 Future Direction

Machine Learning and AI Techniques: Applying machine learning and AI techniques can enhance the accuracy and effectiveness of resource-aware, network-aware, application-aware, and energy-aware VM selection and placement. These techniques can improve prediction models, optimize decision-making algorithms, and adapt to dynamic workload patterns and resource conditions.

Dynamic and Real-Time Adaptation: Future research should focus on developing more dynamic and real-time VM selection and placement approaches. By considering real-time data, monitoring information, and workload dynamics, it becomes possible to make adaptive decisions that can quickly respond to changes in resource availability, network conditions, application demands, and energy profiles.

Hybrid Approaches: Integrating multiple factors and considerations into a unified decision-making framework can lead to more robust and effective VM selection and placement. Hybrid approaches that combine statistical methods, optimization techniques, and heuristic algorithms can provide better trade-offs and balance conflicting objectives.

Resource Optimization at Edge Computing: VM selection and placement in edge environments pose unique challenges as edge computing becomes more prevalent. Future research should address the specific requirements of edge computing, such as limited resources, low-latency communication, and distributed architectures, to optimize VM placement in these contexts.

Security and Privacy Considerations: The future direction of VM selection and placement research should also incorporate security and privacy considerations. Developing algorithms that account for security requirements, isolation constraints, and data privacy concerns can enhance the overall effectiveness and reliability of the placement decisions. By addressing these areas, researchers can overcome the limitations and advance the field of resource-aware, network-aware, application-aware, and energy-aware VM selection and placement, leading to more efficient and optimized cloud resource management.

3.6   CONCLUSION

This chapter collectively reviews VM selection and placement (VMSAP). Here, 89 research papers related to VMSAP from the IEEE, Springer Link, Elsevier, IET, and Google Scholar surveys are assessed. Geometric analysis is provided in this review, which was conducted by taking information from 89 research papers published between 2017 and 2023. It reviews various algorithms used in VM selection and placement and five different organizations based on the objectives of the VMSAP system. Also, it gives an essential background and information about VM selection and placement problems in a CDC. Here, several criteria are analyzed, such as timeline, static and dynamic VMSAP approach analysis, VMSAP algorithm-based classification analysis, VMSAP scheme based on objectives, and analysis of VMSAP algorithm based on VMSAP key factor.

Observations from VM selection and placement surveys have provided valuable insights into the strengths and weaknesses of existing algorithms. While VM placement algorithms have proven beneficial in optimizing resource utilization and improving system performance, several negative aspects have been identified with the help of Enhanced Heterogeneity Awareness, Energy Efficiency Considerations, and Lack of Heterogeneity Consideration. By addressing the limitations of existing algorithms and exploring these possibilities for expansion and improvement, VM selection and placement can be further optimized, leading to more efficient resource utilization, improved system performance, and better support for diverse application requirements.

REFERENCES

1. Ponraj, A. (2019). Optimistic virtual machine placement in cloud data centers using queuing approach. *Future Generation Computer Systems*, 93:338–344.
2. Portaluri, G., Adami, D., Gabbrielli, A., Giordano, S., and Pagano, M. (2017). Power consumption-aware virtual machine placement in cloud data center. *IEEE Transactions on Green Communications and Networking*, 1(4):541–550.
3. Luo, F., Chen, C., and Fuentes, J. (2020). An improved chemical reaction optimization algorithm for solving the shortest common supersequence problem. *Computational Biology and Chemistry*, 88:107327.
4. Mishra, S. K., Puthal, D., Sahoo, B., Jayaraman, P. P., Jun, S., Zomaya, A. Y., and Ranjan, R. (2018). Energy-efficient vm-placement in cloud data center. *Sustainable computing: informatics and systems*, 20:48–55.
5. Ghobaei-Arani, M., Rahmanian, A. A., Shamsi, M., and Rasouli-Kenari, A. (2018). A learning-based approach for virtual machine placement in cloud data centers. *International Journal of Communication Systems*, 31(8):e3537.
6. Verma, C. S., Reddy, V. D., Gangadharan, G., and Negi, A. (2017). Energy efficient virtual machine placement in cloud data centers using modified intelligent water drop algorithm. In *2017 13th International Conference on Signal-Image Technology & Internet-Based Systems (SITIS)*, pages 13–20. IEEE.
7. Mohammadhosseini, M., Toroghi Haghighat, A., and Mahdipour, E. (2019). An efficient energy-aware method for virtual machine placement in cloud data centers using the cultural algorithm. *The Journal of Supercomputing*, 75(10):6904–6933.

8. Alboaneen, D., Tianfield, H., Zhang, Y., and Pranggono, B. (2021). A metaheuristic method for joint task scheduling and virtual machine placement in cloud data centers. *Future Generation Computer Systems*, 115:201–212.

9. Jangiti, S. and VS, S. S. (2018a). Scalable and direct vector bin-packing heuristic based on residual resource ratios for virtual machine placement in cloud data centers. *Computers & Electrical Engineering*, 68:44–61.

10. Tseng, F.-H., Jheng, Y.-M., Chou, L.-D., Chao, H.-C., and Leung, V. C. (2017). Link-aware virtual machine placement for cloud services based on service-oriented architecture. *IEEE Transactions on Cloud Computing*, 8(4):989–1002.

11. Parvizi, E. and Rezvani, M. H. (2020a). Utilization-aware energy-efficient virtual machine placement in cloud networks using nsga-iii meta-heuristic approach. *Cluster computing*, 23(4):2945–2967.

12. Satpathy, A., Addya, S. K., Turuk, A. K., Majhi, B., and Sahoo, G. (2017). A resource aware vm placement strategy in cloud data centers based on crow search algorithm. In *2017 4th International Conference on Advanced Computing and Communication Systems (ICACCS)*, pages 1–6. IEEE.

13. Azizi, S., Zandsalimi, M., and Li, D. (2020b). An energy-efficient algorithm for virtual machine placement optimization in cloud data centers. *Cluster Computing*, 23(4):3421–3434.

14. Ghobaei-Arani, M., Shamsi, M., and Rahmanian, A. A. (2017). An efficient approach for improving virtual machine placement in cloud computing environment. *Journal of Experimental & Theoretical Artificial Intelligence*, 29(6):1149–1171.

15. Xu, M., Tian, W., and Buyya, R. (2017a). A survey on load balancing algorithms for virtual machines placement in cloud computing. *Concurrency and Computation: Practice and Experience*, 29(12):e4123.

16. Shirvani, M. H., Rahmani, A. M., and Sahafi, A. (2020). A survey study on virtual machine migration and server consolidation techniques in dvfs-enabled cloud datacenter: Taxonomy and challenges. *Journal of King Saud University-Computer and Information Sciences*, 32(3):267–286.

17. Silva Filho, M. C., Monteiro, C. C., Inácio, P. R., and Freire, M. M. (2018). Approaches for optimizing virtual machine placement and migration in cloud environments: A survey. *Journal of Parallel and Distributed Computing*, 111:222–250.

18. Xu, M., Tian, W., and Buyya, R. (2017b). A survey on load balancing algorithms for virtual machines placement in cloud computing. *Concurrency and Computation: Practice and Experience*, 29(12):e4123.

19. Alghamdi, M. I. (2022). Optimization of load balancing and task scheduling in cloud computing environments using artificial neural networks-based binary particle swarm optimization (bpso). *Sustainability*, 14(19):11982.

20. Jangra, A. and Mangla, N. (2023). An efficient load balancing framework for deploying resource schedulingin cloud based communication in healthcare. *Measurement: Sensors*, 25:100584.

21. Bermejo, B. and Juiz, C. (2020). Virtual machine consolidation: a systematic review of its overhead influencing factors. *The Journal of Supercomputing*, 76(1):324–361.

22. Le, T. (2020). A survey of live virtual machine migration techniques. *Computer Science Review*, 38:100304.

23. Amri, S., Brahmi, Z., Prado, R. P. d., García-Galán, S., Muñoz-Expósito, J. E., and Marchewka, A. (2018). Interference-aware virtual machine placement: A survey. In *International Conference on Image Processing and Communications*, pages 237–244. Springer.

24. Khan, M. A., Paplinski, A., Khan, A. M., Murshed, M., and Buyya, R. (2018). Dynamic virtual machine consolidation algorithms for energy-efficient cloud resource management: a review. *Sustainable cloud and energy services*, pages 135–165.

25. Zhang, F., Liu, G., Fu, X., and Yahyapour, R. (2018). A survey on virtual machine migration: Challenges, techniques, and open issues. *IEEE Communications Surveys & Tutorials*, 20(2):1206–1243.

26. Cao, G. (2019a). Topology-aware multi-objective virtual machine dynamic consolidation for cloud datacenter. *Sustainable Computing: Informatics and Systems*, 21:179–188.

27. Farzai, S., Shirvani, M. H., and Rabbani, M. (2020a). Multi-objective communication-aware optimization for virtual machine placement in cloud datacenters. *Sustainable Computing: Informatics and Systems*, 28:100374.

28. Shaw, R., Howley, E., and Barrett, E. (2019). An energy efficient anti-correlated virtual machine placement algorithm using resource usage predictions. *Simulation Modelling Practice and Theory*, 93:322–342.

29. Nwe, K. M., Oo, M. K., and Htay, M. M. (2018). Efficient resource management for virtual machine allocation in cloud data centers. In *2018 IEEE 7th Global Conference on Consumer Electronics (GCCE)*, pages 419–420. IEEE.

30. Ashraf, A., Byholm, B., and Porres, I. (2018a). Distributed virtual machine consolidation: A systematic mapping study. *Computer Science Review*, 28:118–130.

31. Calheiros, R. N., Ranjan, R., Beloglazov, A., De Rose, C. A., & Buyya, R. (2011). CloudSim: A toolkit for modeling and simulation of cloud computing environments and evaluation of resource provisioning algorithms. *Software: Practice and Experience*, 41(1), 23–50.

32. Amri, S., Brahmi, Z., Prado, R. P. d., García-Galán, S., Muñoz-Expósito, J. E., and Marchewka, A.(2018). Interference-aware virtual machine placement: A survey. In *International Conference on Image Processing and Communications*, pages 237–244. Springer.

33. Shaw, R., Howley, E., & Barrett, E. (2019). An energy efficient and interference aware virtual machine consolidation algorithm using workload classification. In *Service-Oriented Computing: 17th International Conference, ICSOC 2019, Toulouse, France, October 28–31, 2019, Proceedings 17*, pages 251–266, Springer International Publishing.

34. Paulraj, G. J. L., Francis, S. A. J., Peter, J. D., and Jebadurai, I. J. (2018). A combined forecast-based virtual machine migration in cloud data centers. *Computers & Electrical Engineering*, 69:287–300.

35. Azizi, S., Zandsalimi, M., and Li, D. (2020c). An energy-efficient algorithm for virtual machine placement optimization in cloud data centers. *Cluster Computing*, 23(4):3421–3434.

36. Arroba, P., Moya, J. M., Ayala, J. L., and Buyya, R. (2017). Dynamic voltage and frequency scaling-aware dynamic consolidation of virtual machines for energy efficient cloud data centers. *Concurrency and Computation: Practice and Experience*, 29(10):e4067.

37. Khoshkholghi, M. A., Derahman, M. N., Abdullah, A., Subramaniam, S., and Othman, M. (2017). Energy-efficient algorithms for dynamic virtual machine consolidation in cloud data centers. *IEEE Access*, 5:10709–10722.

38. Ye, X., Yin, Y., and Lan, L. (2017). Energy-efficient many-objective virtual machine placement optimization in a cloud computing environment. *IEEE access*, 5:16006–16020.

39. Zhao, D.-M., Zhou, J.-T., and Li, K. (2019a). An energy-aware algorithm for virtual machine placement in cloud computing. *IEEE Access*, 7:55659–55668.

40. Zhu, T., Wang, H., and Wei, H. (2017). Cost-aware virtual machine allocation for off-grid green data centers. In *2017 IEEE International Conference on Pervasive Computing and Communications Workshops (PerCom Workshops)*, pages 291–295. IEEE.

41. Rawas, S., Zekri, A. S., and El Zaart, A. (2018). Power and cost-aware virtual machine placement in geo-distributed data centers. In *CLOSER*, pages 112–123.

42. Xu, X., Zhang, Q., Maneas, S., Sotiriadis, S., Gavan, C., and Bessis, N. (2019). Vmsage: a virtual machine scheduling algorithm based on the gravitational effect for green cloud computing. *Simulation Modelling Practice and Theory*, 93:87–103.

43. Li, H., Li, W., Wang, H., and Wang, J. (2018a). An optimization of virtual machine selection and placement by using memory content similarity for server consolidation in cloud. *Future Generation Computer Systems*, 84:98–107.

44. Chen, T., Zhu, Y., Gao, X., Kong, L., Chen, G., and Wang, Y. (2018). Improving resource utilization via virtual machine placement in data center networks. *Mobile Networks and Applications*, 23(2):227–238.

45. Mekala, M. S. and Viswanathan, P. (2019). Energy-efficient virtual machine selection based on resource ranking and utilization factor approach in cloud computing for iot. *Computers & Electrical Engineering*, 73:227–244.

46. Li, H., Wang, S., and Ruan, C. (2019). A fast approach of provisioning virtual machines by using image content similarity in cloud. *IEEE Access*, 7:45099–45109.

47. Dabhi, D., & Thakor, D. (2022). Energy and SLA-aware VM placement policy for VM consolidation process in cloud data centers. In Sustainable Technology and Advanced Computing in Electrical Engineering: Proceedings of ICSTACE 2021, pages 351-365. Springer Nature Singapore.

48. Lopez-Pires, F., Baran, B., Benítez, L., Zalimben, S., and Amarilla, A. (2018). Virtual machine placement for elastic infrastructures in overbooked cloud computing datacenters under uncertainty. *Future Generation Computer Systems*, 79:830–848.

49. Xiao, Z. and Ming, Z. (2019b). A state based energy optimization framework for dynamic virtual machine placement. *Data & Knowledge Engineering*, 120:83–99.

50. Selvaraj, A., Patan, R., Gandomi, A. H., Deverajan, G. G., and Pushparaj, M. (2019). Optimal virtual machine selection for anomaly detection using a swarm intelligence approach. *Applied soft computing*, 84:105686.

51. Zhang, T. and Lee, R. B. (2017). Design, implementation and verification of cloud architecture for monitoring a virtual machine's security health. *IEEE Transactions on Computers*, 67(6):799–815.

52. Luo, J., Song, W., and Yin, L. (2018). Reliable virtual machine placement based on multi-objective optimization with traffic-aware algorithm in industrial cloud. *IEEE Access*, 6:23043–23052.

53. Omer, S., Azizi, S., Shojafar, M., and Tafazolli, R. (2021). A priority, power and traffic-aware virtual machine placement of iot applications in cloud data centers. *Journal of systems architecture*, 115:101996.

54. Liu, X., Cheng, B., Yue, Y., Wang, M., Li, B., and Chen, J. (2019). Traffic-aware and reliability-guaranteed virtual machine placement optimization in cloud datacenters. In *2019 IEEE 12th International Conference on Cloud Computing (CLOUD)*, pages 91–98. IEEE.

55. Jobava, A., Yazidi, A., Oommen, B. J., and Begnum, K. (2018). On achieving intelligent traffic-aware consolidation of virtual machines in a data center using learning automata. *Journal of computational science*, 24:290–312.

56. Kanniga Devi, R., Murugaboopathi, G., and Muthukannan, M. (2018). Load monitoring and system-traffic-aware live vm migration-based load balancing in cloud data center using graph theoretic solutions. *Cluster Computing*, 21(3):1623–1638.

57. Hu, B., Chen, S., Chen, J., and Hu, Z. (2016). A mobility-oriented scheme for virtual machine migration in cloud data center network. *IEEE Access*, 4:8327–8337.

58. Donyagard Vahed, N., Ghobaei-Arani, M., and Souri, A. (2019). Multiobjective virtual machine placement mechanisms using nature-inspired metaheuristic algorithms in cloud environments: A comprehensive review. *International Journal of Communication Systems*, 32(14):e4068.

59. Alhammadi, A. S. A., & Vasanthi, V. (2021, July). Multi-objective algorithms for virtual machine selection and placement in cloud data center. In 2021 International Congress of Advanced Technology and Engineering (ICOTEN), pages 1–7. IEEE.

60. Zhang, X., Zhao, Y., Guo, S., and Li, Y. (2017). Performance-aware energy-efficient virtual machine placement in cloud data center. In *2017 IEEE International Conference on Communications (ICC)*, pages 1–7. IEEE.

61. Gharehpasha, S., Masdari, M., and Jafarian, A. (2021). Virtual machine placement in cloud data centers using a hybrid multi-verse optimization algorithm. *Artificial Intelligence Review*, 54(3):2221–2257.

62. Adamuthe, A. C. and Patil, J. T. (2018). Differential evolution algorithm for optimizing virtual machine placement problem in cloud computing. *International Journal of Intelligent Systems & Applications*, 10(7).

63. Braiki, K. and Youssef, H. (2018). Multi-objective virtual machine placement algorithm based on particle swarm optimization. In *2018 14th International Wireless Communications & Mobile Computing Conference (IWCMC)*, pages 279–284. IEEE.

64. Abdel-Basset, M., Abdle-Fatah, L., and Sangaiah, A. K. (2019). An improved lévy based whale optimization algorithm for bandwidth-efficient virtual machine placement in cloud computing environment. *Cluster Computing*, 22(4):8319–8334.

65. Alharbi, F., Tian, Y.-C., Tang, M., Zhang, W.-Z., Peng, C., and Fei, M. (2019). An ant colony system for energy-efficient dynamic virtual machine placement in data centers. *Expert Systems with Applications*, 120:228–238.

66. Farshin, A. and Sharifian, S. (2019). A modified knowledge-based ant colony algorithm for virtual machine placement and simultaneous routing of nfv in distributed cloud architecture. *The Journal of Supercomputing*, 75(8):5520–5550.

67. Jangiti, S. and VS, S. S. (2018b). Scalable and direct vector bin-packing heuristic based on residual resource ratios for virtual machine placement in cloud data centers. *Computers & Electrical Engineering*, 68:44–61.

68. Wei, C., Hu, Z.-H., and Wang, Y.-G. (2020). Exact algorithms for energy-efficient virtual machine placement in data centers. *Future Generation Computer Systems*, 106:77–91.

69. Abdessamia, F., Zhang, W.-Z., and Tian, Y.-C. (2020). Energy-efficiency virtual machine placement based on binary gravitational search algorithm. *Cluster Computing*, 23(3):1577–1588.

70. Azizi, S., Shojafar, M., Abawajy, J., and Buyya, R. (2020a). GRVMP: A greedy randomized algorithm for virtual machine placement in cloud data centers. IEEE Systems Journal, 15(2):2571–2582.

71. Ibrahim, A., Noshy, M., Ali, H. A., and Badawy, M. (2020). Papso: A power-aware vm placement technique based on particle swarm optimization. *IEEE Access*, 8:81747–81764.

72. Zhang, X., Zhao, Y., Guo, S., and Li, Y. (2017). Performance-aware energy-efficient virtual machine placement in cloud data center. In 2017 IEEE International Conference on Communications (ICC), pages 1–7. IEEE.

73. Tian, H., Wu, J., and Shen, H. (2017). Efficient algorithms for vm placement in cloud data centers. In *2017 18th International Conference on Parallel and Distributed Computing, Applications and Technologies (PDCAT)*, pages 75–80. IEEE.

74. Fu, X. and Zhou, C. (2017). Predicted affinity based virtual machine placement in cloud computing environments. *IEEE Transactions on Cloud Computing*, 8(1):246–255.

75. Rawas, S., Zekri, A. S., and El Zaart, A. (2018). Power and cost-aware virtual machine placement in geo-distributed data centers. In *CLOSER*, pages 112–123.

76. Fu, X. and Zhou, C. (2017). Predicted affinity based virtual machine placement in cloud computing environments. *IEEE Transactions on Cloud Computing*, 8(1):246–255.

77. Verma, C. S., Reddy, V. D., Gangadharan, G., and Negi, A. (2017). Energy efficient virtual machine placement in cloud data centers using modified intelligent water drop algorithm. In 2017 13th Inter-national Conference on Signal-Image Technology & Internet-Based Systems (SITIS), pages 13–20. IEEE.

78. Bermejo, B. and Juiz, C. (2020). Virtual machine consolidation: A systematic review of its overhead influencing factors. The Journal of Supercomputing, 76(1):324–361.

4 An Analysis of Security and Privacy in Cloud and IoT

Gayatri K. Chaturvedi, Mahesh Shirole, and Ulka Shirole

4.1 INTRODUCTION

Within the last decade, cloud computing and the Internet of Things (IoT) have become inseparable parts of our everyday lives. In recent years, several use cases have been developed among academia, government agencies, industry, and industry groups across various sectors: transportation, building, manufacturing, the distribution grid, farming, and healthcare. The cloud and IoT computing paradigms utilize a geographically distributed network of smart devices to collect large amounts of data and automate the entire ecosystem of connected devices.

Different technologies for processing and storing data in IoT are integrated through various networking protocols, including WiFi, 5G, etc. Cloud-assisted connected systems can benefit from edge-based computational resources when real-time decisions are important and bandwidth and latency are an issue.

As the adoption and broad deployment of cloud-assisted IoT systems increase, it is critical to ensure the security and privacy of such systems. Threats and attacks against these systems are growing, not only from internal adversaries but also from external entities.

Additionally, integrating clouds and edge computing broadens the attack surface by encompassing cloud, edge, and IoT-related risks and vulnerabilities. There is the risk of security and privacy violations for both data in motion and data at rest transmitted or collected from personal and pervasive smart devices. User's personal information and behavior patterns are highly sensitive data gathered by these devices and shared with other components.

This chapter presents an overview of security frameworks and privacy-preserving approaches for cloud-assisted IoT to ensure IoT data security.

4.1.1 ARCHITECTURE AND APPLICATION OF CLOUD AND IoT

Cloud computing offers a tremendous storage of memory with high computation power for outsourcing data. For example, the devices connected with cloud service upload the sensor data for data processing and provide the desired value-added services to the users. The goal of storing and processing the sensor data by cloud is to reduce the cost of implementation and provide real-time access. On-demand storage and processing services offered by the cloud enable innovation for fewer computation

DOI: 10.1201/9781003390954-4

"

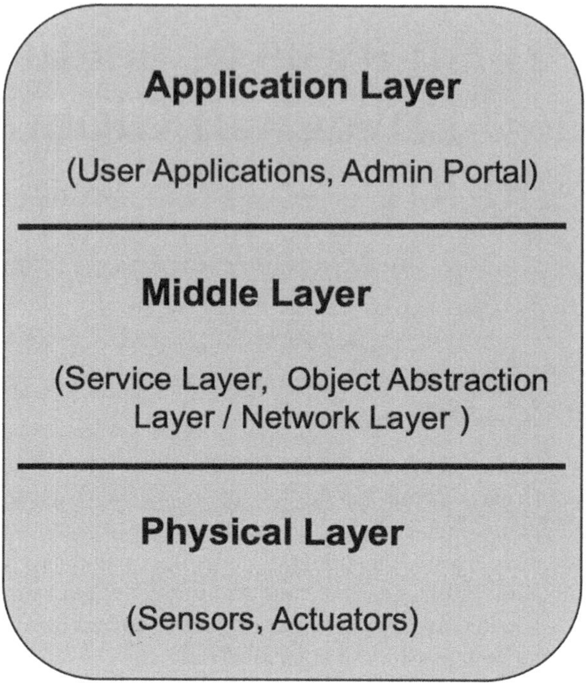

FIGURE 4.1   Three-layered IoT architecture.

devices, such as IoT, which can be used to provide robust processing of sensory data streams and new monitoring services for various applications. For example, cloud-assisted IoT helps to develop smart cities by collecting the sensor data from interior and exterior points that enable to manage systems such as healthcare, water supply systems, crime detection, waste management, and other opportunities. There is no generic IoT architecture available in the literature. Different architectures are proposed based on applications and data analysis to the users. In the literature survey, the authors proposed different IoT architectures based on the configuration model (distributed/centralized/connected). This chapter will discuss the authorization in IoT systems based on a three-layered architecture [1, 2]. As shown in Figure 4.1, components of IoT systems can be divided into three layers: application layer, middle layer, and physical layer.

1. **Application Layer:** This is the topmost layer where the end-user primarily interacts to access the offered services. This layer is responsible for communicating with the middle layer and presenting the information to the end user; for example, in the fitness tracker, the mobile application shows the statistics of calories burnt and steps taken. Apart from this, the admin portal also resides in the application layer. This admin portal allows users to set a security policy, such as in a smart home system granting a temporary guest access to a friend.

2. **Middle Layer:** This layer is subdivided into two layers: the service layer and the object abstraction layer. The service layer receives the data from the physical layer and processes that data, and presents actionable data to the application layer. Also, the middle layer processes commands passed from the user to the application layer. The abstraction layer handles the object transmission to the service layer. In another literature, the network layer replaces the object abstraction layer, which facilitates the information flow between nodes [2]. The service layer is implemented using cloud infrastructure, which receives data, stores it, and performs computations on it.

3. **Physical Layer:** This layer consists of sensors and actuators. Sensors collect information about the environment and share that information with the middle layer to act on it. The middle layer sends a command to actuators to act; for example the middle layer sends a command to the coffee machine to make a cup of coffee.

Similar to architecture layers, various architectural styles have been proposed in the literature. In this chapter, we will primarily consider the three types of architecture described below [3].

1. Centralized: Communication between the application and the physical layers always happens via the middle layer. Physical nodes are completely dependent on the middle layer for computation.

2. Distributed: Physical nodes have the computational capability. Nodes present at the physical layer can directly communicate with application nodes and provide services to the application layer.

3. Connected: Physical nodes can do a computation as well as it has a middle layer with computational capability. Both the middle layer and the physical layer nodes can directly provide services to application layer nodes.

As a part of cloud-assisted IoT, the paper [1] specified that most implementations in the literature refer to three/four essential layers to perform various functionalities, as depicted in Figure 4.2.

Alshehri and Sandhu [1] proposed an access control model as a middleware to support authentication and authorization for IoT architecture. The proposed access control (AC) model for IoT is called AC-Oriented (ACO), where the middle layer is divided into two layers: a virtual object layer and a cloud services layer. In this way, the proposed architecture has four layers: an object layer, a virtual object layer, a cloud services layer, and an application layer. Integrating IoT services with the cloud offers unlimited storage, computation, and analytics capabilities. However, it also increases the security concern of storing, accessing, and securely processing the outsourced data.

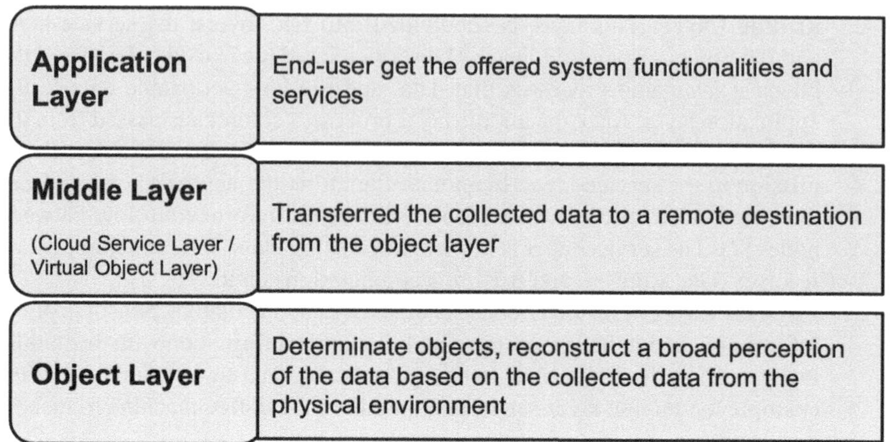

FIGURE 4.2 Layered architecture and its functionalities for cloud-assisted IoT.

4.1.2 Characteristics of Cloud and IoT

Cloud-assisted IoT is characterized by higher connectivity, scalability, and data processing capabilities due to the integration of cloud computing and IoT technologies. Cloud-assisted IoT has the following characteristics:

1. **Interconnection with Integration:** Connecting and integrating IoT systems, sensors, and devices through the cloud is called cloud-assisted IoT. Cloud infrastructure and IoT devices communicate seamlessly with each other.
2. **Dynamic Service Configuration:** IoT deployments can be scaled with cloud architectures. When IoT devices expand, and data volume increases, the cloud can allocate resources dynamically. Without expensive infrastructure investments, organizations can scale IoT systems quickly.
3. **On-Demand Storage and Processing:** Large amounts of IoT data can be stored and processed using cloud-assisted IoT. The cloud can handle large-scale data streams from multiple devices due to its ample storage capacity and computation power. The system is designed to analyze and store data in real time and perform advanced analytics on historical data.
4. **Centralized Management of IoT Devices:** The administrator can manage IoT devices effectively with cloud computing, which provides an efficient method for authentication, dynamic provisioning, and management of IoT devices.
5. **Security and Business Analytics and Insights:** With cloud-assisted IoT, valuable insights can be gained from IoT data through analytics tools and machine learning (ML) algorithms. Using real-time and historical data, we can predict maintenance, detect anomalies, optimize, and make intelligent decisions.
6. **Security and Privacy:** Security mechanisms are embedded in cloud-based IoT architectures. Data and devices connected to IoT can be protected in the

cloud by encrypting messages, controlling access to resources, and authenticating users. The administrator can use a centralized monitoring system for detecting threats, monitoring incidents, and responding to them.

7. **Cost Efficiency Based on Pay-Per-Use Model:** Cloud-assisted IoT is cost efficient because no significant upfront infrastructure investment is required. It offers a pay-per-use model with dynamic computing and storage resources adoption with less maintenance effort.

4.1.3 OUR CONTRIBUTION

This chapter aims to provide a comprehensive overview of the current studies related to the issues and challenges surrounding the security and privacy of cloud-assisted IoT. Following is a summary of the contribution of this chapter:

1. Study a security framework of cloud-assisted IoT as per the security issues related to access control and authorization, risk evaluation, service-to-service interaction, and anomaly detection.
2. Study a privacy-preserving framework with their attack model for cloud-assisted IoT that is related to the encryption-based method, anonymization-based privacy-preserving approaches, and other techniques.
3. Explore the possibilities and research directions for cloud-assisted IoT security and privacy.

This chapter is organized as follows: Section 4.2 provides a security overview of cloud and IoT. Section 4.3 discusses the security framework for cloud and IoT. Section 4.4 deals with the privacy-preserving approaches for cloud and IoT. Section 4.5 presents the future research directions for cloud-assisted IoT security and privacy. Finally, the conclusion is presented in Section 4.6.

4.2 SECURITY OVERVIEW OF CLOUD AND IOT

IoT systems connect smart devices and enable the interconnected relationship between processes, information, devices, and people by exchanging real-time data related to the participating entities for processing, storing, and deriving the decision based on the type of configured service [3]. IoT system has different applications in various domains, such as in the general-purpose use case of smart homes, where it offers a user different services in the home environment, and in the medical health domain, where it monitors patients health. In an IoT system, there is a heavy information flow between entities, posing the risk of information being stolen. If an adversary gets access to smart vehicle data flow, he/she can figure out the user's past global positioning system (GPS) information. Similarly, there is a risk of confidential medical records of patients getting disclosed.

Nodes involved in IoT systems have low computational power. To overcome this drawback, nowadays, most IoT systems rely on cloud infrastructure for computational processing, making it cloud-assisted IoT. Cloud-assisted IoT facilitates interconnected smart devices with outsourced data to the cloud, which may expose critical data to the cloud provider and increase the risk of vulnerability for compromising the security and privacy of the user. Outsourcing data from IoT devices to the cloud requires an effective security control measure to handle the data-in-transit to processing with secure storing of it. Here, cloud infrastructure serves as middleware for IoT systems that solve the problems of computational limitations. However, it also increases the surface of attacks due to the connection of the IoT system to the public Internet. This attack surface includes a physical node, user application, and web admin portal for managing a system. Due to this, another security concern of unauthorized access to the system arises. An adversary with unauthorized access to a system can harm the victim or use it for a lateral movement. Mirai Botnet is one of the prominent examples of an IoT system being used for a lateral movement to mount a distributed denial of service attack [4].

4.3 SECURITY FRAMEWORK FOR CLOUD AND IOT

4.3.1 Access Control and Authorization-Based Framework

IoT devices have limited computation power and storage capacity. To address this limitation now, IoT devices are used along with cloud platforms, making it a cloud-assisted IoT. Big cloud platform service providers like Azure, Amazon Web Services (AWS), and Google Cloud Platform (GCP) provide support for IoT. While this fixes the limited resources issues, But it also raises security and privacy concerns due to outsourcing of data and its computation.

To protect the user's privacy with authorized and controlled access to resources, researchers have proposed several access control models – User determined access control model, for example, Discretionary Access Control (DAC), Mandatory Access Control (MAC), Role-Based Access Control (RBAC), Attribute-Based Access Control (ABAC), and Usage Control (UCON). Researchers have also proposed different architectures for authorization (e.g., policy-based, token-based, and hybrid).

There is no generic model suitable for all types of IoT applications. Different models are suitable for different types of applications, such as RBAC, suitable for home IoT devices. This section performs a comparative study of different authorization architecture and access control models, and then we discuss authorization architecture in place for cloud-assisted IoT with respect to different use cases and applications. This study also focuses on the security and privacy issues of the access control models for cloud-assisted IoT with the suggestion of improving it. The most widely known access control models are the following:

1. User Determined Access Control: The end user or owner of the system acts as the administrator of the system and is responsible for creating an access control policy for the different system users. DAC is an example of user-determined access control model. In a smart home system, an end user or system owner defines what resources a friend of the user can access.
2. Mandatory Access Control: Unlike the DAC, where access is granted or revoked by the system's owner, the MAC admin defines a set of security labels used to grant access. Each user and resource is assigned to a particular security

label, and those labels are used while granting access. MAC is used mainly in government applications where security labels implement several layers of security clearances to restrict unauthorized resources and data access.

3. Role-Based Access Control: In RBAC, access permissions are defined with respect to a role, and each user is assigned a role. This forms a hierarchy of roles and permissions. The user assigned to a role inherits access permissions associated with that role. For example, a user in the hospital environment can be a doctor or a nurse, and based on these roles, access permissions to medical resources and patient data are defined.

4. Attribute-Based Access Control: ABAC provides more fine-grained access control than RBAC. ABAC model considers the attributes of subject, object, and environment while authorizing a requested operation. While the RBAC model will allow access to a patient's file to all other doctors, the ABAC model can restrict this access to only doctors assigned to a patient.

5. Usage Control: UCON is similar to ABAC. It defines policies in terms of attributes of subject and object. In addition to functionalities provided by ABAC, UCON provides two more functionalities, i.e., attribute mutability and continuity of decision. UCON considers more dynamic parameters than ABAC. This access model is based on two characteristics:

   a. Attribute Mutability: It considers the changes in the attributes of the subject and object resulting from its usage.

   b. Continuity of Decision: Access permissions are validated throughout the access time. For instance, a call from a prepaid phone is an example of this model. When the user requests a call, the system checks for the available balance; if the balance is sufficient, the call is successfully placed. The call is charged on a minute basis, so the available balance changes continuously every minute. It also checks for the access permissions after every minute to check if there is sufficient balance to keep the call active. As soon as the balance is zero, the call is dropped, indicating access to resources is denied.

4.3.1.1 Access Control Authorization Architectures

In this section, the three architectures for the implementation of above-mentioned access control models are discussed. Components of access control architecture are [2, 5]:

- Policy Enforcement Point (PEP): Protects the system from unauthorized access by consulting the policies with PDP.
- Policy Decision Point (PDP): Evaluates incoming access requests against defined policies and based on this policies it grants or denies the access.
- Policy Administration Point (PAP): Facilitates policy creation and storage.
- Policy Information Point (PIP): Source of attribute values used for evaluating access request.

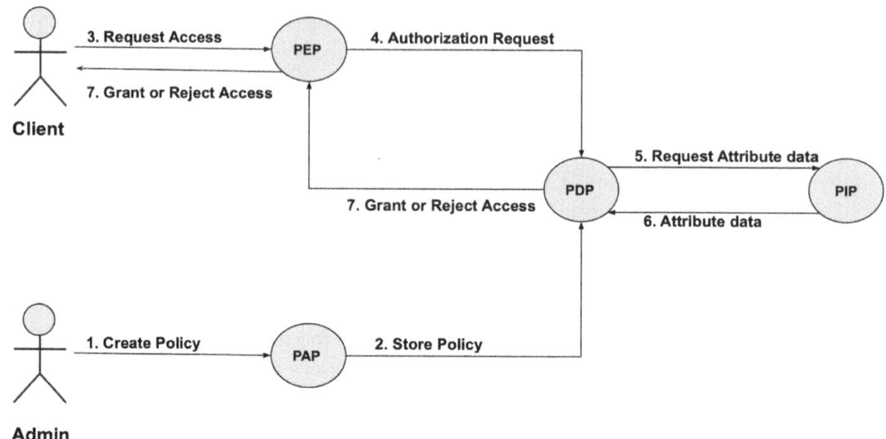

FIGURE 4.3   Policy-based architecture.

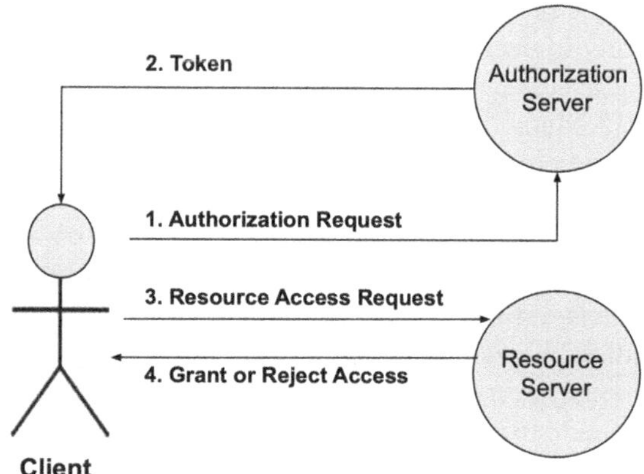

FIGURE 4.4   Token-based architecture.

Access control architecture for cloud-assisted IoT can be configured as:

1. Policy Based: Policy-based architecture [2, 5] proposed by the Extensible Access Control Markup Language (XACML) is most widely used. When PEP receives the request, it forwards the request to PDP. PDP retrieves the policies set by PAP and attributes information available to PIP as shown in Figure 4.3. Using these two pieces of information, it grants or rejects the access request of the client.

2. Token Based: Token-based system [2] is best suited for distributed architecture, allowing easy scalability. OAuth is a widely used example of this architecture. As shown in Figure 4.4, two servers are involved in this architecture. The authorization server authenticates endpoint and provides an access token, whereas the resource server acts as PEP and evaluates access request with the help of access token.

3. Hybrid Based: In token-based system, the user must provide the credential at least once to delegate the access by generating an access token. This is difficult in an IoT environment, considering a large number of devices. To address this issue, Kantara Initiative [6] has proposed User-Managed Access (UMA) which combines the policy-based and token-based architectures. In UMA authorization server can generate tokens without the end user's involvement.

AWS supports infrastructure to integrate with IoT for providing services to the users. To define a flexible fine-grained access control model, Bhatt et al. [7] have discussed ABAC scheme to improve the flexible access control for AWS-enabled IoT and called as AWS-IoTAC model. This study specified that AWS IoT supports limited attributes in access control policy. However, the use of these attributes in access control policies is limited. Therefore, the existing ABAC model cannot provide the fine-grained authorization for IoT to connect with cloud. So, the proposed ABAC is an enhancement to the AWS-IoTAC model by incorporating a complete form of ABAC. The proposed work [7] has contributed to developing a formal access control model for AWS IoT (AWS-IoTAC) and then model authorization using the access control for the smart-home IoT use case.

Google Cloud offers a cloud-assisted IoT deployment to ensure wider adoption by supporting fast data collection and processing from different objects. Gupta et al. [8], proposed a formal access model to formalize the fine-grained access control for the GCP called the GCPAC model. This model offers a mechanism to ensure the authorization of identity among the smart devices while interconnected for sharing the information. In subsequent research, the author included the IoT security criteria and extended the GCPAC model to satisfy the access control requirements for IOT system and called this proposed model as GCP IoT Access Control (GCP-IoTAC) model [8]. As a part of the application, the proposed GCP-IoTAC demonstrated applicability to two use cases for E-health use case and smart home.

4.3.2 Risk Assessment-Based Framework

Cloud computing has evolved from simple private environments to complex ecosystems with multiple coexisting public and hybrid clouds. Several factors must be considered beyond just virtualizing hardware and resources akin to a single cloud infrastructure. Cloud service providers (CSPs) require a well-balanced infrastructure to provide high Quality of Service (QoS) and minimize breaches of their service-level agreements (SLAs). Risk assessment methods using computing are a subset of risk assessment methods that incorporate the combined efforts of CSPs and consumers. Risk model development and implementation are emphasized throughout the different stages of the service life cycle, where risk is assessed. As a result of the experimental evaluation, the risk assessor proved to be successful. In recent years, cloud computing toolboxes have included it. According to its broadest definition, risk consists of the probability and consequences of an event. A lack of trust in cloud services is one of the barriers preventing consumers from adopting cloud services

[9]. The SLA between the CSP and the consumer guarantees quality service. A typical SLA details the agreement between an IoT provider and the customer regarding service levels and performance standards.

A risk assessment provides a framework for managing and anticipating risks and opportunities related to cloud infrastructure. An assessment of risk is divided into four stages:

1. Contributing to the development of a methodology for identifying cloud-related hazards.
2. Assisting in the development of dialogue among many parties.
3. Identifying potential dangers and risks, and applying mitigation measures.
4. Monitoring current risks over time and detecting emerging hazards.

The risk associated with cloud computing must be considered at every stage, in connection with the assets that must be protected. Service providers (SPs) deploy and operate services, while infrastructure providers (IPs) control admission and manage internal operations. Risk can be assessed based on categories that facilitate risk management and mitigation measures. For example, a legal issue requiring legal mitigation regarding SLA risks may be identified. Furthermore, each risk item will be evaluated according to its impact and likelihood, and established risk categories. Depending on the intensity of the risk, a risk level of 1 is extremely low, a risk level of 2 is low, a risk level of 3 is medium, and a risk level of 5 is very high. By managing risk items according to their risk levels, future risk items can be mitigated more effectively. The risk inventory will contain this information. The risk assessment process involves the evaluation of end users during service deployment and operation, internal operations, and IPs throughout the service's life cycle.

1. Service Provider: To meet the end user's requirements, the SP must ensure that the appropriate IPs are matched with the end user's needs. Consequently, the SP needs to rank IPs in accordance with their risk level. Figure 4.5 illustrates the main components of the SP: a confidence service (which can be divided into Risk Assessor and Provider Assessor), a risk inventory, and a historical database [10]. Confidence service will examine the various risks involved in IP management. This functionality will be included in the Service Deployment Optimizer (SDO). To make these judgments, the SDO will use a database of previous interactions with IPs and a risk inventory associated with each asset. Risk inventories are simple databases of risks, threats, and weaknesses associated with assets. Besides risk assessment, risk mitigation techniques are included. SDO will consider all these factors when selecting IPs to run deployed services.
2. Infrastructure Provider: When an SP requests to access the IP service, the IP evaluates the service to be deployed, thus improving the IP's performance and quality. An SLA acceptance will increase the IP's quality of service and reliability by taking fault tolerance mechanisms into account and taking steps to prevent a breach. The SP would send an information manifest to the IP providing details about the feasibility of admitting the new service. This is based

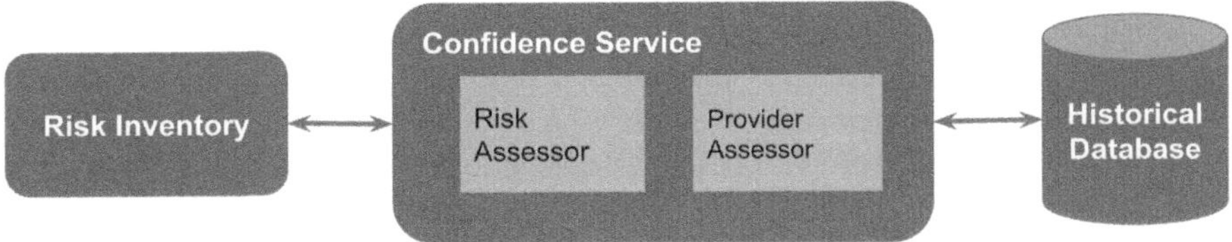

FIGURE 4.5   Service provider – Risk assessment components.

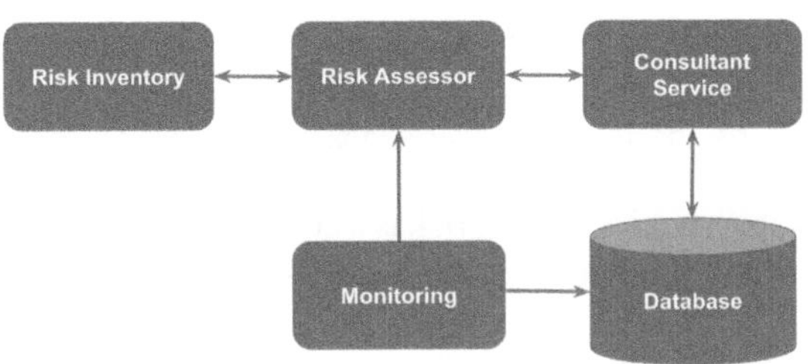

FIGURE 4.6   Infrastructure provider -- Risk assessment components.

on current infrastructure load, future capacity, and risk. Based on IP's local management policy and non functional requirements, virtual machines (VMs) can be placed where they are needed. Figure 4.6 shows how IP risk assessment components are organized [10]. It may involve using data mining techniques to examine the history of events. The consultant service can also access all monitoring data, allowing IPs to stay on top of changes even when running comparable services or working with the same SP. Resources and process data can be static or dynamic. An organization can use this type of information to determine its current workload, fix outages, boost performance for a short period, track network traffic, locate experts, or determine the number of services it offers [9].

3. Risk Inventory: In many research fields, business risk inventories have been developed to manage and evaluate risks. A risk inventory's development and improvement process varies by purpose and context. The following processes are described for developing and maintaining a risk inventory during the implementation of the framework:

   a. Decide what use case scenario to focus on.
   b. Identify the cloud's points of interaction. Several levels of interaction happen between participating entities, such as an SP interacting with end users and an IP interacting with SPs. At each level, an SLA is agreed upon, and a review is conducted to ensure that it is met.
   c. Determining whether assets need protection from external or internal threats depends on their vulnerability and threats.
   d. The risk characteristics of these assets should be determined.
   e. Analyze assets concerning other circumstances or occurrences that may mitigate risk.

In this approach, the decision to reduce or mitigate the risk is determined by the factor of the use case, the assets at risk, and the risk associated with the attack event happened. Risk can also be dynamic, changing with the situation and actions in the cloud. These might include policy changes, trades, etc. This adds a new dimension to risk mitigation techniques, as they may change over time.

4.3.2.1 Risk Assessment Models and Risk Categories

During the cloud service life cycle, the operation time influences risk assessment, allowing the risk level to fluctuate over time. To select appropriate mitigation methods for specific conditions and threats, several risk models are being investigated.

1. In the probabilistic risk model, risk is a combination of likelihood and severity. Previous difficulties determine the likelihood of a problem occurred.
2. A possibilistic risk model predicts failures of physical machines, VMs, or other systems based on stochastic processes like Gamma distributions.
3. In hybrid risk models, different hazards can be quantified by combining the two preceding models. They are designed to forecast and assess the risk of occurrence of events. In spite of the fact that certain components may have a numerical likelihood for a particular event to occur, some events may be dynamic in nature, as exposures may cause a variety of interactions between variables that spread the risk.

The novel risk assessment models are built and developed using a combination of probabilistic, possibilistic, and hybrid models for each risk category identified [10].

4.3.3 SECURE SERVICE-BASED FRAMEWORK

The service type impacts network security. In some cases, lightweight frameworks that do not consume many resources are more beneficial. In contrast, in others, it may be more helpful to use fast processing frameworks requiring higher processing resources and fewer delays. In a typical scenario, such a framework is used to facilitate authentication procedures for managing network security, and such practices consume resources and cause delays during the authentication phase. According to service requirements, these frameworks can be classified as time-constrained, resource-constrained, and hybrid frameworks:

1. Time-constrained: These service-based frameworks employ time as a critical entity in safeguarding network services and users. Such frameworks, as studied in [11], are lightweight and highly efficient in processing and evaluating security regulations. Typically, such frameworks assess the time spent authenticating users and regularly allocate connectivity up-links for data transmissions. In time-bound security frameworks, evaluation, discovery, and authentication time are critical characteristics.
2. Resource-constrained: Most intelligent IoT devices are resource-constrained and face the threat of a short lifetime. In general, the longevity of these devices

is determined by how much memory and energy they require to run the services installed on them; mandatory services consume the most memory and energy [12]. As a result, it is critical to create frameworks that focus on security while maintaining control over IoT resources. This is done with a low load on the device's operational management and activity. Such frameworks employ a checkpoint system to regulate resource usage to ensure IoT application security.

3. Hybrid: Time and resources are the significant constraints on developing intelligent applications today. Consequently, frameworks utilizing all these aspects must be designed to protect smart IoT operations. These frameworks will leverage resource checkpoints and periodic security policy evaluations to establish an integrated hybrid services-based framework for safeguarding the operation in smart IoT. Accessibility and response time are essential factors for assessing such framework performance [13].

4.3.4 Anomaly Detection-Based Framework

An anomaly detection-based framework can be considered a security control measure in an interconnected network of cloud-assisted IoT. However, all the anomaly detection mechanisms cannot be considered a straightforward solution due to the volume, variety, and velocity of data shared from IoT to the cloud. Due to the dynamic nature of IoT data, the anomaly detection mechanism in IoT has the following challenges [14]:

1. The IoT data has various patterns and sources, which causes the distorted pattern with inaccurate and noisy data.
2. The types of relationships among the data points depend on the application, which causes much difficulty to drive the pattern for anomaly detection.

Butun et al. [15] performed a survey on anomaly detection and privacy preserving mechanisms in cloud-assisted IoT. The authors used the wireless sensor network (WSN)-based anomaly detection mechanism based on the configured network, such as distributed, centralized, and hierarchical approaches. As per the literature survey in [15], Ngai et al. [16] have proposed a mechanism based on statistically based anomaly and routing pattern analysis for detecting sinkhole attacks applicable to IoT and cloud. Bhuse and Gupta [17] offer a stand-alone approach with a rule-based anomaly detection mechanism based on the network data (for MAC, routing, and network structure of IoT), which is associated with IoT and cloud.

Anomaly detection is based on analyzing data patterns from the heterogeneous IoT devices connected through Internet require a broad scale computing, storage, and processing resources in the cloud. Cloud computing offers on-demand computing resources on a pay-per-use model with different categories of services, such as Software as a Service (SaaS), Platform as a Service (PaaS), and Infrastructure as a Service (IaaS). Performing anomaly detection on a large volume of data objects, the cloud-assisted IoT must provide an efficient method for storing, processing, and evaluating the data with a combing model of private/public data and public/private

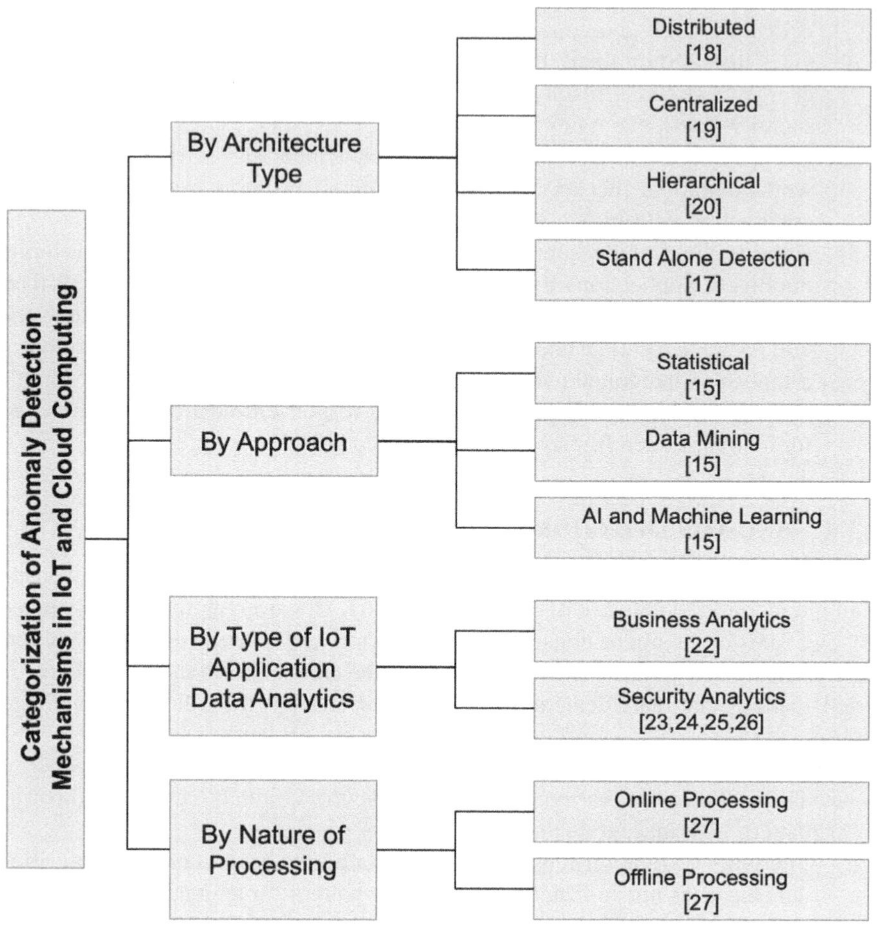

FIGURE 4.7   Categorization of anomaly detection mechanisms in IoT and cloud computing.

analytics. The anomaly detection mechanism in the cloud and IoT are categories based on the type of IoT network, approached solutions to the problem as per data sources agent, a method to implement it, and a class of data storage [15, 14] as shown in Figure 4.7.

1. By Architecture Type: The mechanism can be categorized as per IoT device networking configuration type, such as distributed, centralized, hierarchical, or stand-alone detection method. In a distributed approach, da Silva et al. [18] present rule-based anomaly detection based on the data points of receive power of IoT devices and average packet retrieved from IoT. It applies statistical anomaly to detect any attacks. As compared to distributed approach, the centralized approaches are simple to implement and require less computation cost. Ngai et al. [16] store routing pattern data points in a centralized fashion

and use a statistical detection scheme to detect sinkhole attacks. Wang et al. [19] use the computation-based approach to detect the malicious node in the IoT network by using a heuristic ranking algorithm. Hierarchical approaches support greater scalability for a heterogeneous network of IoT. Chen et al. [20] offer a rule-based hierarchical approach to detect malicious nodes by monitoring routing tables and groups of nodes. A stand-alone detection approach is implemented individually with each node of IoT network, which individually analyzes the dynamic statics of events to detect anomaly [21]. In another work Bhuse and Gupta extended the anomaly detection by extending the data points collected from different sources: MAC, routing, application layer, network layer, and transport layer [17].

2. By Approach: As per the applied approaches of the anomaly detection mechanism, the categories are further divided as statistical-based, data mining-based, and AI and ML based. The statistical-based anomaly detection mechanism uses the analysis method as univariate, multivariate time series. The data mining-based approach uses the concept of expert systems, description languages, finite state machine, data clustering, and outlier detection. AI- and ML-based anomaly use the Bayesian method, Markov method, fuzzy logic, genetic algorithm, neural network, and principal component analysis [15].

3. By Type of IoT Application Data Analytics: As the nature of IoT data, detecting anomalies can be a business function that enhances the criteria for evaluating business behavior [22]. Generally, the data related to IoT applications is collected to detect security anomalies based on connection and network framework [23], data transport security [24], device infrastructure security [25], and user security and privacy framework [26].

4. By Nature of Processing: The anomaly detection mechanism can be applied while after storing the data point in storage or at the data connection stage [27].

4.4 PRIVACY PRESERVING APPROACHES FOR CLOUD AND IOT

The recent development of interconnected IoT smart devices generates a large amount of data outsourced to cloud computing for processing and storing. However, leakage of contextual information while outsourcing sensitive data incurs compromised privacy for cloud-assisted IoT. Privacy protection mechanisms are traditionally insufficient for IoT applications due to the broader scope of data collected from non standard IoT devices [28]. The current schemes of privacy-preserving mechanisms should consider distributed cyber-security controls focusing on IoT device development with cloud vulnerabilities. As a part of current research, learning-based privacy-preserving approaches provide an excellent option to explore the IoT security that improves the detection of cyber-attacks in cloud-assisted IoT. It also provides an opportunity to consider government regulation policies while including privacy protection. As per the current literature on privacy issues of cloud-assisted IoT, Rigaki and Garcia [29] survey privacy attacks for IoT with ML approaches to mitigate them. Recently, Rodríguez et al. [30] analyzed the privacy protection mechanism for IoT.

Broadly, many survey papers focus on privacy-preserving mechanisms based on differential privacy, searchable encryption, homophobic encryption, and anonymization mechanism. The researchers have categorized the privacy-preserving mechanisms based on the groups of:

1. User/service-to-service authentication privacy attack,
2. Provides, attacks on integration with cloud computing,
3. Attacks on anonymization techniques, and
4. Attacks on data processing.

As shown in Figure 4.8, a set of proposed privacy-preserving mechanisms can be categorized as encryption methods with learning approaches, anonymization techniques, and other techniques.

4.4.1 Encryption Methods with Learning Approaches

As a part of configuring architecture with a learning model, the privacy preserving schemes can be categorized as distributed or centralized architecture-based encryption schemes. The encryption techniques such as homomorphic encryption, attribute-based encryption, searchable encryption, and multi-party computation are used widely as privacy protection mechanisms. Jiang et al. [31] proposed a centralized learning model for privacy-preserving processing over data objects of IoT. The distributed architecture with a learning model gains popularity while implementing a secure model for IoT and cloud. Shokri and Shmatikov [32] proposed a privacy-preserving mechanism based on a deep learning approach to protect the privacy of the data. To support privacy preservation, the encryption techniques integrate with distributed configuration to achieve privacy in IoT applications. The most commonly used homomorphic encryption mechanism is one in which data is encrypted and ensures privacy before outsourcing to the cloud. It prevents data leakage from cloud servers without revealing the characteristics of data. Zhang et al. [33] present a secure IoT-enabled healthcare system that uses homomorphic encryption to preserve the privacy of the IoT data.

To provide fine-grained access control over outsourcing IoT data to the cloud, a researcher uses attributed based encryption (ABE) to encrypt, and only authorized entities can use it. Xiong et al. [34] provide privacy-preserving search function with fine-grained data access control based on ABE for the cloud-assisted autonomous transport system. Searchable encryption allows the IoT device to outsource the data to the cloud securely, and keywords search functionalities enable safe search over the encrypted data for analyzing data points without revealing any information about IoT data. Oh et al. [35] suggest key aggregation searchable encryption for the data objects in a fog-enabled IoT environment. To overcome the drawback of high computation in homomorphic encryption, the multi-party computation can be used for privacy preserving in cloud-assisted IoT. Goyal and Saha [36] proposed concurrency-based communication technologies to efficiently utilize multi-party computation based on Shamir's secret sharing to support privacy preserving in resource-constrained IoT systems.

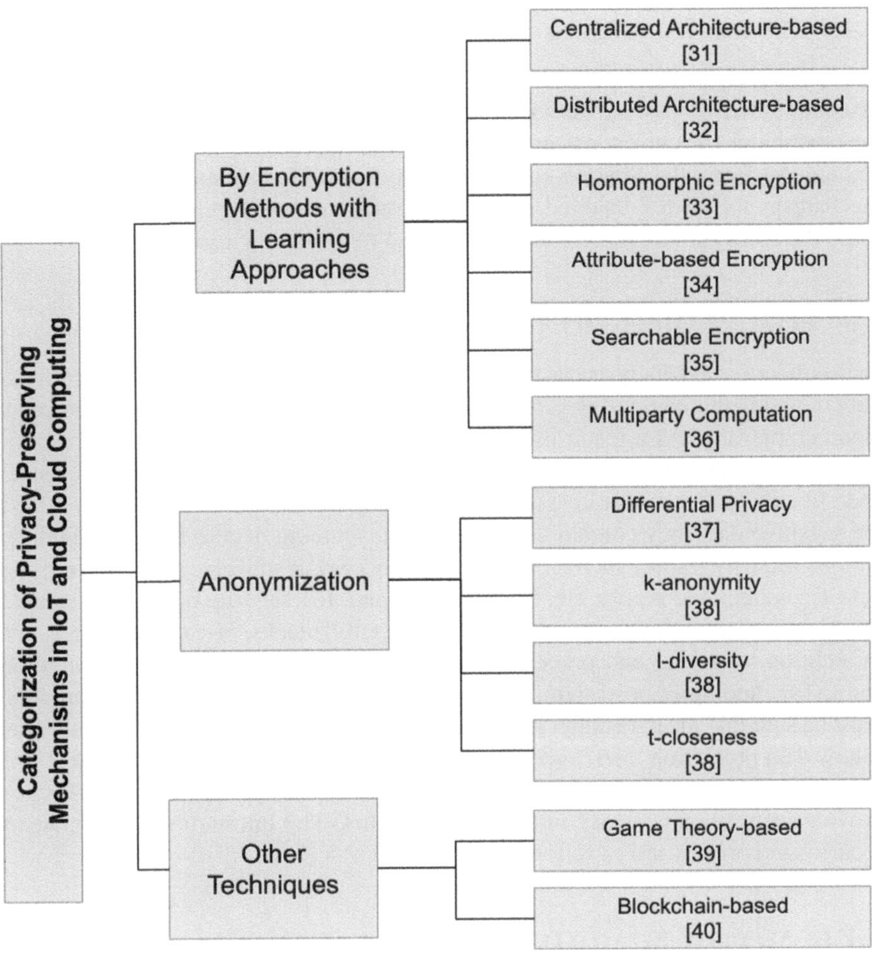

FIGURE 4.8 Categorization of privacy preserving mechanisms in IoT and cloud computing.

4.4.2 ANONYMIZATION-BASED PRIVACY-PRESERVING APPROACHES

The meaning of sensitive information in cloud-assisted models can be changed by using various anonymized techniques such as k-anonymity, l-diversity, t-closeness, and differential privacy. Jiang et al. [37] survey various differential privacy-preserving methods and explore the opportunities and applications of an IoT-enabled cloud. Sangaiah et al. [38] suggest an AI-based k-anonymity model by characterizing the entropy of the class of IoT data and enhancing the privacy preserving for cloud-assisted IoT.

4.4.3 OTHER TECHNIQUES

Li et al. [39] proposed a framework for preserving privacy by analyzing the complex association between SPs, users in the network, and adversaries in the network. On the other hand, the current research is based on blockchain-based privacy-preserving techniques. Recently, Sharma et al. [40] proposed a blockchain privacy-preserving mechanism for an IoT-enabled healthcare system where it generates interfaces to share the records in a secure container without revealing the user's privacy.

4.5 FUTURE DIRECTIONS

Data collection and analysis can be automated in IoT devices and made more efficient with it, and cloud computing provides scalable and flexible computing resources by cloud computing. As a result of technological advancements, privacy and security will become increasingly important to ensure that these technologies are used safely and effectively. It is imperative to ensure security for cloud and IoT services due to the sensitive data they contain and process. Outsourcing data to the cloud included various security issues, such as unauthorized access to sensitive data, uncontrolled data access, confidentiality, etc. On the other hand, IoT security has issues related to transport and communication attacks, data integrity attacks, privacy challenges, etc. In addition to storing and processing vast amounts of personal information, cloud and IoT technologies pose significant privacy concerns. Cloud computing providers must be sure that the data collected, processed, and stored on behalf of the customers follow data protection laws, such as the General Data Protection Regulation.

Based on the literature survey, it is still necessary to explore a variety of areas to provide security and privacy in cloud-assisted IoT. The future directions related to security and privacy are as follows:

4.5.1 SECURITY-RELATED FUTURE DIRECTIONS

Cloud-assisted IoT security is continuously evolving to address emerging threats and challenges due to the continuous development of the technology. In the coming years, the following factors will be considered as an area of security in cloud-assisted IoT:

1. Fog/Edge Cloud Security: IoT deployments increasingly use edge/fog computing for nearby data processing and analytics. Despite that, fog/edge servers and terminal IoT devices near the network's edge are primarily unprotected, making them vulnerable to many cyber-attacks. There are some attacks to which they are more susceptible, for example, stealing their devices, manipulating their devices, stealing their identities, and eavesdropping on their data. Hence, it is essential to ensure that IoT data are secure and private by using fog or edge computing. Various research areas in fog/edge computing-based IoT can be categorized as secure data system integration, secure data transmission, secure data storage, secure data computing, etc. [41].
2. Intelligent Adaptive Approaches in Security: As a result, Artificial Intelligence (AI) and ML are increasingly being recognized as essential security

tools, such as for detecting threats, detecting anomalous behavior, and analyzing human behavior. As IoT security systems become cloud-enabled, AI and ML techniques will be integrated to improve threat analysis, automate security responses, and detect anomalies [42].

3. Distributed Technology for IoT Security: The decentralized and immutable transaction records provided by blockchain technology could improve IoT security. As a result of IoT's cloud-enabled capabilities, it can securely communicate over the cloud, manage authenticated identities, authenticate devices, and ensure data integrity in an environment that enables IoT. An IoT ecosystem can be made more secure, transparent, and trustworthy by using blockchain solutions [42].

4. IoT Hardware Security: For IoT devices to be secured at the hardware level, there needs to be a solid firmware security foundation, along with side-channel attacks and hardware tampering mitigation. Hardware security modules are being integrated into IoT devices to strengthen security from the ground up, including secure hardware architectures and hardware-based encryption mechanisms.

4.5.2 PRIVACY-RELATED FUTURE DIRECTIONS

Cloud computing and IoT are both regarded as significant concerns in terms of privacy. Users may need to be made aware of how their data is being used or who has access to it if they outsource the data in the cloud to a remote server under the CSP's control. In the absence of transparency, CSPs may breach their customer's privacy. CSPs should safeguard user's privacy by implementing transparent, easy-to-understand privacy policies. In the same way, IoT devices can violate privacy as well. In addition to location, health, and behavior data, these devices can collect a wide range of personal information. Users may need to be made aware of or consent to use their data for commercial purposes or sell it to third parties without their knowledge. Privacy by design should be implemented by IoT device manufacturers in order to protect user privacy. Aside from that, users should be informed. These users should have control over how their data is used and shared and understand what is being collected about them.

Individuals need to have control over their data in order to protect their privacy in cloud-assisted IoT. Some key future directions related to privacy security in cloud-assisted IoT are presented here:

1. Privacy-Preserving Requirement for Data Sharing: It is expected that future developments will focus on enabling devices connected to the IoT to share data securely while protecting their privacy. In order to maintain privacy while facilitating data collaboration, techniques such as secure data federation, decentralized data-sharing frameworks, and data usage control mechanisms will be explored [43].

2. Privacy-Enhancing Technologies: To protect sensitive data in cloud-assisted IoT environments while facilitating meaningful analysis and collaboration,

one should continue to explore and refine advances in privacy-enhancing technologies, such as differential privacy, homomorphic encryption, secure multi party computation, and federated learning [43].

3. Data Ownership and Consent: In the future, increased rights for individuals to own their data and improved consent mechanisms for IoT enabled by the cloud will be essential. In order to protect the privacy of users, it is imperative to provide them with clear information about the collection, usage, and retention of their data and to obtain informed consent.

In the context of evolving technology and regulatory environments, these future directions highlight how privacy protection plays an increasingly important role in cloud-assisted IoT.

4.6 CONCLUSION

Cloud computing has emerged as the dominant paradigm for delivering IT services, while IoT is transforming how we interact with the physical world. However, the vast amount of data generated and processed by cloud and IoT systems, along with the complexity and heterogeneity of these systems, poses significant security and privacy challenges.

In this chapter, we have studied the security and privacy issues of cloud-assisted IoT. The security framework of cloud and IoT describes the framework with techniques for authorization and access control models for cloud-assisted IoT. Then, we have focused on a risk assessment framework for evaluating the vulnerabilities of IoT and cloud with mitigation approaches. Next, we explored the layered architecture of IoT and its functionalities. Then, we studied how the ABAC model can be used in Google Cloud and AWS Cloud to configure authorized access and enable secure data transfer between IoT devices and the cloud. We have considered the anonymization detection mechanism that helps to protect the security issues in cloud-assisted IoT.

For IoT applications to be secure, it is essential to protect user's privacy. Several methods of preserving privacy in the IoT are discussed in this chapter. This chapter surveyed various privacy-preserving techniques in IoT, such as the k-anonymity technique, secure multi party computation, attribute-based encryption, and homomorphic encryption techniques.

REFERENCES

1. A. Alshehri and R. Sandhu, "Access control models for cloud-enabled internet of things: A proposed architecture and research agenda," in *2016 IEEE 2nd International Conference on Collaboration and Internet Computing (CIC)*, pp. 530–538, IEEE, 2016.
2. S. Ravidas, A. Lekidis, F. Paci, and N. Zannone, "Access control in internet-of-things: A survey," *Journal of Network and Computer Applications*, vol. 144, pp. 79–101, 2019.
3. O. Logvinov "IEEE standard for an architectural framework for the Internet of Things (IoT)," *IEEE Std 2413-2019*, pp. 1–269, 2020.
4. M. Antonakakis, T. April, M. Bailey, M. Bernhard, E. Bursztein, J. Cochran, Z. Durumeric, J. A. Halderman, L. Invernizzi, M. Kallitsis, *et al.*, "Understanding the

mirai botnet," in *26th USENIX security symposium (USENIX Security 17)*, pp. 1093–1110, USENIX Association, 2017.

5. O. Standard, "extensible access control markup language (XACML) version 3.0," *A:(22 January 2013)*. http://docs. oasis-open. org/xacml/3.0/xacml-3.0-core-spec-os-en. html, 2013.

6. M. Machulak, J. Richer, and E. Maler, "User-managed access (UMA) 2.0 grant for OAuth 2.0 authorization," *Kantara Initiative: Richmond, VA, USA*, 2018.

7. S. Bhatt, F. Patwa, and R. Sandhu, "Access control model for AWS internet of things," in *Network and System Security: 11th International Conference, NSS 2017, Helsinki, Finland, August 21–23, 2017, Proceedings 11*, pp. 721–736, Springer, 2017.

8. D. Gupta, S. Bhatt, M. Gupta, O. Kayode, and A. S. Tosun, "Access control model for Google Cloud IoT," in *2020 IEEE 6th Intl Conference on Big Data Security on Cloud (BigDataSecurity), IEEE Intl Conference on High Performance and Smart Computing (HPSC) and IEEE Intl Conference on Intelligent Data and Security (IDS)*, pp. 198–208, IEEE, 2020.

9. R. Al Attar, J. Al-Nemri, A. Homsi, and A. Qusef, "Risk assessment for emerging domains (IoT, cloud computing, and AI)," in *2021 IEEE Jordan International Joint Conference on Electrical Engineering and Information Technology (JEEIT)*, pp. 120–127, IEEE, 2021.

10. K. Djemame, J. Padgett, I. Gourlay, and D. Armstrong, "Brokering of risk-aware service level agreements in grids," *Concurrency and Computation: Practice and Experience*, vol. 23, no. 13, pp. 1558–1582, 2011.

11. M. Tao, J. Zuo, Z. Liu, A. Castiglione, and F. Palmieri, "Multi-layer cloud architectural model and ontology-based security service framework for IoT-based smart homes," *Future Generation Computer Systems*, vol. 78, pp. 1040–1051, 2018.

12. A. F. A. Rahman, M. Daud, and M. Z. Mohamad, "Securing sensor to cloud ecosystem using Internet of Things (IoT) security framework," in *Proceedings of the International Conference on Internet of Things and Cloud Computing*, pp. 1–5, Association for Computing Machinery, 2016.

13. L. Seitz, G. Selander, and C. Gehrmann, "Authorization framework for the Internet-of-Things," in *2013 IEEE 14th International Symposium on "A World of Wireless, Mobile and Multimedia Networks" (WoWMoM)*, pp. 1–6, IEEE, 2013.

14. A. A. Cook, G. Mısırlı, and Z. Fan, "Anomaly detection for IoT time-series data: A survey," *IEEE Internet of Things Journal*, vol. 7, no. 7, pp. 6481–6494, 2019.

15. I. Butun, B. Kantarci, and M. Erol-Kantarci, "Anomaly detection and privacy preservation in cloud-centric Internet of Things," in *2015 IEEE International Conference on Communication Workshop (ICCW)*, pp. 2610–2615, IEEE, 2015.

16. E. C. Ngai, J. Liu, and M. R. Lyu, "On the intruder detection for sinkhole attack in wireless sensor networks," in *2006 IEEE International Conference on Communications*, vol. 8, pp. 3383–3389, IEEE, 2006.

17. V. Bhuse and A. Gupta, "Anomaly intrusion detection in wireless sensor networks," *Journal of High Speed Networks*, vol. 15, no. 1, pp. 33–51, 2006.

18. A. P. R. da Silva, M. H. T. Martins, B. P. S. Rocha, A. A. F. Loureiro, L. B. Ruiz, and H. C. Wong, "Decentralized intrusion detection in wireless sensor networks," in *Proceedings of the 1st ACM International Workshop on Quality of Service & Security in Wireless and Mobile Networks*, Q2SWinet '05 (New York, NY, USA), pp. 16–23, Association for Computing Machinery, 2005.

19. C. Wang, T. Feng, J. Kim, G. Wang, and W. Zhang, "Catching packet droppers and modifiers in wireless sensor networks," *IEEE Transactions on Parallel and Distributed Systems*, vol. 23, no. 5, pp. 835–843, 2012.

20. R.-C. Chen, C.-F. Hsieh, and Y.-F. Huang, "A new method for intrusion detection on hierarchical wireless sensor networks," in *Proceedings of the 3rd International Conference on Ubiquitous Information Management and Communication*, pp. 238–245, Association for Computing Machinery, 2009.

21. I. Onat and A. Miri, "A real-time node-based traffic anomaly detection algorithm for wireless sensor networks," in *2005 Systems Communications (ICW'05, ICHSN'05, ICMCS'05, SENET'05)*, pp. 422–427, IEEE, 2005.

22. T. Asakura, W. Yashima, K. Suzuki, and M. Shimotou, "Anomaly detection in a logistic operating system using the mahalanobis–taguchi method," *Applied Sciences*, vol. 10, no. 12, pp. 4376, 2020.

23. Y. An, F. R. Yu, J. Li, J. Chen, and V. C. Leung, "Edge intelligence (Ei)-enabled http anomaly detection framework for the Internet of Things (IoT)," *IEEE Internet of Things Journal*, vol. 8, no. 5, pp. 3554–3566, 2020.

24. W. Song, M. Beshley, K. Przystupa, H. Beshley, O. Kochan, A. Pryslupskyi, D. Pieniak, and J. Su, "A software deep packet inspection system for network traffic analysis and anomaly detection," *Sensors*, vol. 20, no. 6, 2020.

25. A. M. Said, A. Yahyaoui, and T. Abdellatif, "Efficient anomaly detection for smart hospital IoT systems," *Sensors*, vol. 21, no. 4, p. 1026, 2021.

26. F. Cauteruccio, L. Cinelli, E. Corradini, G. Terracina, D. Ursino, L. Virgili, C. Savaglio, A. Liotta, and G. Fortino, "A framework for anomaly detection and classification in multiple IoT scenarios," *Future Generation Computer Systems*, vol. 114, pp. 322–335, 2021.

27. I. Hafeez, M. Antikainen, A. Y. Ding, and S. Tarkoma, "IoT-keeper: Detecting malicious IoT network activity using online traffic analysis at the edge," *IEEE Transactions on Network and Service Management*, vol. 17, no. 1, pp. 45–59, 2020.

28. J. Sen and S. Dasgupta, "Data privacy preservation on the internet of things," *arXiv preprint arXiv:2304.00258*, 2023.

29. M. Rigaki and S. Garcia, "A survey of privacy attacks in machine learning," *arXiv preprint arXiv:2007.07646*, 2020.

30. E. Rodríguez, B. Otero, and R. Canal, "A survey of machine and deep learning methods for privacy protection in the Internet of Things," *Sensors*, vol. 23, no. 3, p. 1252, 2023.

31. L. Jiang, L. Chen, T. Giannetsos, B. Luo, K. Liang, and J. Han, "Toward practical privacy-preserving processing over encrypted data in IoT: an assistive healthcare use case," *IEEE Internet of Things Journal*, vol. 6, no. 6, pp. 10177–10190, 2019.

32. R. Shokri and V. Shmatikov, "Privacy-preserving deep learning," in *Proceedings of the 22nd ACM SIGSAC Conference on Computer and Communications Security*, CCS '15 (New York, NY, USA), pp. 1310–1321, Association for Computing Machinery, 2015.

33. L. Zhang, J. Xu, P. Vijayakumar, P. K. Sharma, and U. Ghosh, "Homomorphic encryption-based privacy-preserving federated learning in IoT-enabled healthcare system," *IEEE Transactions on Network Science and Engineering*, vol. 10, no. 5, pp. 2864–2880, 2022.

34. H. Xiong, H. Wang, W. Meng, and K.-H. YehSenior Member, "Attribute-based data sharing scheme with flexible search functionality for cloud assisted autonomous transportation system," *IEEE Transactions on Industrial Informatics*, vol. 19, no. 4, pp. 10977–10986, 2023.

35. J. Oh, J. Lee, M. Kim, Y. Park, K. Park, and S. Noh, "A secure data sharing based on key aggregate searchable encryption in fog-enabled IoT environment," *IEEE Transactions on Network Science and Engineering*, vol. 9, no. 6, pp. 4468–4481, 2022.

36. H. Goyal and S. Saha, "Multi-party computation in IoT for privacy-preservation," in *2022 IEEE 42nd International Conference on Distributed Computing Systems (ICDCS)*, pp. 1280–1281, IEEE, 2022.

37. B. Jiang, J. Li, G. Yue, and H. Song, "Differential privacy for industrial Internet of Things: Opportunities, applications, and challenges," *IEEE Internet of Things Journal*, vol. 8, no. 13, pp. 10430–10451, 2021.

38. A. K. Sangaiah, A. Javadpour, F. Ja'fari, P. Pinto, and H.-M. Chuang, "Privacy-aware and AI techniques for healthcare based on k-anonymity model in Internet of Things," *IEEE Transactions on Engineering Management*, pp. 1–15, 2023. doi:10.1109/TEM.2023.3271591

39. K. Li, L. Tian, W. Li, G. Luo, and Z. Cai, "Incorporating social interaction into three-party game towards privacy protection in IoT," *Computer Networks*, vol. 150, pp. 90–101, 2019.

40. P. Sharma, S. Namasudra, N. Chilamkurti, B.-G. Kim, and R. Gonzalez Crespo, "Blockchain-based privacy preservation for IoT-enabled healthcare system," *ACM Transactions on Sensor Networks*, vol. 19, no. 3, pp. 1–17, 2023.

41. A. Hazra, P. Rana, M. Adhikari, and T. Amgoth, "Fog computing for next-generation Internet of Things: Fundamental, state-of-the-art and research challenges," *Computer Science Review*, vol. 48, p. 100549, 2023.

42. A. K. Tyagi, S. Dananjayan, D. Agarwal, and H. F. Thariq Ahmed, "Blockchain Internet of Things applications: Opportunities and challenges for industry 4.0 and society 5.0," *Sensors*, vol. 23, no. 2, p. 947, 2023.

43. L. Malina, G. Srivastava, P. Dzurenda, J. Hajny, and S. Ricci, "A privacy-enhancing framework for Internet of Things services," in *Network and System Security: 13th International Conference, NSS 2019, Sapporo, Japan, December 15–18, 2019, Proceedings 13*, pp. 77–97, Springer, 2019.

5 CoT in Smart Cities

Ramprasad Ohnu Ganeshbabu,
Mothiram Rajasekaran, and
Chitra Sabapathy Ranganathan

5.1 INTRODUCTION

In the midst of the rapid growth of the world's urban population, municipal governments have a lot of opportunities and the same level of challenges to meet their basic needs. So, the city services should improve the way to handle critical issues such as traffic congestion, crime, and climate change. Smart cities are one such technological solution that is transforming the way cities operate (see Table 5.1). Cities depend on technology to provide more efficient and effective services in response to challenges. To improve infrastructure and services, smart cities use data and insights. An urban area is defined as a smart city when it collects data and offers insights. Sensor-driven IoT (IoT) technologies allow smart cities to generate vast amounts of data and analyze the data to improve service quality. Also it provide access to real time data like traffic congestion, air quality, and energy consumption of the smart cities. Using this data, urban infrastructure can be optimized for efficiency and sustainability. For instance, data of traffic flow can be used to optimize traffic lights and reduce congestion, whereas data about energy usage can help optimize streetlight use and save energy. Smart cities bring significant challenges and take a substantial investment in information technology (IT) infrastructure and skilled personnel for municipal governments to gather, store, and analyze the sheer amount of data generated by technologies. Data security and privacy are also challenges associated with smart cities. With so much data generated and collected, sensitive information may be compromised or misused. To ensure citizen data remains secure, municipalities must invest in robust cybersecurity measures and data protection policies. Smart cities offer both opportunities and challenges to municipal governments. A smart city can improve its services by leveraging data insights. However, the stream of data creates significant challenges, including the need for investment in IT infrastructure and cybersecurity measures. Smart city initiatives must be navigated carefully by municipal governments to ensure success. The role of data in enabling IT-driven urban services and the rise of smart cities are explored in this chapter.

Municipal governments are turning to technology to streamline operations and boost efficiency. In recent years, cities have tapped new data sources to drive innovation and improve services. And 5G capabilities will make this trend stronger. Using a variety of sources of information, including cars, parking meters, surveillance cameras, traffic lights, and utility infrastructure, data is disrupting and revolutionizing civic government. Smart cities can understand their citizens' needs, prepare for future

DOI: 10.1201/9781003390954-5

TABLE 5.1
19% CAGR for smart city market size worth $3110.58 billion, globally, by 2028 [4]

Global Smart City Market	Report Scope
Market Size Value (2028)	US $3110.58 billions by 2028
Market Size Value (2022)	UD $1094.23 billions by 2022
Growth Rate	CARG of 19% from 2022 to 2028
Forecast Period	2022 to 2028
Base Year	2022

challenges, and develop feasible solutions by harnessing this data. Globally, two-thirds of people live in cities[1]. Smart technology is being adopted by cities. As of 2023, the global smart city market is expected to reach $223.3 billion according to forecasts [2]. Smart cities collect, manage, and analyze vast amounts of data generated by IoT devices, sensors, and municipal infrastructure. Artificial intelligence (AI) and 5G networks support a data-driven environment. Monitoring traffic and parking in real time can help civic leaders identify key population trends, improve utilities, enhance security, and promote economic development. With IoT and AI-based solutions, supported by 5G connectivity, this data will grow exponentially in average cities. Cities need an IT infrastructure to handle the volume and variety of the data generated from various sources to extract value. To meet future challenges, cities are embracing IoT, AI, and other data-based solutions today. To better organize and manage resources, they analyze multiple large data sets and analyze their relationships. In addition, they aim to add context to the data by combining sensor data from devices with other internal and external data.

5.1.1 What the Future Holds: Cities Embracing IoT, AI, and Other Data-Based Solutions

As cities around the world embrace technology to enhance the quality of life of their citizens, many of them are adopting it to improve the quality of life of their citizens. Data-driven solutions, including the IoT and AI, are being used in order to create smart cities that can monitor traffic, manage utilities, enhance security, and promote economic development. As cities become more efficient and responsive to the needs of their residents, the investments made today are likely to pay dividends in the future by fulfilling their residents' needs. A total value of $2.57 trillion will be reached by 2025 for smart city products and services, with a growth rate of 18.4% on average [3].

- **High-Volume Areas: Monitor Traffic, Parking, and Other** – Many cities experience significant problems with traffic congestion and parking, causing commuters frustration and time loss. By using IoT and AI solutions, cities can monitor traffic patterns and identify congestion areas in real time, adjust traffic signals, and reroute traffic based on those findings. In addition to alerting drivers about available parking spots, IoT sensors can guide them to their destination, reducing the time spent looking for parking.

- Improve Utility Infrastructure Management: Power and Water – The efficient management of utilities is crucial to the success of cities. Cities can use IoT sensors to monitor energy consumption and identify waste areas, which will allow them to make adjustments and save money. A smart water system can also detect leaks and monitor water consumption, reducing waste and saving resources. Smart systems also provide more flexibility, as they can easily be adjusted and controlled remotely, allowing cities to better respond to sudden changes in demand. By doing so, cities can save money and resources while providing services that are cost-effective and efficient.

- Surveillance and Monitoring in Real Time Enhance Security – IoT and AI solutions can be used to improve security by providing real time monitoring and surveillance capabilities to cities. Smart cameras can detect suspicious behavior and alert authorities, and facial recognition technology can identify potential threats as well. Security cameras with AI can be used for surveillance. In addition to facial recognition and real time monitoring, they have behavior detection features. A real time alert can also be sent using IoT sensors if an emergency occurs.

- Economic Development by Identifying Population Trends – By analyzing consumer behavior and foot traffic, cities can identify areas of high demand for services and adjust their services accordingly. By tracking population demographics, cities can also identify areas of growth, which in turn can help them invest in infrastructure and services to support that growth. Having this detailed map shows the landscape, roads, and population centers of a country or region; the government can then decide where new infrastructure should be built or where more resources should be allocated.

5.2 BUILDING INFRASTRUCTURE TO SUPPORT TECHNOLOGY IN CITIES: COLLECTING, MANAGING, AND ANALYZING DATA

Smart cities generate vast amount of data using IoT devices, AI, and 5G connectivity. It is crucial that cities have a robust IT infrastructure that can collect, manage, and analyze this data. As a result, they would be able to fully utilize this data and make smarter, more efficient cities. We explore the challenges that smart cities face with regard to dealing with the sheer volume of data generated by infrastructure, as well as to examine the key elements of a successful IT system, throughout this chapter. The average city already has to deal with terabytes of data from traffic sensors, cameras, and municipal infrastructure such as water meters. It is expected that IoT adoption and AI implementations continue to accelerate, which exponentially increase the volume of data produced by cities, and in order for these cities to glean value from all of this structured and unstructured data, they will need an IT infrastructure that can perform this task.

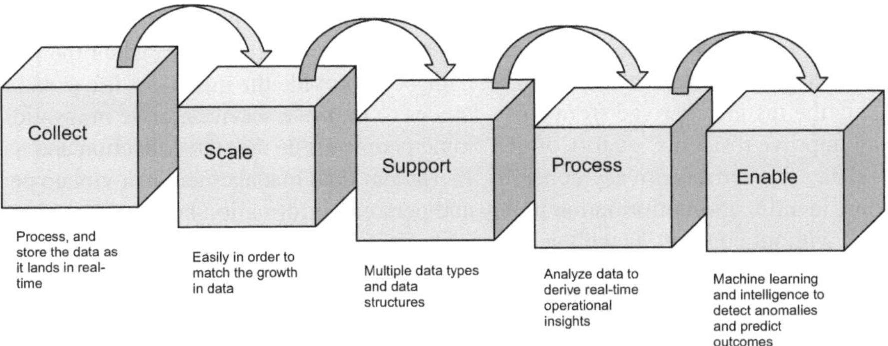

FIGURE 5.1 Collecting, managing, and analyzing data is core to modernization: building it infrastructure to support IoT, AI, and 5G connectivity in cities.

5.2.1 COLLECT DATA FROM VARIOUS SOURCES

Collect data from a variety of sources to gain comprehensive information on smart cities infrastructure and identify areas that require improvement. Collected data from sensors, cameras, traffic lights, waste management systems, and other municipal infrastructure provides valuable insights into traffic patterns, air quality, waste management, etc. The process should be standardized and easily accessible to utilize the data fully. Moreover, in addition to fixed air quality monitoring stations, vehicles can also contribute air quality data, which also needs to be standardized and readily accessible to effectively address air quality concerns. Analysts can identify patterns and trends, streamlining their work through the analysis and comparison of data from multiple sources. Without proper standardization, data becomes difficult to utilize, making it challenging to gain a comprehensive view of the city. Cities can make well-informed decisions with better infrastructure management, improved services, enhanced quality of life for the public, and a more sustainable future by collecting and standardizing data from various sources.

5.2.2 MANAGE DATA EFFECTIVELY

The massive amount of data generated by the IoT devices, sensors, and other connected devices and the cities faces the majority of challenges in data management. The ultimate key is to handle the massive amount of data generated using robust, secure, and scalable data management systems with proper data privacy and security. This means that effective data management must also include data quality and accuracy. Cities should follow established protocols and standards to ensure consistency in collecting and managing reliable and accurate data. It requires huge investment in the right infrastructure like cloud computing and big data analytics tools to store and process data effectively and it is mandatory to label, categorize, and classify the data to facilitate analysis and interpretation. With the right infrastructure, cities can store and process massive amounts of data in real time and it can allow to respond more quickly and effectively to adapt to any conditions and make informed decisions.

Cities that wish to leverage the IoT and other connected devices need effective data management. Data privacy and accuracy are guaranteed at all times when the right technologies and systems are in place. Cities can provide the quality of life possible using the insights derived from these devices to improve services, drive innovation, and improve residents' quality of life. Some people argue that the collection and use of data raise serious privacy concerns. Inefficient data management can violate personal identification information policy and personal information being accessed and used without consent, as well as inaccurate data being used to make decisions that negatively affect people.

5.2.3　ANALYZE DATA FOR INSIGHTS

To benefit from the collected data, cities should analyze it in real time. They need analytical tools to help them find trends, patterns, and actions to take. A city's ability to analyze this data effectively is also essential for leveraging its insights. A good analytical tool will allow you to mine data for trends, patterns, and actionable insights, so municipal governments can identify trends and patterns, and take action on them. As cities react to changing conditions, real time data analytics is especially important because it allows the municipal government to make informed decisions quickly. As an example, cities can identify traffic congestion patterns and adjust traffic flow accordingly, or monitor air quality levels and issue alerts and advisories. To analyze data effectively, cities may need to use advanced analytics tools and technologies. City officials can identify patterns and relationships in large data sets, which are otherwise difficult to discern using analytical tools. Getting the most value from the data that cities collect requires not only technology, but also skilled analysts and data scientists who can interpret the data and extract insights relevant to their needs. Increasing demand for data science jobs indicates economic growth. As businesses continue to rely on data-driven insights to stay competitive, the need for professionals in this field is only going to grow, creating new jobs and contributing to the economy [5]. It is possible for cities to unlock the full potential of their data by combining advanced analytics tools with skilled human analysis in order to create innovative solutions, improve services, and improve the quality of life of their citizens.

5.2.4　MAKE DATA ACCESSIBLE

All city officials, citizens, and private organizations must have access to their data if a city wants to maximize the benefits of IoT and connected devices. The data must be used to drive innovation, improve services, and improve the quality of life and it should be properly stored, organized, and secured before cities can make it accessible to the public. Data sharing is the process of the data resources available to multiple applications, users, or organizations. Therefore, data-sharing platforms must be implemented to make it simple and easy to access and share the data. With this kind of platform, an authorized user can access data in real time, collaborate on data analysis, and share insights across departments. Using different teams collective knowledge and insights, organizations can make faster, more informed decisions.

Additionally, the data-driven decisions can make better organizations. The cities can use data-sharing platforms to establish data governance and to ensure data privacy and security. Some techniques like data anonymization or other methods are required to protect personal privacy while the departments allow the shared data for public benefit. The data from the IoT and other connected devices can provide insights for private organizations to accelerate the development of new products and services and it could lead to a more sustainable and prosperous future for the city and its residents [6].

5.2.5 INCORPORATE EMERGING TECHNOLOGIES

The city should stay ahead with the latest technologies and in the advancement of AI and 5G connectivity to maximize the value of its data. In addition to providing powerful new tools for analyzing data, it can also improve the city's operations, and quality of life. Enterprise hybrid big data platform can analyze large amounts of data quickly and efficiently and it allows the cities to identify patterns, trends, and insights. Without right platforms with AI and machine learning (large language models) capabilities, the data insights will go unnoticed. AI with leveraging machine learning algorithms can make predictions for future events, identify potential problems before they occur, and optimize the operations and services of the city. Cities can achieve the process of collecting and analyzing the data in real time using 5G connectivity to provide faster, more reliable data transmission. Smart traffic management benefits the cities but if there is a delay in data transmission, then it can have significant impacts on traffic flow and safety. To take advantage of these technologies, cities must do right investment in the infrastructure and stay up to date with the technologies. Regular upgrade is required for IT systems to support new features of AI and 5G, new hardware and software must be purchased, and employees should be trained on how to use these technologies efficiently.

5.3 OVERCOMING DATA SILOS AND ENSURING SECURITY AND GOVERNANCE IN SMART CITIES

In today's data-driven world, different departments and systems are becoming increasingly reliant on the ability to share data and collaborate, and cities are becoming increasingly required to do so. While critical information is often not readily available across departments in many cities, data silos still persist. Data investments can only be fully leveraged if they integrate information from a variety of sources, such as real time traffic data, traffic pollution information, and utility and security camera data. Modernizing cities involves ensuring data security, compliance, and governance. The increasing number of devices and systems as well as the sensitive nature of some data may lead to a number of potential vulnerabilities and liability risks. Cyber-attacks can cripple a city's IT infrastructure for weeks, disrupting essential services and putting citizens' safety at risk.

Data needs to be handled, accessed, and shared efficiently and effectively by cities that have well-defined protocols and rules, as well as effective security measures and controls. In order to ensure data integrity, cities must ensure strict accountability and transparency. Established data standards, metadata management, and data monitor quality are all part of this process. To address these challenges, cities need to adopt a unified data platform that works anywhere and avoids vendor lock-in, at the same time. A city should consider where data will land and be analyzed before implementing devices and systems. A city's ability to identify opportunities for improvement and take advantage of its data depends on its ability to overcome data silos and ensure governance and security. An effective smart city transformation requires a holistic approach, including a unified data platform, effective security measures, and robust governance processes.

As cities invest in data-driven solutions, they need to integrate collected data and efficiently get rid of silos and collected data should be dispersed across diverse disconnected systems in order to realize the full potential of their investments.

5.3.1   Real Time Traffic Data

Collected real time traffic data improves the flow of traffic for cars, bike riders, and pedestrians with the help of in-road sensors and smart cameras. Data insights gained from the traffic data help to analyze the traffic patterns, congestion hotspots, and other traffic-related issues. It can optimize traffic management strategies and make a city's traffic flow more efficient. Real-time traffic data manages traffic in a timely and efficient manner by calculating the traffic signal timings minutely, rerouting traffic to less congested routes, and implementing other traffic management measures. As a result of the improvement, travel times of vehicles will be reduced, pedestrians and

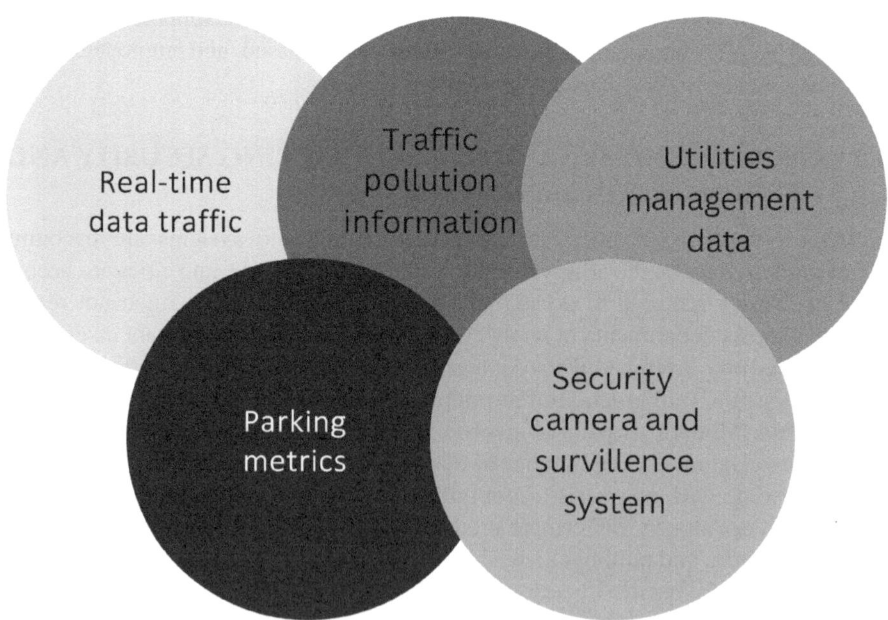

FIGURE 5.2   Smart cities data driven investments.

cyclists will be safer, and smoke emissions will be identified from the vehicles' idle state and moving state. Real-time traffic data provides real time maps like Google Maps to drivers with up-to-date information on traffic conditions like road closures, accidents, and construction, and it can help them plan their routes accordingly and make better informed decisions. As a result of the data, the drivers are able to reduce frustration and stress when they understand the traffic conditions better and learn how to navigate. Overall, real time traffic data collected from in-road sensors and smart cameras provide a valuable resource to improve traffic management strategies and the quality of life. The cities are able to create a safer, more efficient, and more sustainable transportation system for all with this data effectively.

5.3.2 Traffic Pollution Information

To reduce the carbon footprint and improve air quality, it is imperative that cities provide traffic pollution information to planners. With traffic pollution information like traffic patterns, the planners can identify times and areas where the air pollution level is high and how they can reduce it. When it comes to public transportation, traffic pollution information is particularly useful for cities to consider converting their fleet of buses and other vehicles to electric power. With the proper analysis of pollution data, planners can identify routes and areas and prioritize the deployment of electric buses and other low-emission vehicles in these areas. In addition, traffic pollution information develops targeted interventions to reduce emissions from other vehicles on the road, such as cars and trucks. This could include measures such as carpooling incentives, congestion pricing, or targeted emissions regulations for high-polluting vehicles. In general, real time traffic data and traffic pollution information mutually benefit the cities that aim to reduce their carbon footprints and improve their air quality, and create a healthier, cleaner environment for the citizens by reducing emissions.

5.3.3 Parking Metrics

Using parking metrics as a tool can provide valuable insights into how cities can allocate parking resources better with mobile applications and kiosks. With data gathered from parking metrics, demand and availability of the cities can be identified, and reallocation of resources can be done more efficiently. One of the primary applications of parking metrics is aligning parking facilities with public transportation facilities. Cities can gain an in-depth understanding of how parking can support public transportation, such as by providing limited parking for commuters. If public transportation is used in peak times like weekdays, then parking usage will be less. So cities can increase the parking availability only on weekends. This process reduces traffic congestion and promotes sustainable modes of transportation. Moreover, parking metrics can improve alignment between parking supply and daily requirements. The cities can increase parking rates and enforce policies in high-demand areas to increase the use of public transportation, or by allocating more parking resources to support specific events or activities with the parking metrics. In addition, ensuring parking resources are utilized efficiently and effectively can also reduce environmental impact by reducing the amount of time drivers spend circling for parking.

5.3.4 Data from Utilities

Data from utilities becomes a powerful tool for the cities and customers, such as knowing when water or electric usage spikes. The cities can have more control over the services, and thus become a proactive part of a smart city solution, rather than waiting for the unknown when services are either not optimized, or worse–out of commission. Data from utilities, both private and city-owned, can provide critical insights of energy usage patterns and plans for high demand during peak hours or possible supply shortages when it occurs. As a result of the data, cities can determine areas where energy consumption is particularly high and develop strategies to reduce energy usage during peak times with valuable insights. Both consumers and utilities will be able to reduce overall energy consumption and lower costs. Additionally, utility data can also be used for predicting potential outages or brownouts by analyzing weather patterns, energy demand, and utility supply. During these scenarios, cities can develop contingency plans to minimize their impact on residents and businesses with proper planning ahead. In general, utility data can be a powerful tool for cities that are seeking to optimize their energy usage and prepare for potential energy shortages. The cities can improve the resilience of their energy infrastructure with the data by reducing energy costs and environmental impact.

5.3.5 Security Cameras and Surveillance Systems

Security cameras and surveillance systems become increasingly important as part of city policing and providing law enforcement a powerful tool to prevent and identify suspects after a crime. More than capturing video footage of criminal activity, surveillance systems can also monitor public areas like streets, parks, and buildings. Security cameras can be effective tools for preventing crime, but there is the possibility that the public can raise privacy concerns. A city's use of surveillance technology must be balanced with the protection of individual rights if it is to be transparent. Developing clear policies and procedures for the use of surveillance systems and security cameras is crucial to addressing their concerns. The guidelines should specify when and where the cameras can be installed, who has the privilege to see the footage, and how long it can be retained before the deletion. The city must also provide oversight so that law enforcement officers do not violate these systems, with proper training. Security cameras and surveillance systems should be secure and not susceptible to hacking. To ensure security, strong cybersecurity measures should be implemented and systems should be regularly monitored for potential threats. While surveillance systems and security cameras can be valuable tools for city policing, privacy and civil liberties must be considered. Using this technology effectively while protecting individual rights requires clear policies and procedures, as well as ensuring that these systems are secure.

A city's data-driven investment will be fully maximized if data collected from multiple sources is integrated and made accessible across departments. Thus, vendor lock-in must be minimized by using a unified stack approach that can operate anywhere, whether on-premises or in the cloud. Data silos can result from implementing modern devices and systems without considering where data will be analyzed. As cities add devices and sensors to their infrastructure, the risks of vulnerabilities increase. In addition to integration challenges, cities must deal with data security, compliance, and governance issues. Data handling rules, such as who has access to sensitive data, when, and how, should be established by cities and public entities as a way to reduce risk. Security measures are as important as data governance in a data-centric urban model. By prioritizing data governance, cities can leverage data for civic improvement and minimize liability risks. The risk of data breaches, fines, and reputational damage is high without well-defined data handling rules. Cities can reduce liability risk and protect data by establishing clear data handling rules. Furthermore, data governance ensures that data is collected, stored, and used ethically and in accordance with privacy laws. In addition to protecting themselves from liability, cities can ensure that data is used responsibly and ethically by establishing data handling rules.

5.4 BIG DATA ANALYTICS AND MACHINE LEARNING HELP SMART CITIES

Despite their potential to revolutionize our way of living and working, smart cities will always need to evolve in response to changing technological advancements, urban challenges, and the evolving needs and preferences of their residents.

A growing amount of data generated by cities is making it increasingly difficult to gather and interpret. As urban areas add more sensors, it is harder to organize and use data gathered for specific scenarios or individuals. Through the combination of structured and unstructured data, we can find new efficiencies and solutions in real time. As a result, many other areas can be improved, including urban planning and traffic management. To maximize the benefits of big data and machine learning in an era of digitalization, cities must embrace them. With rapid urbanization and the proliferation of connected devices, the amount of data generated each day has increased exponentially. It is predicted that by 2025, 175 trillion gigabytes of data will be created, captured, and replicated, up from 33 trillion gigabytes in 2018 IDC-Seagate-DataAge-Whitepaper. In order to effectively manage cities, it is crucial to collect and analyze vast amounts of data. Organizing data and using it in certain situations can be challenging. Whenever sensors collect data, it is crucial to analyze and combine it to improve citizens' lives, find new solutions, and be more efficient. It is important for urban planners to know traffic flow, but not during peak hours. Both structured and unstructured data sets are required to make data-driven decisions that benefit citizens. Video footage or an estimate of how long emergency services will take to respond is more useful.

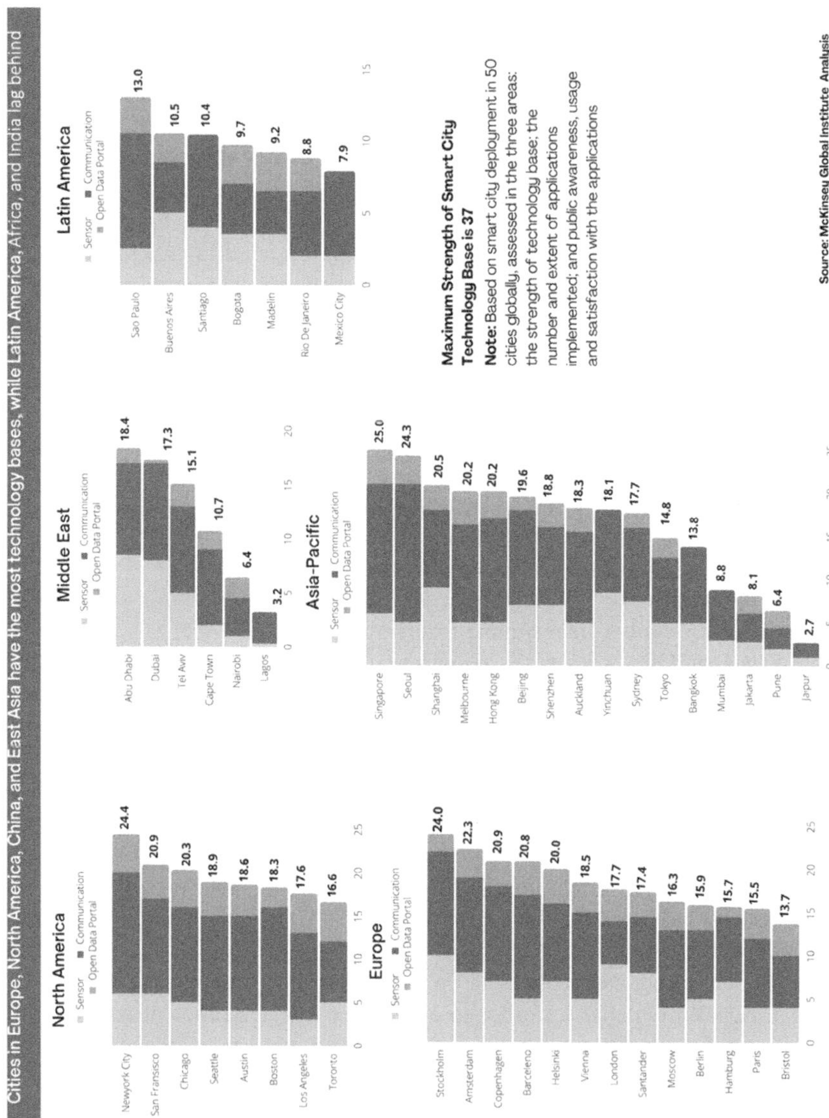

FIGURE 5.3 Cities with most developed technology bases [7].

5.4.1 Smart City Dubai

By 2025, it is estimated that smart cities will create business opportunities worth $2.46 trillion [10].

Dubai, with an ambitious goal of becoming the happiest city on earth, has spearheaded an initiative to harness the power of data to power a smarter city. By empowering, delivering, and promoting the city's best, Smart Dubai aims to create a seamless, efficient, safe, and personalized city experience for residents and visitors. By modernizing the city's infrastructure and providing innovative experiences through efficient, seamless, safe, and personalized experiences, Smart Dubai has developed a comprehensive data strategy. As part of its initial vision to create a repository of data that can be shared between public and private entities, Smart Dubai has developed a platform for data gathering and computing – Dubai Pulse – that allows more data to be ingested and computed to be able to gain valuable insights for the city as a whole. The Smart Dubai initiative has been able to deliver a wide range of use cases across planning, development, social services, and energy for the local population by using analytics and open data. The electricity and water consumption dashboard created in collaboration with Dubai Electricity and Water Authority (DEWA) [11] is a good example of such a use case. With the dashboard, DEWA consumption data, geographic information system (GIS) data, and other district cooling operators' data can be analyzed to provide a better understanding of the way water is used throughout the city and to plan for it. By using DEWA's open data records, Smart Dubai has also been able to generate a 360-degree picture of Dubai's citizens. This live dashboard has proven to be an extremely valuable resource for city planners, government entities, real estate developers, and even new start-up companies, as it allows them to track community population trends and to visualize how the city's residents are moving over time and space.

Smart Dubai's data platform has already identified over 2,000 data sets from participating Dubai government and semi-government entities, of which more than 550 have already been ingested. Due to the growing importance of city data, the platform received more than 2 million hits in 2019 [12]. As we strive toward a brighter future, Smart Dubai has launched its Smart Cities Global Network, bringing 500 members from around the world to share knowledge and expertise in smart cities. By engaging with and participating in international bodies, the organization is committed to developing and using international standards, sharing best practices, and planning for cities' future. By leveraging analytics and open data, Dubai has become the happiest city on earth by transforming itself into a smart city. Using a single strategy and connecting the efforts of a citywide network of partners, the Smart Dubai initiative should achieve real time dashboards based on available information, resulting in a stronger economy, better quality of life, and more effective governance.

5.4.2 A Smart City Powered by AI and Data Using Cloudera Data Platform

As cities grow rapidly in an urbanized world, they face a number of challenges, including becoming more sustainable, resilient, and efficient. Due to the increased

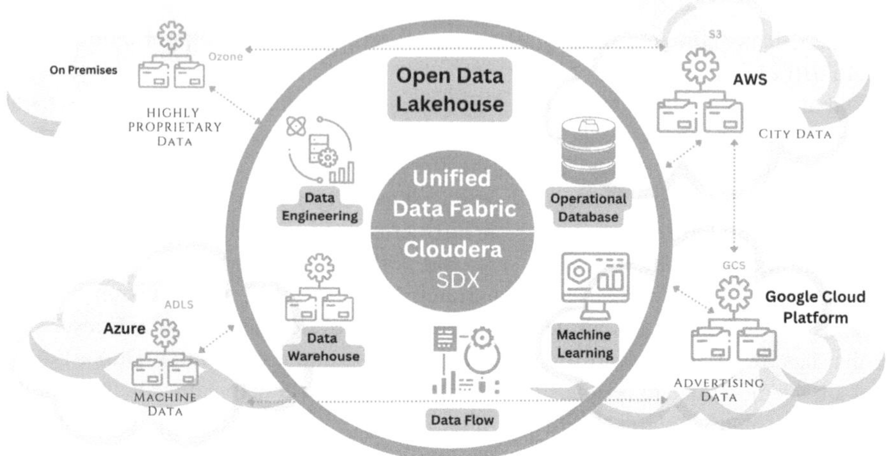

FIGURE 5.4 Cloudera Data Platform: Making hybrid data strategy a reality [14].

availability of data, new technologies and the increasing availability of data provide the opportunity to create smarter cities that are better able to deal with these
challenges. AI and data will power cities in the future. The Cloudera Data Platform
(CDP) and edge computing will also be used to revolutionize our lives and businesses.
According to the United Nations, 68% of the world's population will live in urban
areas by 2050 [13]. Climate change, traffic congestion, transportation management,
public safety, healthcare access, and education are some of the challenges arising
from this growth.

The city of the future must be designed to be resilient to unexpected events, such
as pandemics and natural disasters. In order to address this issue holistically, it is
necessary to take into account the interconnection of different systems, such as transportation, energy, and public safety. Within the context of these challenges, the city
must leverage data to gain insights and make better decisions. A more comprehensive
picture of the city can be formed with sensors, cameras, and other data sources, allowing real time decisions to be made. In cities, AI allows them to extract insights from
their data, automate decision-making, and identify patterns and trends. Furthermore,
AI can enable predictive maintenance of critical systems and enable more efficient
and effective use of energy and transportation infrastructure.

Data and AI will be leveraged more effectively by cities leveraging edge computing
and 5G networks. With edge computing, data can be processed closer to the source,
reducing latency and enabling real time decision-making. 5G networks offer faster
speed and lower latency than previous generations of wireless networks, making real
time data transfers possible and enabling new applications like autonomous vehicles
and remote healthcare. By combining these technologies, a truly data-driven city can
be built where data can be used to optimize resource use, improve public safety, and
improve citizens' quality of life. CDP offers a flexible and scalable data platform to
unlock the value of city data. In addition to sensors, cameras, and other IoT devices,

CDP enables cities to collect, process, and analyze data. In order to optimize the use of resources like energy and transportation infrastructure, CDP can help cities gain insights from their data. CDP also allows cities to experiment and iterate on new applications without committing to a fixed workflow. Using CDP's Private, Hybrid, and Multi-Cloud approaches, a city's data infrastructure can also be controlled, analyzed, and experimented with where it is located, increasing its agility and scalability. Data fuels the future city, and AI runs it. Smarter, more efficient, and more resilient cities can be created with the increasing availability of data and technologies like edge computing and 5G. With the CDP, cities can unlock the value of their data and experiment with new applications.

5.4.3 COLLABORATION AND PARTNERSHIPS

A truly intelligent city of the future will be fueled by data and powered by AI if we work together and form partnerships. There are many stakeholders in our cities, including government agencies and private businesses, who seek to help each other by achieving their own objectives. Collaboration is important especially for smart city projects to work together for betterment rather than pursue their own interests. Collaboration ensures stakeholders pool resources and share knowledge which can lead to more efficient and effective solutions. Collaboration brings all stakeholders to a single forum to solve complex problems together, as each stakeholder has their own unique perspective. In the coming sections, we can see some of the successful collaborations between stakeholders and city governments and what is the key factor in driving successful outcomes and ensuring long-term sustainability.

One of the successful examples of collaboration is the partnership between the city of Amsterdam and the Amsterdam Institute for Advanced Metropolitan Solutions [15]. Together, these two organizations have created a smart city initiative called "City-zen." The primary benefit of this collaboration is to make Amsterdam a more sustainable city through the use of smart technologies. Collaboration involves the city government, universities, and the private sector as part of the project (it is a multi-multi stakeholder project). Each and every party is able to share their resources, and expertise, and develop innovative solutions to the challenges that the city faces. The collaboration initiative revolutionizes the city management of Amsterdam and aids in the development of smarter, more efficient cities that are greener, more accessible, and more sustainable.

A collaboration between the city of San Diego [16] and General Electric (GE) [17] is an another example of a successful collaboration. In partnership, they have created the "Current" initiative, a smart city initiative designed to make San Diego more energy-efficient and sustainable. Several stakeholders like city government officials, utility companies, and private businesses are involved in the smart initiative project. The collaboration leads to the development of new technologies and services that benefit the entire city by sharing of data and resources. A key component of creating a smart city is collaboration, but we should not forget that citizen participation is also crucial. Citizens can participate in a variety of ways, including public meetings, online surveys, and feedback mechanisms. As a result, the project received good feedback from citizens and their involvement in the planning and decision-making process that leads to a smarter, more sustainable city.

One of the cities in the Netherlands participating in "My Smart City District" [18] is Eindhoven, a city with a vast amount of citizen input. Its goal is to engage residents and involve them in the process to create a sustainable neighborhood. Everything from the layout of the streets to the energy systems is decided by the residents. It proves that it is possible for the city to create smart cities which are tailored to meet the needs and desires by directly involving the residents in the planning process. The goal of smart cities is to benefit the entire population, not just a few wealthy or privileged members of society. As a last point, the project is exemplary for collaboration with a focus on inclusivity and equity. Toward this success, the city ensures that everyone has access to the technologies and services and no one is left behind when it comes to using them.

In the same way, Barcelona prioritized inclusivity and equity in the city planning. Many initiatives are taken by the city to ensure all citizens benefit from the advantages of the smart city, regardless of their socio-economic status. For low-income families, the city has developed a "Digital Inclusion Plan" [19] that provides free computers and Internet access. As part of the efforts, there are many initiatives taken to make public transport more accessible and affordable. By leveraging data and AI, creating a future city requires collaboration, citizen participation, and equity and inclusivity. With the right approach and partnerships, cities create smarter, more sustainable, and more resilient cities which are fit for the 21st century with innovation and technology.

5.5　CONCLUSION

The use of big data and machine learning revolutionizes the humans, lives and works in cities and it quickly changes the way we live and work. CoT technologies improve urban services and infrastructure in departments like transportation, public safety, waste management, and energy efficiency. CoT opens new economic opportunities by providing access to more accurate and real time information. It connects every human to urban services and harnesses data to inform decision-making, thereby enabling cities to drive innovation and create new opportunities for citizens to thrive. In the future, cities can provide more efficient and public-centric services as data sets are abundant and new technologies are being developed. The success of smart city initiatives is determined by their ability to integrate data from multiple sources, create a good data governance system, and operate seamlessly across departments. The cities should adopt flexible big data platforms that can scale up in order to meet future demands as cities become more connected and data becomes richer, and as cities become more connected, new services and opportunities will arise, making the adoption of such platforms crucial. CoT in smart cities inspires and excites us through delivering services that enhance our quality of life, thereby improving it.

ACKNOWLEDGMENTS

The Cloudera Resource Library greatly enhanced authors' knowledge and assisted them in writing this chapter. As a team, the authors would like to thank Cloudera for providing them with the opportunity and experience to work on this project. They are grateful to Cloudera for their generous support, which enabled them to create a high-quality chapter for this book.

REFERENCES

1. Boretti, A., Rosa, L. (2019) "Reassessing the projections of the World Water Development Report." *NPJ Clean Water* 2, 15. https://doi.org/10.1038/s41545-019-0039-9

2. Singh, S. (2019) Smart city platforms market worth $223.3 billion by 2023 - exclusive report by MarketsandMarkets™, PR Newswire: Press release distribution, targeting, monitoring, and marketing. Available at: www.prnewswire.com/news-releases/smart-city-platforms-market-worth-223-3-billion-by-2023—exclusive-report-by-marketsandmarkets-300800263.html (Accessed: April 18, 2023).

3. Li, B. (2022) "Effective energy utilization through economic development for sustainable management in smart cities," Energy Reports, 8, pp. 4975–4987. Available at: https://doi.org/10.1016/j.egyr.2022.02.303

4. Joshi, S. (2023) 19% CAGR for smart city market size worth $3110.58 billion, globally, by 2028 – comprehensive study by the insight partners. GlobeNewswire News Room. The Insight Partners. Available at: www.globenewswire.com/news-release/2023/02/20/2611452/0/en/19-CAGR-for-Smart-City-Market-Size-Worth-3110-58-Billion-Globally-by-2028-Comprehensive-Study-by-The-Insight-Partners.html (Accessed: April 18, 2023).

5. C. O. Fantoni, A. Mero, and C. Vaca (2020) "Tech for Hire: Data science-related jobs signal economic growth," *2020 Seventh International Conference on eDemocracy & eGovernment (ICEDEG)*, Buenos Aires, Argentina, pp. 151–156. doi: 10.1109/ICEDEG48599.2020.9096765

6. Kresl, P.K. (2020) *Towards a competitive, sustainable modern city*. Cheltenham, UK, Gloucestershire: Edward Elgar Publishing Limited.

7. Supra, J. D. Accelerating change: From smart city to smart society. (2020, May 28) Retrieved March 17, 2023, from www.jdsupra.com/legalnews/accelerating-change-from-smart-city-to-83944/

8. M. Hazara and V. Kyrki (2016) "Reinforcement learning for improving imitated in-contact skills," *2016 IEEE-RAS 16th International Conference on Humanoid Robots (Humanoids)*, Cancun, Mexico, pp. 194–201. doi: 10.1109/HUMANOIDS.2016.7803277

9. David Reinsel Global Datasphere Expansion is Never-ending David Reinsel, John Gantz, John Rydning The Digitization of the World – From Edge to Core – November 2018 Doc#US44413318. www.seagate.com/files/www-content/our-story/trends/files/idc-seagate-dataage-whitepaper.pdf.

10. "Smart cities predicted to create $2.46 trillion worth of business" (2020) Smart Cities [Preprint]. https://bacpress.com. Available at: https://bacpress.com/smart-cities-predicted-to-create-2-46-trillion-worth-of-business-opportunities-by-2025/ (Accessed: April 18, 2023).

11. M. S. Hussin, M. S. Al-Mehairi, and H. Al-Madhani (2016) "DEWA distribution power asset management system in view of ISO 55000 Standard," *Asset Management Conference (AM 2016)*, London, pp. 1–5. doi: 10.1049/cp.2016.1424

12. Sabri, S. (2021) "Smart Dubai IOT strategy: Aspiring to the promotion of happiness for residents and visitors through a continuous commitment to innovation,"

Smart Cities for Technological and Social Innovation, pp. 181–193. Available at: https://doi.org/10.1016/b978-0-12-818886-6.00010-1

13. Hoornweg, D., and Pope, K. (2017) Population predictions for the world's largest cities in the 21st century. *Environment and Urbanization*, 29(1), 195–216. https://doi.org/10.1177/0956247816663557

14. Moxey, D. (2022) Cloudera Data Platform (CDP) Private and Public, Cloudera Data Platform (CDP). Cloudera. Available at: www.cloudera.com/products/cloudera-data-platform.html (Accessed: April 18, 2023).

15. Amsterdam, and Amsterdam Institute for Advanced Metropolitan Solutions. (2019) About Smart Cities®. Retrieved from www.amsterdamsmartcity.com/about/about-smart-cities/

16. San Diego (n.d.) Smart Cities San Diego. Retrieved from www.smartcitysdk.com/san-diego/

17. General Electric (n.d.) Current Retrieved from www.currentbyge.com/current

18. Nijland I.C.M. (2020) Smart City Start-ups: Performance after business incubation in Utrecht and Gothenburg [Unpublished Master Thesis]. U.S.E.

19. Parteka, Eloisa, and Alcides Rezende, Denis. (2018) Digital planning of the city of Barcelona and its relations with the strategic digital city. *Journal of Technology Management & Innovation*, 13(4), 54–60. https://dx.doi.org/10.4067/S0718-27242018000400054

6 BGLMC

A Bidirectional GRU-Based Learning Model for Cloud-Computing Resource Workloads Forecasting

Ananya Kaim, Surjit Singh, and Yashwant Singh Patel

6.1 INTRODUCTION

The architectural move toward cloud computing is taking place in data centers worldwide. Each physical machine is exclusively available to one client as part of the conventional hosting model. The more effective concept of cloud computing is replacing the conventional model of hosting data centers. Cloud computing allows services to be layered into virtual machines (VMs), which increases operational efficiency for the hosting provider (VM) [1]. Cloud computing is becoming increasingly popular because of its improved scalability, availability, and promise of lower infrastructure costs [2]. As the Internet has grown, more businesses have adopted cloud-based online services. An e-common system can leverage service components, for instance, servers from Alibaba, Microsoft, and Amazon, without software creation or reworking, due to the power of cloud computing to give on-demand network connectivity [3]. Yet, maintaining such a high level of service while using the fewest amount of resources is difficult [4]. As a result, workload forecasting assists cloud-end cluster maintainers in determining if the existing resource allocation approach is adequate or not [5].

The virtualization method is crucial to cloud computing because it enables users to build processing nodes in accordance with their software needs and financial constraints [6]. It also significantly contributes to assuring the highest level of service and decreases energy usage because the workload of virtual computers is continually changing based on the time of day [7]. As an example, when workloads are low, more resources than necessary are supplied, and when workloads are high, few resources are made available [8]. Thus, reliable workload prediction is necessary to optimize resource consumption. The service provider can allocate or de-allocate resources in advance with the accurate forecast of future demands, which will allow for more efficient and logical resource planning [9].

DOI: 10.1201/9781003390954-6

With the aid of multivariate time series analysis, it is possible to uncover hidden temporal relationships, gain a deeper comprehension of the system, and make superior predictions. It is difficult to adequately describe multivariate time series data using conventional statistical methods because of its elevated spatio-temporal dependency and huge dimensionality. Furthermore, noisy data makes modeling difficult [10]. Several fields, including computer vision and speech recognition, have made use of artificial intelligence, particularly deep learning [11]. While time series data can be modeled using the conventional statistical approach, estimating data from time series with deep learning is gaining popularity [12]. Therefore, we suggest a multivariate BiGRU model in our work for assessing and forecasting workload in cloud environments.

6.2 RELATED WORK

Several time series prediction systems have lately incorporated a variety of deep learning methodologies [13]. Deep learning algorithms may be used to extract both geographical and temporal information from workload history data without the requirement of pre processing.

In order to forecast future CPU utilization requirements and deal with workload fluctuations for future time intervals, Al-Asaly et al. [14] proposed an effective diffusion convolutional Recurrent Neural Network (RNN) model. The main novel feature of the proposed method is its ability to efficiently capture key topological properties of the network that have a significant, positive influence on forecasting performance. For their research, they used the Planetlab data set. As evaluation measures, they used root mean square error (RMSE) and mean absolute percentage error (MAPE).

Based on historical data, a VM's allocation of resources can be predicted with greater accuracy, for which Shu et al. [15] suggested a gated recurrent unit-attention mechanism (GRU-AM). Azure and Alibaba were two known workload data sets that were utilized. RMSE, MAPE, and R_2 were employed as evaluation metrics. They assessed their suggested model in three different ways. Firstly, they compared their forecasted model, i.e., GRU without the use of attention mechanism (AM). Secondly, they compared their forecasted model with Long Short Term Memory (LSTM)-AM model, and, lastly, they compared the window sizes to confirm its impact on the outcomes.

A one-step LSTM prediction model was proposed by Nguyen et al. [16] that predicts a point in the future by using both the expected and actual values of the prior time series. Their research focused on CPU usage for workload prediction. Each data set they used for the trials had a span of 2 weeks and had over 20K data records. They combined two actual data sets with three hypothetical data sets situations. They used mean squared error (MSE), RMSE, mean absolute error (MAE), and MAPE as evaluation metrics.

For the purpose of predicting the resource demand for cloud computing, Li et al. [17] presented the hybrid forecasting BiLSTM-GRU model. Their model can successfully forecast how the cloud platform will use its resources, address the issue of uneven resource usage, and enhance the functionality and dependability of the application. With the Google cluster data set, they examined and confirmed numerous classical forecasting techniques. They have used MSE, RMSE, MAE, and MAPE as evaluation metrics.

Liu et al. [18] proposed a Tr-Predictor algorithm to forecast cloud workload. They integrated the LSTM model, which is employed as a regression tool, with the TrAdaBoost.R2 algorithm, which helps to enhance the model. To determine the ultimate effect, the suggested method combines the weights with the weak learner's prediction outcomes. Google cluster and Alibaba data sets were two data sets that were utilized. MSE, MAE, MAPE, and R_2 were employed as evaluation metrics.

A cloud workload forecasting approach using Adaptive Pattern Mining was suggested by Bao et al. [19]. To capture the low- and high-frequency aspects of highly fluctuating workloads, they developed a multi-model two-step strategy. Then, in order to combine the forecasts from several models into a single result that could be used to forecast future workloads, we developed an error-based weights aggregation approach for a specific workload. Google cluster and Alibaba data sets were two data sets that were utilized. They used MAE and ATT as evaluation metrics.

A workload forecasting method based on Particle Swarm Optimization (PSO) was proposed by Leka et al. [20]. They combined three RNNs using an ensemble learning technique, followed by a layer of dense neural networks. They used deep learning models (LSTM, BiLSTM, and GRU) to evaluate their suggested model. The data sets from PlanetLab and Bitbrains were used in their work. RMSE, AE, and MAPE were used as evaluation metrics.

An RCP-CL model that combines the LSTM layer with several 1D CNN layers was proposed by Patel and Kushwaha [21]. While LSTM generates temporal dependencies, 1D-CNN layers are concatenated into an SPE block to grasp spatial patterns. They employed the Google cluster and the Alibaba data set in their research. As evaluation measures, they used MSE, RMSE, and MAE.

Following a review of earlier publications, we have found that the majority of cloud evaluations are limited to the unidirectional future resource demands, which only capture past resource use data. Moreover, there hasn't been much research acknowledgment for BiGRU. Moreover, BiGRU networks had a lighter architecture and more reliable forecasting performance than LSTM since they used fewer training parameters, less memory, and faster execution.

6.3 PROPOSED MODEL

An RNN may process a series of inputs in order using its memory. Unfortunately, the gradient diminishing makes it ineffective for RNNs in learning long-term memory requirements. Some improved RNNs have been proposed to overcome this restriction, such as LSTM and GRU, that have indicated a remarkable ability in learning long-term memory requirements. GRU is a particular type of RNN introduced by Kyunghyeon Cho et al. [22]. It tries to resolve the vanishing gradient problem that is present in conventional recurrent neural networks. Because of their comparable designs, GRU can also be seen as a variant of LSTM. GRU controls the data flow with two gates as opposed to LSTM's three [23]. Update and reset gates are two different kinds of gates as shown in the right corner of Figure 6.1 The short-term memory of the network is controlled by the reset gate (rg_t) and

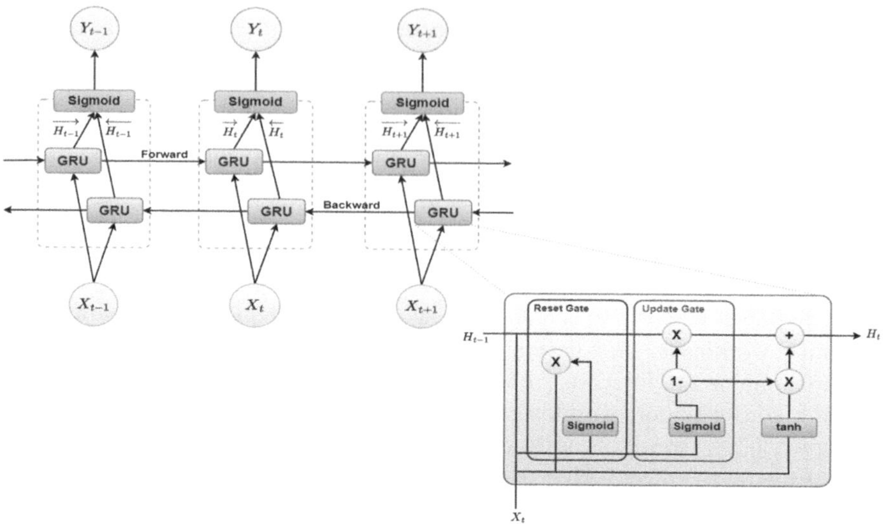

FIGURE 6.1 Proposed Architecture.

it can be calculated using Equation 6.1. It chooses how much historical data to throw out and the reset gate also helps in calculating a candidate's hidden state as shown in Equation 6.3. The update gate (zg_t) controls the long-term memory of the network and it can be calculated using Equation 6.2. It decides the amount of previous data that should be delivered in the future [24]. The reset gate and candidate's hidden state help in calculating the hidden state as shown in Equation 6.4.

$$rg_t = \sigma(B_r + H_{t-1} * Wt_{hr} + Z_t * Wt_{xr}) \tag{6.1}$$

$$zg_t = \sigma(B_z + H_{t-1} * Wt_{hz} + Z_t * Wt_{xz}) \tag{6.2}$$

$$\widetilde{H}_t = tanh(Z_t * Wt_{xh} + Wt_{hh}(hl_{t-1} * rg_t) + B_z) \tag{6.3}$$

$$H_t = (H_{t-1} * zg_t) + \widetilde{H}_t * (1 - zg_t) \tag{6.4}$$

where σ indicates the sigmoid, B signifies the bias of respective gate unit, and Wt indicates the weight of respective gate unit.

Unfortunately, GRU has a limitation in that it can only understand the historical conditions of time series. Hence, we propose using BiGRU in this research. Just integrating two separate GRUs into one is what BiGRU does. Processing the input sequence both forward and backward occurs with each time step as shown in Figure 6.1. The recurrent pattern is used by the GRU to maintain and access information across extended timespans. However, the network's efficiency might not be as good in fact as it is in principle because it only accesses old data [25]. In a subsequent layer of the BiGRU network, the data sequence flows in the other direction to bypass this issue. These characteristics enable the bidirectional technique to assist the RNNs in retrieving new data, which improves the execution of the learning process. It has been demonstrated that BiGRU performs better in feature extraction than unidirectional GRUs [26, 27]. The output for the forward layer as shown in Equation 6.5 and backward layer as shown in Equation 6.6 is determined as follows:

$$\overrightarrow{H_t} = \sigma(Wt_{\overrightarrow{H_t}} * \overrightarrow{H_{t-1}} + X_t * Wt_{\overrightarrow{x_t}}) \tag{6.5}$$

$$\overleftarrow{H_t} = \sigma(Wt_{\overleftarrow{H_t}} * \overleftarrow{H_{t-1}} + X_t * Wt_{\overleftarrow{x_t}}) \tag{6.6}$$

Equation 6.7 calculates the final output vector:

$$Y_t = (\overrightarrow{H_t}, \overleftarrow{H_t}) \tag{6.7}$$

where $\overrightarrow{H_t}$ indicates the result of forward GRU, $\overleftarrow{H_t}$ indicates the result of backward GRU, σ signifies the sigmoid, and Wt signifies the weight of forward and backward GRU.

When forecasting multivariate data, multiple variable values are typically present at each observational time step. In some cases, the other variables' importance also affects each variable's dependence on multivariate data and the past values of the variable in question. For modeling multivariate data, these correlations are crucial. Due to these factors, a multivariate BiGRU prediction model is suggested in this work.

6.4 RESULTS ANALYSIS

6.4.1 DATA SET

We used the Grid Workload Archive (GWA)-T-13 Materna data set [28]. It has three distributed data center traces, each comprising data from 520, 527, and 547 VMs. Resources (such as CPU, disk, and memory) given to VMs and the resources those VMs actually used are displayed in this trace data, which is bundled in CSV files. First trace data from 520 VMs were used in our research. In Table 6.1, a list of resource attributes is provided.

6.4.2 SIMULATION ENVIRONMENT

An Intel(R) Core(TM) i3-6006U CPU, 8 GB of RAM, a 512 GB SSD, and Windows 10 Pro OS were used to execute the simulations. The Keras library and Python 3.10.9 were used to execute the time series forecasting model. It is crucial to set the hyper-parameters for the suggested strategy. Table 6.2 represents the hyper-parameter table of our model.

6.4.3 PERFORMANCE MEASURES

For the analysis of the forecasting models, we used performance indicators such as RMSE and MAE.

TABLE 6.1
List of resource attributes

Resource Attribute	Description
CU [%]	CPU Usage in %
MU [%]	Memory Usage in %
MU [Mhz]	Memory Usage in Mhz
CU [Mhz]	CPU Usage in Mhz
CCP	CPU Capacity Provisioned in Mhz
MCP	Memory Capacity Provisioned in Mhz
DWT	Disk Write Throughput in KB/s
DS	Disk Space in GB
DRT	Disk Read Throughput in KB/s
NTT	Network Transmitted Throughput in KB/s
NRT	Network Received Throughput in KB/s

TABLE 6.2
Hyper-Parameters of the suggested method

Parameters	Value
Default Hidden Layer	3
Epochs in Training	30
Batch Size	128
Loss Function	MSE
Optimizer	Adam

- RMSE: Root mean square error
- MAE: Mean absolute error

RMSE is the sqrt of the difference between the real resource utilization and the expected resource utilization with respect to n observations.

$$RMSE = \sqrt{\frac{\sum_{i=1}^{n}(p_i - \hat{p}_i)^2}{n}} \qquad (6.8)$$

MAE calculates the average absolute deviation from the real resource utilization and the expected resource utilization.

$$MAE = \frac{\sum_{i=1}^{n}\|p_i - \hat{p}_i\|}{n} \qquad (6.9)$$

TABLE 6.3
RMSE analysis for CPU usage

	CPU Usage			
RMSE	**90**	**180**	**270**	**Overall**
LSTM	0.00849	0.00856	0.00859	0.00886
GRU	0.00869	0.00874	0.00877	0.00909
BiLSTM	0.00843	0.00850	0.00853	0.00878
Proposed	0.00671	0.00677	0.00681	0.00712

TABLE 6.4
MAE analysis for CPU usage

	CPU Usage			
MAE	**90**	**180**	**270**	**Overall**
LSTM	0.00803	0.00802	0.00802	0.00803
GRU	0.00825	0.00825	0.00825	0.00825
BiLSTM	0.00798	0.00798	0.00798	0.00798
Proposed	0.00623	0.00622	0.00622	0.00623

6.4.4 DISCUSSION OF RESULTS

6.4.4.1 Performance Prediction for CPU Usage

The experiment findings are shown in Tables 6.3 and 6.4 for predictions of CPU usage at various time steps, such as 90 (15 minutes), 180 (30 minutes), and 270 steps (45 min). We can see from both evaluation metrics that our suggested work has reduced the amount of errors compared to other models. We discovered through experimental evaluation that the presented model decreases RMSE by 21.6% and MAE by 24.4% especially in comparison to the forecasting methods in use today. Additionally, we created the visualization of prediction results using various deep-learning models from Figure 6.2 to 6.5.

6.4.4.2 Performance Prediction for Memory Usage

The experiment findings are shown in Tables 6.5 and 6.6 for predictions of memory usage at various time steps, such as 90 (15 minutes), 180 (30 minutes), and 270 steps (45 minutes). We can see from both evaluation metrics that our suggested work has reduced the amount of errors compared to other models. We discovered through experimental evaluation that the presented model RMSE by 17.4% and decreases MAE by 29.6% especially in comparison to the forecasting methods in use today. Additionally, we created the visualization of prediction results using various deep-learning models from Figure 6.6 to 6.9.

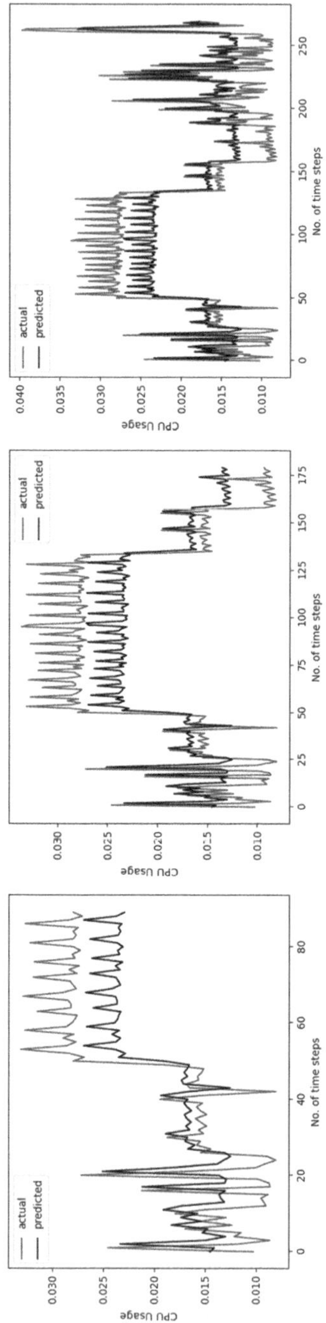

FIGURE 6.2 Representation of LSTM-based multi-step CPU utilization prediction.

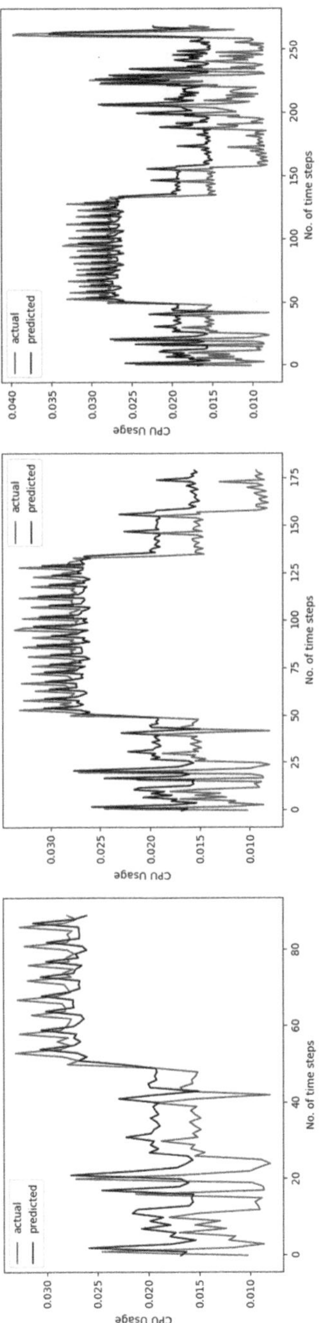

FIGURE 6.3 Representation of GRU-based multi-step CPU utilization prediction.

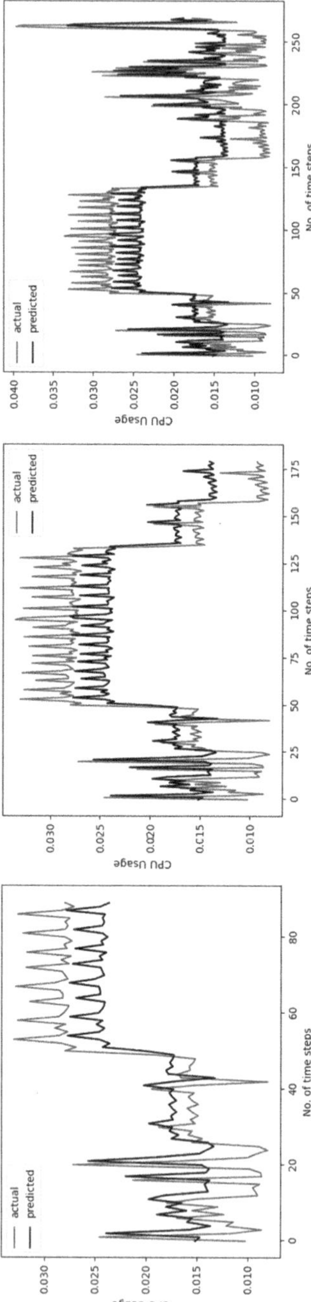

FIGURE 6.4 Representation of BiLSTM-based multi-step CPU utilization prediction.

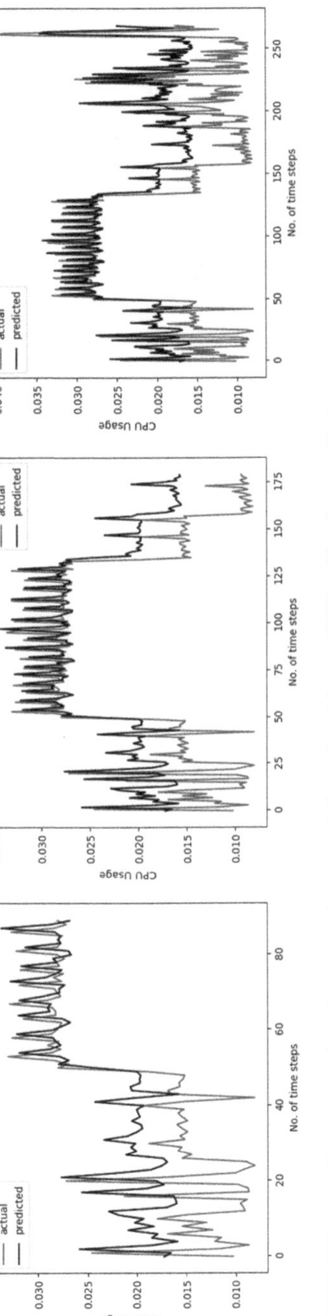

FIGURE 6.5 Representation of proposed approach-based multi-step CPU utilization prediction.

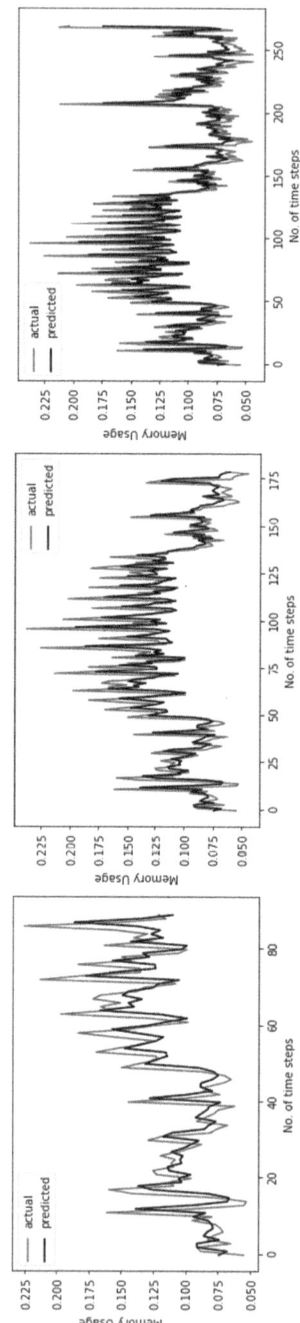

FIGURE 6.6 Representation of LSTM-based multi-step memory utilization prediction.

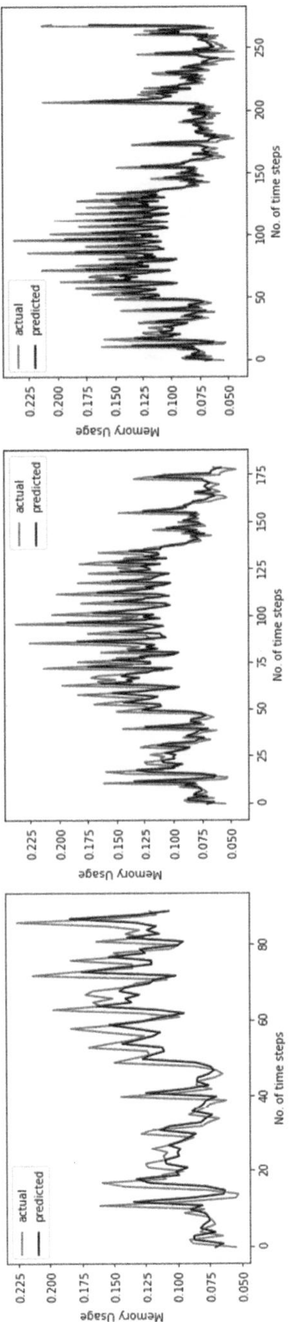

FIGURE 6.7 Representation of GRU-based multi-step memory utilization prediction.

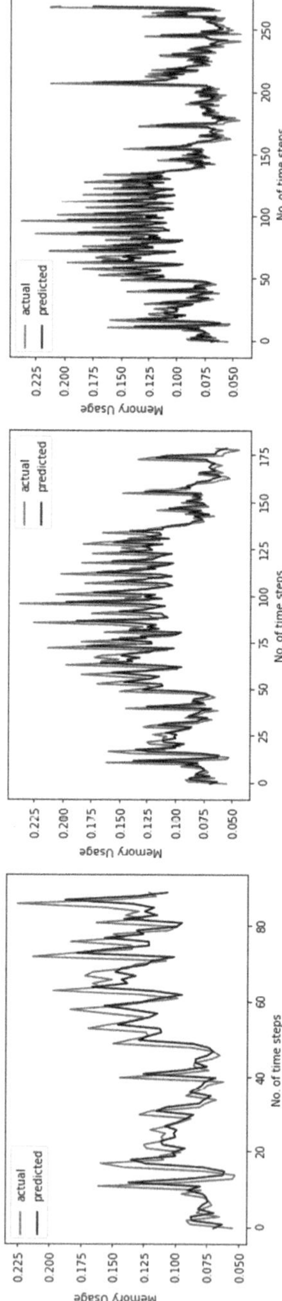

FIGURE 6.8 Representation of BiLSTM-based multi-step memory utilization prediction.

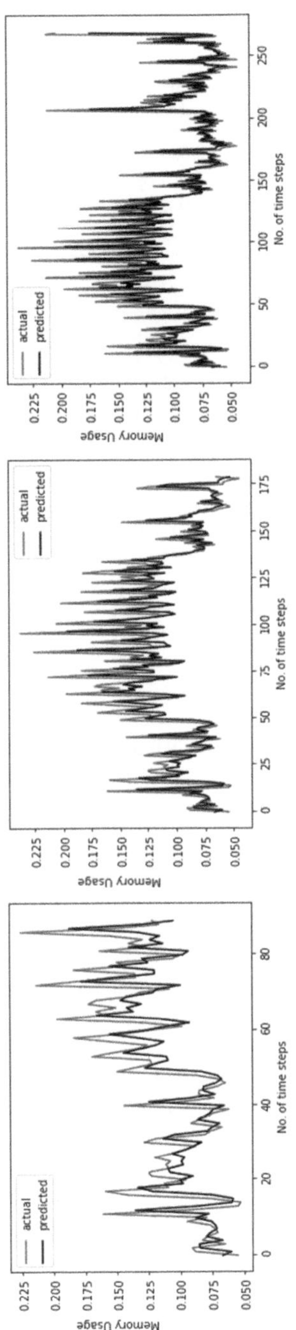

FIGURE 6.9 Representation of proposed approach-based multi-step memory utilization prediction

TABLE 6.5
RMSE analysis for memory usage

	Memory Usage			
RMSE	**90**	**180**	**270**	**Overall**
LSTM	0.02922	0.02934	0.02939	0.03245
GRU	0.02765	0.02778	0.02784	0.03117
BiLSTM	0.02207	0.02232	0.02243	0.02649
Proposed	0.02206	0.02233	0.02245	0.02680

TABLE 6.6
MAE analysis for memory usage

	Memory Usage			
MAE	**90**	**180**	**270**	**Overall**
LSTM	0.02537	0.02534	0.02532	0.02540
GRU	0.02366	0.02363	0.02361	0.02369
BiLSTM	0.01781	0.01777	0.01774	0.01784
Proposed	0.01783	0.01779	0.01776	0.01787

6.5 CONCLUSION

An investigation has been done on a multivariate BiGRU prediction model. Also, this work uses the GWA T-13 Materna trace data, yielding notable gains in prediction accuracy over existing methods. Our suggested approach has been compared to a few other deep-learning models. We discovered through experimental evaluation that the presented model decreases RMSE by 21.6% and MAE by 24.4% for CPU usage and decreases RMSE by 17.4% and MAE by 29.6% for memory usage especially in comparison to the forecasting methods in use today. We intend to expand the work to autoscaling in the future.

REFERENCES

1. Prevost, J. J., Nagothu, K., Jamshidi, M., & Kelley, B. (2014, August). Optimal calculation overhead for energy efficient cloud workload prediction. *In 2014 World Automation Congress (WAC)* (pp. 741–747). IEEE.
2. Kumar, J., & Singh, A. K. (2018). Workload prediction in cloud using artificial neural network and adaptive differential evolution. *Future Generation Computer Systems, 81,* 41–52.
3. Zhu, Y., Zhang, W., Chen, Y., & Gao, H. (2019). A novel approach to workload prediction using attention-based LSTM encoder-decoder network in cloud environment. *EURASIP Journal on Wireless Communications and Networking, 2019,* 1–18.
4. Armbrust, M., Fox, A., Griffith, R., Joseph, A. D., Katz, R., Konwinski, A., ...& Zaharia, M. (2010). A view of cloud computing. *Communications of the ACM, 53*(4), 50–58.

5. Qi, L., Yu, J., & Zhou, Z. (2017). An invocation cost optimization method for web services in cloud environment. *Scientific Programming, 2017*, 1–9.

6. Alouane, M., & El Bakkali, H. (2016, May). Virtualization in Cloud Computing: Existing solutions and new approach. In *2016 2nd International Conference on Cloud Computing Technologies and Applications (CloudTech)* (pp. 116–123). IEEE.

7. Qiu, F., Zhang, B., & Guo, J. (2016, May). A deep learning approach for VM workload prediction in the cloud. In *2016 17th IEEE/ACIS International Conference on Software Engineering, Artificial Intelligence, Networking and Parallel/Distributed Computing (SNPD)* (pp. 319–324). IEEE.

8. Wang, S., Li, X., & Ruiz, R. (2019). Performance analysis for heterogeneous cloud servers using queueing theory. *IEEE Transactions on Computers, 69*(4), 563–576.

9. Chen, Z., Hu, J., Min, G., Zomaya, A. Y., & El-Ghazawi, T. (2019). Towards accurate prediction for high-dimensional and highly-variable cloud workloads with deep learning. *IEEE Transactions on Parallel and Distributed Systems, 31*(4), 923–934.

10. Fu, T. C. (2011). A review on time series data mining. *Engineering Applications of Artificial Intelligence, 24*(1), 164–181.

11. Dang-Quang, N. M., & Yoo, M. (2021). Deep learning-based autoscaling using bidirectional long short-term memory for kubernetes. *Applied Sciences, 11*(9), 3835.

12. Imdoukh, M., Ahmad, I., & Alfailakawi, M. G. (2020). Machine learning-based autoscaling for containerized applications. *Neural Computing and Applications, 32*, 9745–9760.

13. Singh, S., & Mohan Sharma, R. (Eds.). (2019). *Handbook of Research on the IoT, Cloud Computing, and Wireless Network Optimization*. IGI Global.

14. Al-Asaly, M. S., Bencherif, M. A., Alsanad, A., & Hassan, M. M. (2022). A deep learning-based resource usage prediction model for resource provisioning in an autonomic cloud computing environment. *Neural Computing and Applications, 34*(13), 10211–10228.

15. Shu, W., Zeng, F., Ling, Z., Liu, J., Lu, T., & Chen, G. (2021, December). Resource demand prediction of cloud workloads using an attention-based GRU model. In *2021 17th International Conference on Mobility, Sensing and Networking (MSN)* (pp. 428–437). IEEE.

16. Nguyen, T., Do, T., Le, K., Go, S., Na, S., Kim, D., & Tran, D. (2022, December). An LSTM-based Approach for predicting resource utilization in cloud computing. In *Proceedings of the 11th International Symposium on Information and Communication Technology* (pp. 173–179). ACM.

17. Li, X., Wang, H., Xiu, P., Zhou, X., & Meng, F. (2022, August). Resource usage prediction based on BILSTM-GRU combination model. In *2022 IEEE International Conference on Joint Cloud Computing (JCC)* (pp. 9–16). IEEE.

18. Liu, C., Jiao, J., Li, W., Wang, J., & Zhang, J. (2022). Tr-Predictior: An ensemble transfer learning model for small-sample cloud workload prediction. *Entropy, 24*(12), 1770.

19. Bao, L., Yang, J., Zhang, Z., Liu, W., Chen, J., & Wu, C. (2023). On accurate prediction of cloud workloads with adaptive pattern mining. *The Journal of Supercomputing, 79*(1), 160–187.

20. Leka, H. L., Fengli, Z., Kenea, A. T., Hundera, N. W., Tohye, T. G., & Tegene, A. T. (2023). PSO-based ensemble meta-learning approach for cloud virtual machine resource usage prediction. *Symmetry, 15*(3), 613.

21. Patel, E., & Kushwaha, D. S. (2023). An integrated deep learning prediction approach for efficient modelling of host load patterns in cloud computing. *Journal of Grid Computing, 21*(1), 5.

22. Cho, K., Van Merriënboer, B., Gulcehre, C., Bahdanau, D., Bougares, F., Schwenk, H., & Bengio, Y. (2014). Learning phrase representations using RNN encoder-decoder for statistical machine translation. *arXiv preprint arXiv:1406.1078*.
23. Chung, J., Gulcehre, C., Cho, K., & Bengio, Y. (2014). Empirical evaluation of gated recurrent neural networks on sequence modeling. *arXiv preprint arXiv:1412.3555*.
24. Qin, H. (2019). Comparison of deep learning models on time series forecasting: A case study of dissolved oxygen prediction. *arXiv preprint arXiv:1911.08414*.
25. Deng, Y., Jia, H., Li, P., Tong, X., Qiu, X., & Li, F. (2019, June). A deep learning methodology based on bidirectional gated recurrent unit for wind power prediction. In *2019 14th IEEE Conference on Industrial Electronics and Applications (ICIEA)* (pp. 591–595). IEEE.
26. Jinbao, T., Weiwei, K., Yidan, C., Qiaoxin, T., Chenyuan, S., & Long, L. (2021, September). Text classification method based on BiGRU-attention and CNN hybrid model. In *2021 4th International Conference on Artificial Intelligence and Pattern Recognition* (pp. 614–622). ACM.
27. Kaim, A., Singh, S., & Patel, Y. S. (2023, January). Ensemble CNN attention-based BiLSTM deep learning architecture for multivariate cloud workload prediction. In *24th International Conference on Distributed Computing and Networking* (pp. 342–348). ACM Proceedings.
28. *GWA-T-13 Materna*. (n.d.). http://gwa.ewi.tudelft.nl/data sets/gwa-t-13-materna

7 Study on Cost-Effective Solutions for Big Data Processing in the Cloud of Things

Fredrick Ishengoma

7.1 INTRODUCTION

Big Data Analytics (BDA) has revolutionized decision-making across diverse domains, including finance, healthcare, retail, and marketing (Ranjan & Foropon, 2021). The technology has spurred this transformative approach to decision-making from the volume, velocity, and variety of data generated by various sources, such as the Internet of Things (IoT) devices, social media, and mobile apps (Naeem et al., 2022). Diverse sources like IoT devices, social media platforms, and mobile apps are generating enormous amounts of data which support the scrutiny of extensive data sets containing invaluable insights that enhance our decision-making processes for business expansion within an information system (Karupusamy et al., 2021).

Within the information systems domain, BDA has emerged as a critical tool that helps process huge and heterogeneous data sets. Given its capabilities, it becomes possible for organizations to gather valuable insights to drive both strategy and operations (Vats & Sagar, 2019). When leveraged correctly by organizations, BDA leads businesses toward understanding their customers much better. Moreover, achieving this clarity allows businesses to optimize performance and make informed decisions that help accomplish their desired goals (Vitari & Raguseo, 2020).

With real-time data analysis becoming an operational standard across diverse sectors today, BDA is helping organizations respond rapidly as trends shift or issues emerge. As a result of this agility-driven culture within the business world, coupled with analytical capabilities offered by implementing BDA technology, companies can enhance competitiveness significantly while improving overall performance across all areas of operation.

In the meantime, with the advent of smart devices and IoT technologies, there has been a significant increase in the volume of data generated from various sources (Dai et al., 2019). This data is efficiently managed on large-scale platforms like Cloud of Things (CoT) ecosystems. The ecosystems encompass cloud resources with edge processing capabilities alongside smart devices, making them ideal for processing vast amounts of real-time data (Zuo et al., 2018). However, effective management is not enough, as the costs incurred while handling this deluge can pose challenges to firms that cannot afford expensive services, directly impacting their overall efficacy

DOI: 10.1201/9781003390954-7

(Caiza et al., 2020). Achieving a balance between cost-effectiveness and efficient data processing within CoT ecosystems is crucial, and for information systems, it entails ensuring optimal results based on available budgets.

Cloud computing can be expensive, particularly for organizations with significant storage and processing needs. Given this challenge, finding cost-effective ways to process big data in CoT ecosystems has become a critical research focus. One promising method is utilizing edge processing capabilities – the process by which devices located nearby data sources handle information before transmitting it to cloud resources (Cao et al., 2020). This approach has several benefits, including faster response times and greater reliability since it reduces reliance on distant network connectivity. Businesses can unlock major savings while enhancing their overall performance by striking an optimal balance between efficient data processing and affordable costs within CoT ecosystems.

Cloud computing resources are expensive, and the situation is intensifying for firms requiring substantial information system storage and processing power (Sandhu, 2021). Hence, a cost-cutting solution concerning handling big data in CoT ecosystems must be researched. Leveraging edge processing ability to process the data on the devices closer to the source within an information system could resolve this problem. Consequently, less volume of data that needs transmission to the cloud will decrease overheads associated with information systems dramatically. Besides reducing costs, edge processing reduces response times substantially; additionally, it is free from connectivity issues providing superior reliability and enhancing the performance of information systems (Zhang et al., 2021). For businesses looking for effective and inexpensive answers, efficient big data handling solutions are crucial when using CoT ecosystems.

Efficient use of cloud resources can help organizations achieve cost-effective big data processing within CoT ecosystems in information systems. Optimizing the utilization of cloud storage and processing resources is a good place to start, including strategies such as reducing data transfer frequency and implementing data compression techniques before transmission. Another alternative solution is adopting cost-effective cloud services like serverless computing. Companies only pay for actual resource usage in this model instead of committing to fixed amounts (Saeed et al., 2018).

The optimal cost-effective solutions for big data processing in CoT ecosystems are critical factors in efficiently managing information systems. Simulation experiments play an essential role in evaluating different CoT approaches' cost-effectiveness. By considering elements such as the number of devices on networks, data volume, device performance capacity, and associated costs with potential cloud computing options, researchers are better placed to identify viable long-term solutions.

This chapter provides an informative simulation study that delves into information systems, focusing on developing economical solutions for processing big data in CoT ecosystems. To replicate a CoT ecosystem, a simulation model is suggested to encapsulate smart devices, cloud resources, and edge processing capabilities. Experiments were performed to test diverse, viable alternatives to evaluate cost-effectiveness across various scenarios. The experiments help determine the most sustainable economic solution for big data processing within various conditions. Statistical analysis was applied to understand further comparative results gathered from experimentation. By scrutinizing this compiled research through statistical analysis, the study identifies the optimal economic solutions under different working conditions.

This chapter proposes solutions to address the significant cost challenges of processing extensive data volumes in CoT ecosystems. The contribution furnishes necessary tactics for organizations' adoption, enabling them to manage and mitigate costs associated with large volume processing demands, particularly within a CoT ecosystem. The structure of this chapter is as follows: Section 7.2 presents an overview of related works that provide cost-effective solutions for big data processing in CoT ecosystems. Section 7.3 describes the simulation model and the experimental setup Section 7.4 presents the experiments and a detailed analysis of the results. Section 7.5 then discusses the implications of the findings and suggests future research directions worth investigating. Finally, Section 7.6 presents the conclusion and future work.

7.2 RELATED WORKS

CoT ecosystems are grappling with processing big data effectively without draining resources. To overcome this issue, Nan et al., (2016) developed an efficient analysis solution that offers reliability while balancing costs against response time in fog computing (Muniswamaiah et al., 2021). The study found that using lower-capability fog nodes instead of cloud computing reduces average cost but results in longer response times. To mitigate this trade-off between time lag and monetary value, the authors proposed "unit-slot optimization," a compelling online algorithm tool which leverages Lyapunov optimization technology (Qiu et al., 2018) to strike a balance between cost-effective measures for big data processes through weighted decision-making based on currency consumption needs. Experimental outcomes demonstrated that the suggested method ensures sound spectral analysis solutions while satisfying users' requirements for low waiting durations. Hence, these findings offer developers additional solutions to achieve high-quality, inexpensive IoT outcome management systems within fog computing frameworks.

Bao et al. (2017) conducted a simulation study to address the fog-integrated IoT ecosystem device processing and data offloading. The Distributed Weighted Backpressure (DWB) policy was proposed to optimize online control systems for effective reductions in the cost of IoT devices. The DWB policy developed provided a three-way trade-off between communication costs, computation costs, and queue backlogs. It makes use of the Lyapunov optimization technique. A simulation study proved vital in validating this strategy's usefulness and correctness. According to findings from their analysis, Bao et al.'s approach tended toward minimizing IoT devices' overall cost within Fog-integrated IoT ecosystems quite effectively by leveraging on optimizing both distributed computation architecture and network transmission efficiency via their proposed DWB policy inversion techniques.

According to a study by Dhirani et al. (2017), the merger of IoT with cloud computing can pose cost management challenges. Although cloud computing provides added benefits such as better storage capacity and computational resources than IoT alone could offer, using the combined technology can result in increased costs which must be carefully managed by industries looking at adopting this innovation. The study recommends smartly managing expenses related to combining these technologies since achieving successful implementation necessitates identifying cost-effective alternatives for various activities such as processing data, data storage, and hiring cloud resources.

The study by Zhou et al. (2020) discusses the challenges of providing mobile-edge computing and cloud resources that cater to a diverse range of IoT applications. Different IoT applications have varying resource costs, including strict deadlines for multi-access edge computing (MEC) provisioning and flexible soft deadlines, which can be efficiently met by cloud computing. To tackle this challenge and provide operational efficiency, the study introduces an online cloud-edge resource provisioning framework based on Lyapunov optimization designed with delay awareness in mind. The proposed approach allows for optimal allocation withouts any assumptions made regarding system statistics to satisfy both deadlines while maintaining minimal operational cost-effectiveness. The analysis proves that the proposed framework accelerates deadline requirements meeting while minimizing operational costs; empirical evaluation under realistic traffic scenarios establishes its effectiveness further.

Xiang et al. (2022) conducted a study on allocating edge resources toward service composition-based IoT applications to tackle the challenge of resource allocation in a resource-limited environment. The researchers proposed an efficient approach to optimize the cost-effectiveness and performance robustness of IoT applications in MEC environments by balancing their performance metrics in consideration with other factors such as time and cost constraints. The study was built upon mathematical models for optimizing robustness in similar cloud-edge application systems, but it introduced some modifications that set it apart. To evaluate the proposed approach, experiments were conducted to compare the results obtained from two baseline methods against the proposed model across several measures, such as robustness, performance, and total costs incurred. The data-driven optimization approaches effectively managed the performance-related challenges faced by many companies adopting MEC over traditional cloud architectures for commercial projects.

7.3 METHODOLOGY

The methodology for this study to determine cost-effective solutions for big data processing in CoT environments followed several essential phases. This section outlines these phases, including developing a simulation model, experimental setup, and simulation metrics and settings specification.

7.3.1 SIMULATION MODEL

The AnyLogic simulation software created a discrete event simulation model that replicated the CoT ecosystem for this study. Table 7.1 shows the essential components utilized in the development of the simulation model.

TABLE 7.1
Key components of the simulation model

Component	Description
Smart devices	These are the devices that generate data in the CoT ecosystem. The study modeled these devices as sensors that generate data at a fixed rate.
Edge devices	Data processing is mainly done locally by edge devices, reducing the flow of information or data that needs cloud transfer. To minimize cloud dependence, edge devices receive smart device data and process it with reduced algorithms. These edge devices act as a gateway that accepts and curtails these details before moving on toward the cloud system.
Cloud resources	These are the resources used for processing data in the cloud. The study modeled cloud resources as virtual machines (VMs) running on a cloud provider (Amazon Web Services).
Cloud processing	Data processing in the cloud was modeled as a queueing system where first-in-first-out (FIFO) principle is used. In this concept, data are processed based on the order they arrive at the queue.

TABLE 7.2
Processing solutions

Component	Description
Cloud processing only	This solution involves processing all data in the cloud.
Edge processing only	This solution involves processing all data locally on edge devices.
Hybrid processing	This solution involves processing data on both edge devices and the cloud.

TABLE 7.3
Experiment parameters

Parameter	Description
Number of smart devices	Varied the number of smart devices from 10 to 100.
Volume of data generated	Varied the volume of data generated by each smart device from 10 MB to 100 MB per hour.
Processing capabilities of edge sevices	Varied the processing capabilities of edge devices from 0.5 GHz to 2 GHz.
Cost of cloud resources	Varied the cost of cloud resources from \$0.10 to \$0.20 per hour per VM.

7.3.2 Experimental Setup

Table 7.2 displays the processing solutions tested in this study to determine the most cost-effective option within the CoT ecosystem. The experiments conducted evaluated a range of alternatives with varying results.

The study varied several parameters to evaluate the cost-effectiveness of each processing solution. These parameters are presented in Table 7.3:

TABLE 7.4
Simulation metrics

Metric	Description
Total Cost	This is the total cost of processing data in the CoT ecosystem, including the cost of edge devices and cloud resources.
Average Latency	This is the average time it takes for data to be processed in the CoT ecosystem.
Data Transferred	This is the amount of data that needs to be transferred from smart devices to edge devices and from edge devices to the cloud.
Cloud Resource Utilization	This is the percentage of time that cloud resources are used for processing data.

TABLE 7.5
Simulation settings

Parameter	Value
Number of smart devices	10, 50, 100
Volume of data generated	10 MB/hr, 50 MB/hr, 100 MB/hr
Processing capabilities of edge devices	0.5 GHz, 1 GHz, 2 GHz
Cost of cloud resources	$0.10/hr, $0.15/hr, $0.20/hr
Time step	1 second
Simulation Time	24 hours
Replications	10
Cloud provider	Amazon Web Services
VM configuration	2 vCPUs, 8 GB RAM, 100 GB SSD storage

7.3.3 SIMULATION METRICS

In an attempt to analyze processing solutions' cost-effectiveness, the study employed various metrics. These metrics are presented in Table 7.4.

7.3.4 SETTINGS

To simulate the CoT ecosystem, the study used the following settings as shown in Table 7.5.

The study analyzed a spectrum of variables, including the number of smart devices ranging from 10 to 100. The volume of data generated by each smart device varied between 10 MB/hr to 100 MB/hr. In contrast, processing capabilities varied from as little as 0.5 GHz up to 2 GHz and cloud resources varied between the cost of $0.10/hr and $0.20/hr depending on usage, type, or complexity requirements.

The study ran ten replications of its experiments to ensure statistically significant results. Each replication featured the simulation model running with a different random seed, varying the input data every time. The results were then averaged across all replications for greater accuracy and precision in predicting future simulation metrics. To maintain consistency throughout each trial, the study implemented a time step of one second within their CoT ecosystem simulation setup; this allowed them to focus on how events unfolded over 24 hours without compromising system performance or robustness. Amazon Web Services functioned as the cloud provider. Furthermore, virtual machines processing was configured with two virtual central processing units (VCPUs), 8 GB of RAM for memory utilization efficiency, and 100 GB limiting SSD storage usage while maintaining effectiveness.

7.4 FINDINGS

This simulation study examined three processing solutions (cloud processing only, edge processing only, and hybrid processing) to determine which has the lowest average latency. This was done by varying two factors: the number of smart devices in a CoT ecosystem (from 10 to 100) and the data volume generated by each device per hour (from 10 MB/hr to 100 MB/hr). The results are presented in Table 7.6, and they show that edge processing had the lowest average latency among all three solutions, followed by hybrid and solely cloud-based operations. Since edge devices operate closer to devices that generate raw data, it is faster than cloud-based computing resources, which are physically farther from these sources, delaying their response times. Hybrid achieved lower latency than strictly cloud processing because some of the load is shared with edge devices as part of distributed computing structure, reducing overall centralization function obligations and making it efficient compared to stand-alone cloud schema, as shown in Table 7.6.

Every processing solution in the CoT ecosystem has been analyzed under several parameter combinations. Table 7.7 represents the detailed summary of data transmission for each processing solution. Cloud processing, edge processing, and hybrid processing are three solutions considered. Notably, the number of smart devices installed in the ecosystem was exclusively 10, with a fixed volume of data generation of 10 MB/hr from every smart device. Moreover, edge computing processors

TABLE 7.6
Average latency for processing solutions

Processing Solution	Number of smart devices	Volume of data generated	Processing capabilities of edge sevices	Cost of Cloud Resources	Average Latency
Cloud processing only	10	10 MB/hr	0.5 GHz	$0.10/hr	12.87 seconds
Edge processing only	10	10 MB/hr	0.5 GHz	N/A	6.14 seconds
Hybrid processing	10	10 MB/hr	0.5 GHz	$0.10/hr	9.72 seconds

TABLE 7.7
Data transferred for processing solutions

Processing Solution	Number of smart devices	Volume of data generated	Processing capabilities of edge sevices	Cost of Cloud Resources	Data Transferred
Cloud processing only	10	10 MB/hr	0.5 GHz	$0.10/hr	10.8 GB
Edge processing only	10	10 MB/hr	0.5 GHz	N/A	3.6 GB
Hybrid processing	10	10 MB/hr	0.5 GHz	$0.10/hr	6.4 GB

TABLE 7.8
Cloud resource utilization for processing solutions

Processing Solution	Number of smart devices	Volume of data generated	Processing capabilities of edge sevices	Cost of Cloud Resources	Cloud Resource Utilization
Cloud processing only	10	10 MB/hr	0.5 GHz	$0.10/hr	87.2%
Edge processing only	10	10 MB/hr	0.5 GHz	N/A	N/A
Hybrid processing	10	10 MB/hr	0.5 GHz	$0.10/hr	46.9%

were operating at the rate of 0.5 GHz while fixing cloud resources cost to $0.10/hr as per the analysis standardization policy followed during experiments.

Three methods were employed for data processing: cloud processing only, edge processing only, and hybrid cloud-edge processing. The amount of data transferred varied depending on the type of method chosen. Data processed purely in the cloud resulted in 10.8 GB total transfer, while having it at the edge saved a larger chunk of that with 3.6 GB total transfer. Hybrid is an interesting choice since some processes were completed at the immediate location before being sent over to the cloud for finalization. Thus, the total data transfer was 6.4 GB in this case, as shown in Table 7.7.

Different parameter combinations were used to analyze the utilization of cloud resources for processing solutions, summarized in Table 7.8. The three processing solutions considered were cloud-only, edge-only, and hybrid processing. For this analysis, it was assumed that there were only ten smart devices in the CoT ecosystem and each device generated data volume with a fixed throttling limit of 10 MB/hr. Further, it is important to add that all edge devices had a fixed processing capability of half a GHz while costs associated with cloud resources were set at $0.10/hr to arrive at the information presented in Table 7.8.

Findings show that cloud processing had an 87.2% utilization of cloud resources by sending all data directly to the cloud for processing. In contrast, no cloud resources were utilized in edge processing as all the work was processed locally at the network's edge. Hybrid processing combined these two methods; some data went through local edge networks before transferring to the cloud, resulting in only 46.9% utilization. Cloud processing relies solely on using entire cloud resources. It is marginally better than hybrid processing. Edge-only processing used none of it, suggesting an edge experience-oriented solution with significant savings could significantly reduce data transfer and resources expenses. The studies' conclusion demonstrated how organizations can cut their operating costs drastically by opting for edge computing routing when having limited cloud resources or looking to reduce their usage costs, as shown in Table 7.8.

The cost of processing data within the CoT ecosystem in each processing setting and parameter combination is summed up in Table 7.9. For a standardized comparison across all settings, it was assumed that there were ten smart devices within the CoT ecosystem, each with an equal volume of data generation – 10MB/hr. Furthermore, during experimental runs, all edge computing nodes were maintained at a uniform rate of 0.5 GHz power capacity. At the same time, costs for cloud computation remained consistent at $0.10/hour spent on operation throughout every test simulation window analyzed.

The cost of cloud processing, where data is only sent to be processed outside of the edge device, was $14.40. This includes a fee for using the particular cloud node's processing power and duration trading off. Using edge computing only instead, which does not require any external resources outside that of an edge device, results in a cost of $10.80 due to no additional fees except those associated with the device used. In hybrid processing, however, some data is processed on-site at the edge location before being transmitted to remote storage; this option led to a total processing expense of $11.20 due to combined costs from both local devices, including their ability and any utilization by remote resources.

All three options were analyzed in terms of their overall value alongside performance results, whereby purely edge computing remained relatively cheap largely

---

TABLE 7.9
Total cost for processing solutions

Processing Solution	Number of smart devices	Volume of data generated	Processing capabilities of edge sevices	Cost of Cloud Resources	Total Cost
Cloud processing only	10	10 MB/hr	0.5 GHz	$0.10/hr	$14.40
Edge processing only	10	10 MB/hr	0.5 GHz	N/A	$10.80
Hybrid processing	10	10 MB/hr	0.5 GHz	$0.10/hr	$11.20

due to its lack of dependence on costly remote processing capabilities used solely for cloud processing. Blended solutions like hybrid processing provide balance but edge out cloud processing among economic choices because it minimized extraneous bandwidth charges while optimizing price-performance enhancements whenever possible, as shown in Table 7.9.

7.5 DISCUSSION

An information systems perspective study evaluated the cloud's cost-effectiveness for processing big data in CoT solutions. This study analyzed three processing solutions – the hybrid, edge and cloud techniques – used within CoT ecosystems. The findings showed that the edge method with IoT devices integrated into information systems displayed minimal latency compared to hybrid or entirely cloud-based solutions. These results suggest that when data sources are closer to IoT devices, they can be processed much faster than relying on remote servers. Hybrid methods also fared well due to distributing processing tasks, reducing system latency. However, all three processing methods demonstrated effective management of big data in CoT ecosystems despite some differences in performance efficiency.

Data transferred by cloud-based information systems was significantly higher than those utilizing hybrid and edge processing. This is because cloud processing involves handling all data exclusively in the cloud. In contrast, hybrid and edge processing use a mix of local computing with some or no communication to the cloud. Nevertheless, it is worth noting that the variance of data transmitted between each method was small. Cloud-based information systems saw their highest resource utilization when using dedicated cloud resources. The integration of cloud services remained essential even for hybrid ones, despite featuring on-premises computing elements. Edge technology only focuses on locally performing computations without relying on external network utilization or response time. It highlights how cost-effective and efficient offloading tasks to low-power devices can be for organizations looking to benefit from new technologies with defined budgets or improving efficiency within complex ecosystems like managing information systems.

This study indicates that edge processing and hybrid processing in information systems can have a significant advantage over cloud processing, especially in cost-effectiveness, data transfer speed, and latency reduction within the CoT ecosystem. However, not all information systems are made equal under these circumstances; thus, considering requirements and constraints specific to the CoT system is essential when selecting a processing system.

Edge processing has an added benefit by reducing data transfer amounts to clouds and lowering network traffic around devices. At the same time, reducing costs resulting from bigger bandwidth associated with high-speed connectivity feeds that Internet service providers require. This shows that cost-effectiveness is vital for edge processing in its practical use cases. Nevertheless, one must recognize that there is no one-size-fits-all solution. Not all use cases are simple enough for edge platforms – those requiring computational power or real-time decision-making capabilities may need more refined alternatives like hybrid processing merging benefitting aspects from both types of architecture and striking a balance effectively between them as per CoT demands.

Hybrid processing provides two benefits to data processing: combining edge and cloud resources for information transfer. By distributing the workload across edge devices, organizations can reduce their dependency on expensive cloud services, which directly results in cost-cutting. The effective functioning of hybrid computing depends largely on determining how much/less to portion the workload between processors at both ends. Although this may sometimes be challenging, it massively reduces the costs of relying solely on cloud resources. It should be noted, however, that our simulation study findings were based on specific parameter settings and may not necessarily apply to all scenarios. Further research is required to investigate different combination methods across various use cases within information systems.

This study indicates that edge processing and hybrid processing present several advantages over relying solely on cloud processing, particularly in latency, data transfer, and cost-effectiveness. However, selecting an appropriate processing solution is contingent upon the specific demands and limitations of the CoT ecosystem, and a combination of processing approaches is the most efficient strategy in certain scenarios. The outcomes of our study offer valuable insights for organizations seeking to adopt economical solutions for processing large volumes of data in the CoT ecosystem.

7.6 CONCLUSION AND FUTURE WORK

Information systems have become increasingly vital in today's world. This chapter delves into this topic, presenting a simulation study focusing on the best cost-effective approaches to handle Big data processing in the CoT ecosystem. To achieve this goal, the chapter developed a simulation model replicating the CoT environment and examined various simulations looking at viable techniques for processing data while minimizing costs under different scenarios.

This study's findings indicate the benefits of combining edge processing with cloud technologies in large volume big data solutions, especially within CoT frameworks. In addition to substantial cost reductions, this method is effective for coping with all computational needs. The simulation experiment findings further highlight its superiority compared to centralized and decentralized approaches concerning resource usage levels, response time, and throughput. Moreover, CoT technology presents scalability and flexibility advantages while managing data storage capabilities efficiently. Interoperability concerns and security issues must be addressed to successfully implement optimization techniques regarding CoT-based resource allocation in large-scale IoT deployment scenarios.

The policy implications of the findings suggest that policymakers play a vital role in promoting CoT technology and IoT deployment optimization. Policymakers can use incentives to encourage businesses to adopt CoT-based techniques for resource allocation optimization, review factors influencing resource allocation via CoT-based systems, and promote the development of interoperable and secure CoT-based devices.

Future research directions should explore potential cooperative things (CoTs) uses with various large-scale IoT deployment scenarios. Extending initial simulation experiments can facilitate an inclusive assessment of performance analysis regarding diverse facets such as multi-cloud environments or edge computing. Subsequently, addressing security concerns while preserving individual privacy remains a significant hurdle concerning CoT-based IoT networks. Secure protocols need development alongside policies that guarantee regulating authorities' protection over sensitive data generated from various IoT sources.

In addition, it is essential to focus research on CoT-based IoT systems' interoperability challenges within information systems. This will require the development of open standards and protocols that allow for seamless interaction among IoT platforms and devices for efficient integration and management. Furthermore, specific research should be conducted on applying CoT technology to improve resource allocation in various IoT domains such as healthcare and smart cities. Such focused investigations into these domains help tailor the resource distribution strategy based on their unique requirements, potentially enhancing effectiveness. CoT technology has exciting potential to make large-scale IoT deployments more productive while enabling efficient resource allocation. Policymakers also play an important role in advocating its implementation within information systems. Moreover, researching additional areas related to CoT technologies can uncover its full potential affecting improvements across all aspects of life indirectly or directly.

REFERENCES

1. Bao, W., Li, W., Delicato, F. C., Pires, P. F., Yuan, D., Zhou, B. B., & Zomaya, A. Y. (2017). Cost-effective processing in fog-integrated internet of things ecosystems. In *Proceedings of the 20th ACM International Conference on Modelling, Analysis and Simulation of Wireless and Mobile Systems* (pp. 99–108). ACM. DOI: 10.1145/3127540.3127547

2. Caiza, G., Saeteros, M., Oñate, W., & Garcia, M. V. (2020). Fog computing at industrial level, architecture, latency, energy, and security: A review. *Heliyon*, 6(??), e03706. DOI: 10.1016/j.heliyon. 2020.e03706

3. Cao, K., Liu, Y., Meng, G., & Sun, Q. (2020). An overview on edge computing research. *IEEE Access*, 8, 85714–85728.

4. Curry, E. (2016). The big data value chain: Definitions, concepts, and theoretical approaches. In J. M. Cavanillas, E. Curry, & W. Wahlster (Eds.), *New Horizons for a Data-Driven Economy* (pp. 41–52). Springer. DOI: 10.1007/978-3-319-21569-3_3

5. Dai, H. N., Wang, H., Xu, G., Wan, J., & Imran, M. (2019). Big data analytics for manufacturing internet of things: Opportunities, challenges and enabling technologies. *Enterprise Information Systems*, 14(9–10), 1279–1303. DOI: 10.1080/17517575.2019.1633689

6. Dhirani, L. L., Newe, T., Lewis, E., & Nizamani, S. (2017). Cloud computing and internet of things fusion: Cost issues. In *ICST 2017 International Conference on Sensing Technology Proceedings* (pp. 1–6). IEEE. DOI: 10.1109/ICSensT.2017.8304426

7. Karupusamy, S., Ramalingam, M., & Kumar Eswaran, B. (2021). Big data analytics and management in internet of things. *Journal of Information Technology Management*, 13(Special Issue: Big Data Analytics and Management in Internet of Things), 1-5.

8.   Muniswamaiah, M., Agerwala, T., & Tappert, C. C. (2021). Fog computing and the Internet of Things (IoT): A review. In *2021 8th IEEE International Conference on Cyber Security and Cloud Computing (CSCloud)/2021 7th IEEE International Conference on Edge Computing and Scalable Cloud (EdgeCom)* (pp. 10–12). Washington, DC, USA. DOI: 10.1109/CSCloud-EdgeCom52276.2021.00012

9.   Naeem, M., Jamal, T., Diaz-Martinez, J., Butt, S. A., Montesano, N., Tariq, M. I., ... & De-La-Hoz-Valdiris, E. (2022). Trends and future perspective challenges in big data. In *Advances in Intelligent Data Analysis and Applications: Proceeding of the Sixth Euro-China Conference on Intelligent Data Analysis and Applications*, 15–18 October 2019, Arad, Romania (pp. 309-325). Springer Singapore

10. Nan, Y., Li, W., Bao, W., Delicato, F. C., Pires, P. F., & Zomaya, A. Y. (2016). Cost-effective processing for delay-sensitive applications in Cloud of Things systems. In *2016 IEEE 15th International Symposium on Network Computing and Applications (NCA)* (pp. 162–169). Cambridge, MA, USA: IEEE. DOI: 10.1109/NCA.2016.7778612

11. Qiu, C., Hu, Y., Chen, Y., & Zeng, B. (2018). Lyapunov optimization for energy harvesting wireless sensor communications. *IEEE Internet of Things Journal*, 5(??), 1947-1956. DOI: 10.1109/JIOT.2018.2817590

12. Ranjan, J., & Foropon, C. (2021). Big data analytics in building the competitive intelligence of organizations. *International Journal of Information Management*, 56, 102231.

13. Saeed, S., Jhanjhi, N., Abdullah, A., & Naqvi, M. (2018). Current trends and issues legacy application of the serverless architecture. *International Journal of Computing & Network Technology*, 6, 100–108. DOI: 10.12785/IJCNT/060304

14. Sandhu, A. K. (2021). Big data with cloud computing: Discussions and challenges. *Big Data Mining and Analytics*, 5(??), 32-40.

15. Vats, S., & Sagar, B. B. (2019). Data lake: A plausible big data science for business intelligence. In S. Bhatia & R. Singh (Eds.), *Communication and Computing Systems* (pp. 7). CRC Press. Doi: 10.1201/9780429444272

16. Vitari, C., & Raguseo, E. (2020). Big data analytics business value and firm performance: linking with environmental context. *International Journal of Production Research*, 58(??), 5456-5476.

17. Xiang, Z., Zheng, Y., Wang, D., et al. (2022). Robust and cost-effective resource allocation for complex iot applications in edge-cloud collaboration. *Mobile Networks and Applications*, 27(??), 1506-1519. DOI: 10.1007/s11036-022-01977-9

18. Zhang, T., Li, Y., & Chen, C. L. P. (2021). Edge computing and its role in industrial internet: Methodologies, applications, and future directions. *Information Sciences*, 557, 34–65. DOI: 10.1016/j.ins.2021.02.027

19. Zhou, Z., Yu, S., Chen, W., & Chen, X. (2020). CE-IoT: Cost-effective cloud-edge resource provisioning for heterogeneous IoT applications. *IEEE Internet of Things Journal*, 1–1. DOI: 10.1109/JIOT.2020.2994308

20. Zuo, Y., Tao, F., & Nee, A. Y. C. (2018). An Internet of Things and cloud-based approach for energy consumption evaluation and analysis for a product. *International Journal of Computer Integrated Manufacturing*, 31(4–5), 337–348. DOI: 10.1080/0951192X.2017.12854

8 Phishing Attack Detection for Cloud-Based Environment Using Hybrid Machine Learning Model

Md Kamar Raza and Kunwar Pal

8.1 INTRODUCTION

The number of companies has increased exponentially, and new technologies are always being investigated for wider applications. Millions of new websites are created every day. That collects user credentials through login gateways. Due to the sheer number of these websites, it becomes harder to confirm their legitimacy. According to data made public by Dofo, till March 2023, around 400 million domain names and 2 billion subdomains were registered [1]. It is simpler for attackers to entice more users as the user base grows.

One of the most hazardous types of cybercrime is phishing, in which cybercriminals obtain personal data including passwords, bank account details, and credit card numbers. Phishing sounds like "Fishing." Bait is used to capture fish. A phisher targets everyday Internet users. Cybercriminals misuse this data for profit. Phishing is a major cybercrime.

According to research released recently by the Anti-Phishing Working Group [2], there were 1,270,883 phishing sites found in the third quarter of 2022. The proportion of phishing attacks directed at the financial sector was 23.2%. Phishing attacks reached an all-time high during this quarter, making it the worst one the APWG has ever seen. As shown in Figure 8.1, economic sector has been affected by phishing attempts.

When someone gets phished, they are tricked into clicking on a link or opening an attachment that appears to come from a legitimate source. The plot to carry out the attack is now being planned by phishers or attackers. When someone is attacked, they become the victim or target of the attack. If you've ever received an email that seems suspicious, it may be a phishing attempt designed to trick you into divulging personal information or downloading malware.

In its most common form, phishing includes an adversary creating a false website and distributing it to a victim by email or another type of electronic communication, as depicted in Figure 8.2.

DOI: 10.1201/9781003390954-8

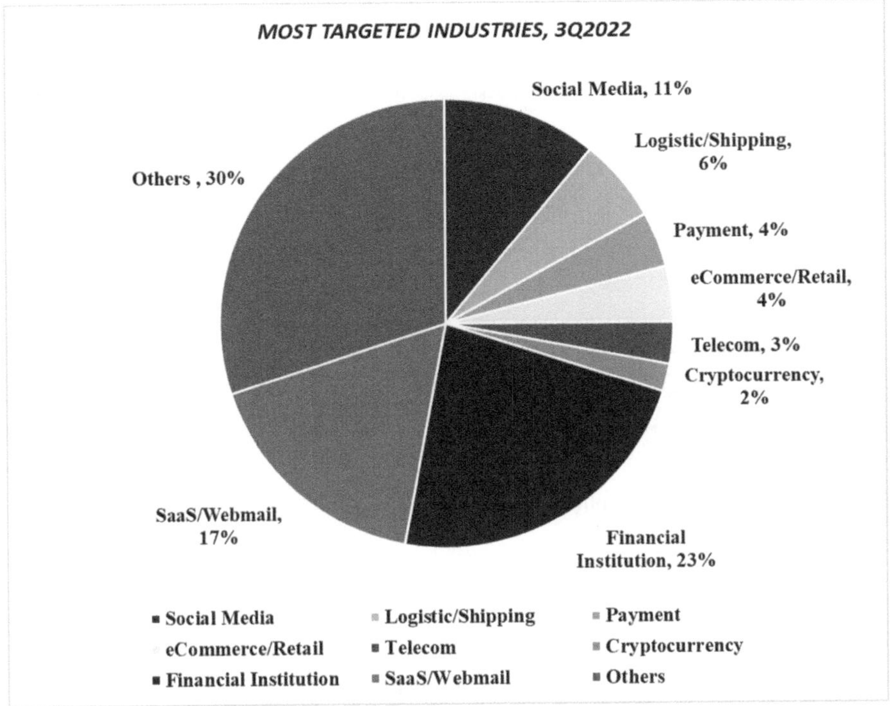

FIGURE 8.1   Most targeted industries.

Using a network or the Internet, users may access and use computer resources and services, such as storage, processing power, and software applications, without the need for local hardware or infrastructure. This is known as cloud computing. In other words, it entails the use of remote servers located on the Internet in place of local servers or individual computers to store, manage, and process data. Businesses and people may benefit from cloud computing in many ways, including increased productivity, cost savings, flexibility, and scalability. Additionally, it offers a variety of deployment options, each with its own capabilities and use cases, such as public, private, and hybrid clouds [3].

The main contributions of this chapter are the following:

- This study presents a ML (ML)-based approach for detecting phishing attacks. We have created a unique data set of phishing URLs and extracted 15 lexical features of the URLs using a novel feature extraction approach.
- We used several feature selection algorithms to identify the ten most effective lexical features of URLs.
- The model is evaluated using hybrid ML algorithms, and the results show that the system outperforms other approaches for phishing detection, with an accuracy rate of 97.26%.
- We also analyzed the significance of the features to show how well the method works in identifying essential aspects of phishing URLs.

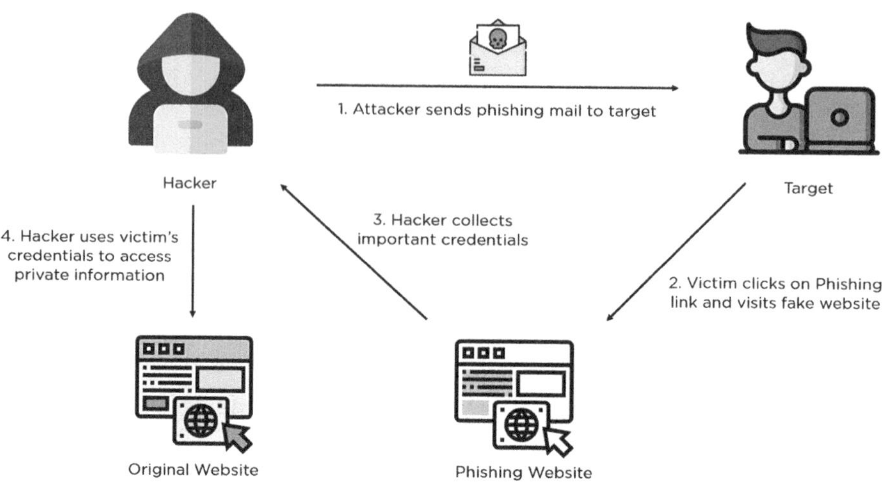

FIGURE 8.2   Conventional phishing attack.

- We suggest a method for phishing detection in cloud environments as an Application Programming Interface [API] of our hybrid model.
- The findings have significant implications for enhancing online communication security and mitigating the effects of phishing attacks on people and companies, offering a more precise and reliable mechanism for detecting phishing URLs.

The chapter is structured as follows: different phishing defense mechanisms are covered in section 8.2, related research is covered in section 8.3, and the proposed technique is covered in section 8.4. Section 8.5 contains details on the implementation, an analysis of the findings, and discussions. Section 8.6 concludes the chapter and discusses some potential future work.

8.2 PHISHING DEFENSE MECHANISM

There are a variety of different approaches that can be used to identify phishing scams. The numerous safeguards against phishing are illustrated in Figure 8.3. The detection of phishing attacks can be done in two ways.

1. User awareness
2. Software-based mechanism

8.2.1 USER AWARENESS

Many Internet users are unaware of online security and threats. Nowadays, hackers use victims' lack of knowledge to phish them. At this stage, users must be educated about online threats and how to protect their data. However, according to the authors of [4], teaching individuals how to recognize a trustworthy website takes time and effort, hence a strong phishing defense system is required.

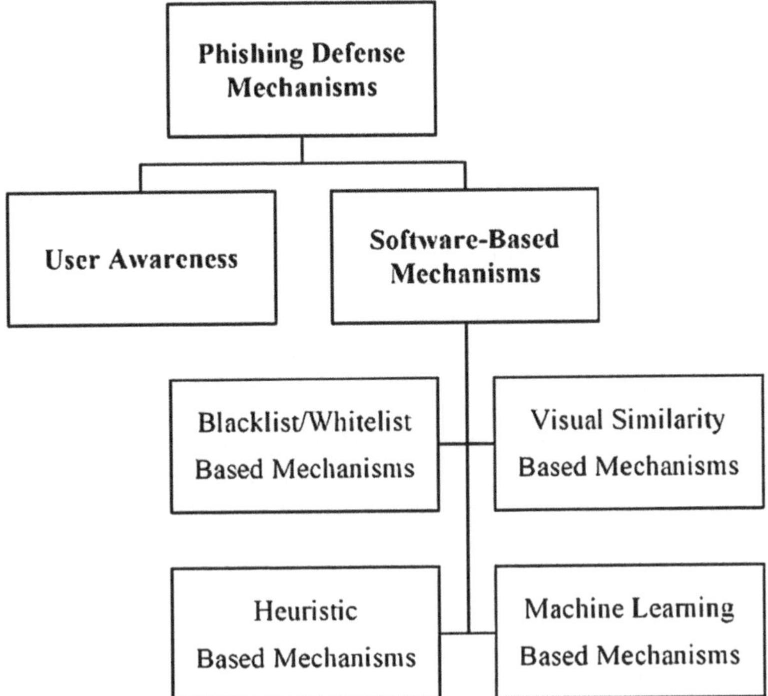

FIGURE 8.3 Phishing defense mechanism.

8.2.2 Software-Based Mechanism

Software-based mechanisms can be classified into subcategories such as blacklist- and whitelist-based mechanisms, heuristic-based mechanisms, visual similarity-based mechanisms, and ML or deep-learning-based processes.

8.2.2.1 Blacklist- and Whitelist-Based Mechanism

Blacklisting suspicious websites is a tried-and-true strategy. A blacklist contains all hazardous domains, URLs, and IP addresses. Since thousands of fake websites are established daily, the blacklist's source updates it frequently. These methods are faster than ML and visual similarity algorithms. Most list-based solutions do not guard against zero-hour attacks, resulting in poor detection rates [5]. To solve this problem, whitelist and blacklist methods have been combined with ML to develop a system that can automatically recognize phishing websites or real websites not on this list.

8.2.2.2 Visual Similarity-Based Mechanism

Visual similarity-based phishing detection can identify and classify phishing websites by comparing their visual similarities to legitimate websites. The technique involves assessing the degree of similarity between the visual components of a suspected phishing website and those of legitimately recognized websites, such as the layout, color scheme, and logos. If the similarity score exceeds a set level, a website is classified as a phishing website. This approach is particularly useful for identifying sophisticated phishing scams that trick visitors into supplying personal information by employing convincing graphic spoofs of reliable websites [6].

8.2.2.3 Heuristic-Based Mechanism

Using a heuristic-based technique the phishing websites are matched with a comprehensive feature set (which may include keywords, URL characteristics, IP addresses, an SSL certificate, Pop-up windows, and other similar elements). The following are some examples of these types of features: it is possible to uncover zero-day attacks by employing detection systems that are heuristically based. In order to identify whether or not the websites in question are phishing sites, one method that is based on heuristics examines the features of malicious websites. These characteristics include things like login forms, URLs, and online traffic. When designing new websites, attackers always make it a point to pay careful attention to heuristic-based characteristics in order to avoid being discovered by phishing detection systems [7].

8.2.2.4 Machine-Learning/Deep-Learning-Based Mechanism

Over the last few years, a number of researchers have developed ML/deep learning approaches for detecting phishing websites. These methods use a big database of legitimate and fake websites. The URL, page content, Domain Name Space (DNS), and other features are extracted, and a new data set containing the specified feature set is created. After that, the data set is preprocessed before ML or deep learning algorithms are deployed.

8.3 RELATED WORK

A study by Alwanain [8] looked into and studied the effect of user awareness on the detection of phishing attacks. Real-time experiments were conducted on a group of users to track their reactions to various cyber threats and gather data on those reactions. The testing revealed that in order to increase people's levels of phishing awareness, there is a major need for hands-on teaching.

An effective method was developed by Ankit et al. [9] and may be used to check if a webpage is real or not using hyperlink-based features and to defend against phishing attacks using an automatically updated whitelist of trustworthy websites that each user visits. The proposed approach is quite effective, with an 86.25% true positive rate (TPR) and a very low false positive rate (FPR).

An improved blacklist strategy was developed by Rao et al. [10] to identify phishing websites that are copies of valid websites already in existence. The proposed approach incorporates significant aspects that were taken directly from the source code of the authentic websites. The proposed method has a phishing detection rate of 84.39%.

An approach based on visual resemblance was suggested in [11]. The authors automatically extracted 224 unique web pages' visual mimics from 2,262 rogue sites without any prior information and achieved an 80% phishing detection rate.

Using a data set with 2,129 phishing webpages and 1,367 genuine webpages, Zhou et al. [12] suggested a unique technique focused on the pure picture level and achieved a 90% TPR.

Chiew et al. [13] proposed a hybrid technique using an ML algorithm to extract a logo from a webpage a person is browsing. The obtained logo is then uploaded to Google's image search engine in order to validate the integrity of the target website. When a website is visited, its URL is compared to the results of several search engines; if the URL matches, the website is considered to be legitimate; otherwise, it is considered to be phishing. If a phishing website has no logo, an algorithm that extracts logos will fail.

Ali et al. [14] has devised the wrapper-based features selection approach that is used for selecting the most significant features to employ in effectively predicting the phishing website. This method can be found in the referenced article. The author incorporated a number of different supervised ML algorithms, including Back-Propagation Neural Network, Support Vector Machine (SVM), Naive Bayes, Decision Tree (DT), Random Forest (RF), K-nearest neighbor (KNN). In addition, wrapper-based feature selection has the potential to be combined with ensemble learning in order to boost the effectiveness of intelligent phishing website detection strategies.

Mahajan et al. [15] used DT, RF, and SVM to categorize data in their research work. Each of 17,058 safe and 19,653 phishing URLs contained 16 features. The data set was collected from [16] and [17]. The data set was split 50:50, 70:30, and 90:10 into training and testing sets. Accuracy, false negative, and false positive rates determined performance. RF has the lowest false negative rate at 97.14%. The study found that increasing training data improves accuracy.

Jain et al. [18] proposed a method for phishing detection in which they acquired 19 features from the client side in order to distinguish between phishing links and legal ones. They used 1,918 lawful web pages and 2,141 phishing pages in order to construct a data set. They succeeded in achieving a 99.39% accuracy as a result. They have created a technique that can be applied to client-side desktop software and has the potential to be used for real-time phishing detection.

Gupta et al. [19] proposed a method for phishing detection in which they acquired nine lexical URL features from the URL in order to distinguish between phishing links and legal ones. The data set was collected from [20]. They used 10,000 legitimate web pages and 9,964 phishing pages in order to construct a data set. They succeeded in achieving a 99.57% accuracy as a result. They have created a technique that has the potential to be used for real-time phishing detection.

8.4 PROPOSED METHODOLOGY

The principal aim of our proposed approach is to distinguish phishing websites from legitimate ones. The proposed architecture is shown in Figure 8.4.

The proposed system is divided into two phases: the pre-deployment phase and the deployment phase. In the pre-deployment phase, the data set goes through a pre-processing phase in which the data is prepared. The pre processing output is fed into a feature extraction and selection process in which a relatively small number of

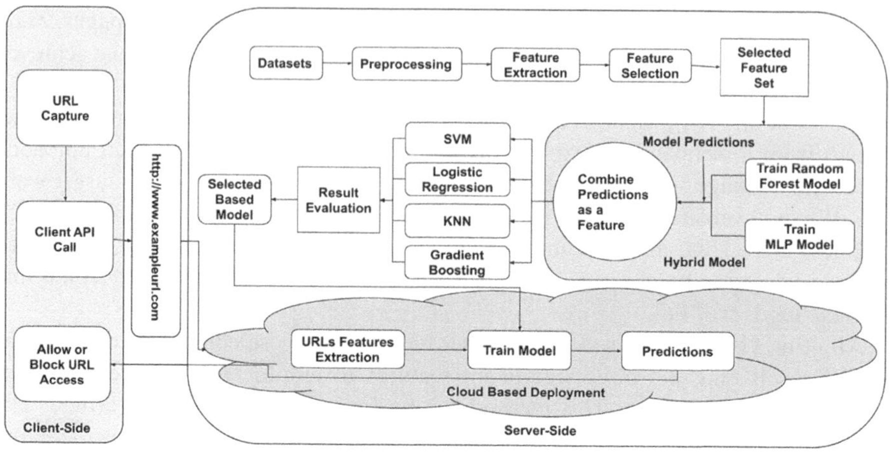

FIGURE 8.4 Proposed methodology.

features are selected. The resulting data set is then used to train and test hybrid ML classifiers. After the initial testing phase, the best-performing hybrid model is tested further to ensure that it can be generalized well beyond the training data set. After systematic testing, the model is archived for further use in the cloud environment.

A hybrid model and a voting classifier are both utilized in ML, but they have distinct characteristics and purposes.

A hybrid model is designed to leverage the strengths of multiple ML algorithms or models to enhance overall performance. It integrates different models that are often diverse, such as RF, SVM, KNN, multi-layer perceptron (MLP), Gradient Boosting (GB), etc. These models work together cooperatively, with specialized roles based on their strengths, to improve the final predictions. A hybrid model aims to combine the unique capabilities of each model and overcome their weaknesses. The final prediction is obtained by fusing the predictions from the constituent models using techniques like averaging or stacking.

On the other hand, a voting classifier is a specific type of hybrid model focused on classification tasks. It combines the predictions of multiple individual classifiers to make a final prediction. Unlike a hybrid model, the constituent models in a voting classifier are typical of the same type, using the same algorithm or learning approach. Each model operates independently and makes predictions on its own. The voting classifier employs a voting strategy, such as majority voting or weighted voting, to combine the predictions and reach a consensus. This approach benefits from the collective wisdom of the models and provides a more robust and accurate prediction.

To summarize, a hybrid model is a broader concept that involves integrating different models to solve a task, whereas a voting classifier is a specific type of hybrid model focused on combining predictions from classifiers of the same type for classification tasks.

8.4.1 Data Set and Preprocessing

We used a combined data set of phishing and legitimate URLs collected from Kaggle and SpyCloud to assess our proposed methodology. Kaggle's URL collection contains both legitimate and phishing URLs, but SpyCloud solely contains phishing URLs during the COVID-19 timeframe. Both [21] and [22] URL-data set have updated URLs through 2021. There are a total of 450,176 instances in Kaggle, of which 345,738 are legitimate and 104,438 are phishing URLs. There are 133,724 instances of Spycloud, and we are using 130,000 phishing URLs. Hence, there are 5801,75 instances in our combined data set, out of which 345,738 are legitimate and 234,437 are phishing URLs.

Our data set only contains URL and label columns; therefore, after preprocessing we simply removed any occurrences that were unrelated to our experiment. We use a feature extraction approach to extract features from the URLs once the data set has been cleaned.

8.4.2 Features Extraction

URLs are characterized by a number of traits and patterns that can be categorized as features of the format. The important components of a typical URL are categorized and displayed in Figure 8.5.

There are 15 features extracted from the URLs in total that are discussed below.

8.4.2.1 Steps to Extract Features from URLs

8.4.2.1.1 Domain Token Count

The domain token count of a URL refers to the number of segments or parts that make up the domain name. In a URL, the domain name is the part that comes after the protocol (e.g., "http://" or "https://") and before the first slash ("/").

For example, in the URL "https://www.exmaple.com/blog/article", the domain name is "www.exmaple.com", and the domain token count is 3, because it has three segments: "www", "example", and "com".

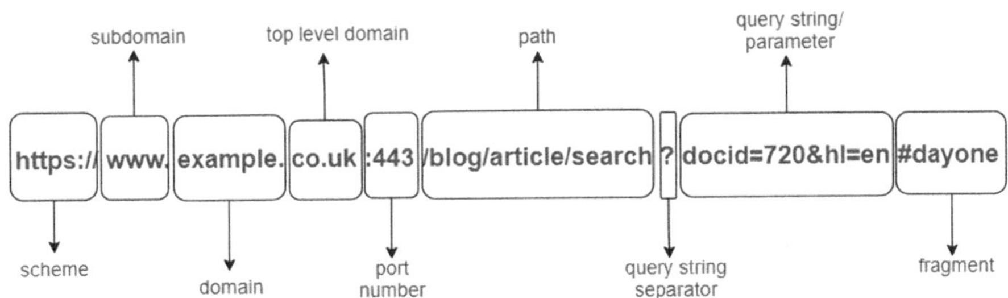

FIGURE 8.5   Anatomy of URLs.

Algorithm 8.1 Feature extraction algorithm.

1: **procedure** PHISHDETECTION(*inputURL*)
2:     Define a function called extract_features that takes a URL as input.
3:     Initialize variables for the various features to be extracted.
4:     Use the urlparse() function from the urllib.parse module to parse the URL
       and extract the domain and path components.
5:     Count the number of tokens in the domain and path.
6:     Count the number of symbols in the domain and path.
7:     Calculate the ratio of the number of tokens to the number of symbols in the
       domain and path.
8:     Check if the URL contains an IP address, HTTPS, prefixes/suffixes, hyphens,
       or TinyURLs.
9:     return a dictionary object containing the extracted features.
10: **end procedure**

8.4.2.1.2 Symbol count in URL

The symbol count in a URL refers to the number of special characters or symbols used in the URL. These can include characters such as ".", "/", "?", "=", "&", and others.

For example, in the URL "https://www.exmaple.com/blog/article?id=123", the symbol count is 4, because there are four special characters: "." (dot) after "com", "/" (slash) after "blog", "?" (question mark) after "article", and "=" (equals sign) and "&" (ampersand) in the query string "id=123".

8.4.2.1.3 Ratio of Path Token and URL Token

The number of segments or parts in the domain name to the number of segments or parts in the URL path is known as the path token to URL token ratio.

For example, in the URL "https://www.exmaple.com/blog/article", the domain token count is 3 ("www", "exmaple", and "com"), and the path token count is 2 ("blog" and "article"). Therefore, the ratio of URL token and path token is 2:3.

8.4.2.1.4 Ratio of Path Length and URL Length

The proportion of the URL's overall character count to its path's character count is known as the path-to-URL length ratio. The portion of the URL that comes after the domain name, including any subdirectories or filenames, is known as the URL path.

For example, in the URL "https://www.exmaple.com/blog/article", the URL length is 31 (including the protocol, domain name, and path), and the path length is 13 ("blog/article"). Therefore, the ratio of URL length and path length is approximately 13:31.

8.4.2.1.5   Ratio of Domain Token and Symbol Count in Doamin

The number of segments or sections in a domain name divided by the number of special characters or symbols used in the domain name is known as the domain token and symbol count ratio.

For example, in "www.exmaple.com", the domain token count is 3 ("www", "exmaple", and "com"), and the symbol count is 2 (".com"). Therefore, the ratio of domain token and symbol count is 3:2.

8.4.2.1.6   Ratio of Path Token and Symbol Count in Path

The number of tokens or sections in a URL path divided by the number of special characters or symbols used in the URL path is known as the path token and symbol count ratio.

For example, in the URL "https://www.exmaple.com/blog/article", the path token count is 2 ("blog" and "article"), and the symbol count is 1 ("/" between "blog" and "article"). Therefore, the ratio of path token and symbol count is 2:1.

8.4.2.1.7   Ratio of URL Token and Symbol Count in URL

The ratio of the number of segments or sections to the number of special characters or symbols used in a URL is known as the URL token and symbol count.

For example, in the URL "https://www.exmaple.com/blog/article", the URL token count is 5 ("https", "www", "exmaple", "com", "blog", and "article"), and the symbol count is 3 (":", "//", and "/"). Therefore, the ratio of URL token and symbol count is approximately 5:3.

8.4.2.1.8   Ratio of Domain Length and URL Length

The amount of characters in the domain name to the total number of characters in the URL, which includes the protocol, domain name, and path, is known as the domain length to URL length ratio.

For example, in the URL "https://www.exmaple.com/blog/article", the domain length is 11 ("exmaple.com"), and the URL length is 31. Therefore, the ratio of domain length and URL length is approximately 11:31.

8.4.2.1.9   Path Symbol Count

The count of path symbols in a URL is the number of special characters or symbols used to separate different segments or parts of the path in the URL.

For example, in the URL "https://www.exmaple.com/blog/article", the count of path symbols is 1 (the forward slash "/" between "blog" and "article").

8.4.2.1.10   Contain IP Address or Not

To check if a URL contains an IP address or not, you can look for a string of digits separated by periods (e.g. "168.105.5.2") in the domain part of the URL.

For example, in the URL "https://168.105.0.1/index.html", the domain part is "168.105.0.1", which is an IP address.

In contrast, in the URL "https://www.exmaple.com/index.html", the domain part is "www.exmaple.com", which is a domain name.

It is important to note that some domain names may contain digits and periods (e.g. "exmaple123.com"), but they are still treated as domain names and not IP addresses.

8.4.2.1.11   Contain https Token or Not

To check if a URL contains "https" or not, you can look for the protocol or scheme part of the URL.

In contrast, in the URL "http://www.exmaple.com/index.html", the protocol or scheme part is "http".

8.4.2.1.12   Contain Prefix or Suffix ("-") or Not

To check if a URL contains a hyphen ("-") or not, you can scan the entire URL for the presence of hyphens.

Hyphens are often used in URLs to separate words in the domain name, path, or query parameters. For example, a URL like "https://www.pay-pal.com/homepage" contains a hyphen in the path segment "pay-pal".

8.4.2.1.13   Contain Tiny URL or Not

To check if a URL contains a TinyURL or not, you can look for a specific pattern in the URL. TinyURLs are a type of URL shortener that allows users to create short, easy-to-remember URLs that redirect to a longer, more complex URL.

TinyURLs typically have a specific format, which includes a domain name such as "tinyurl.com" followed by a series of random characters or numbers. For example, a TinyURL may look something like "https://tinyurl.com/yj7ztd2w".

8.4.2.1.14   Count Dots in URL

The protocol, domain name, and path are just a few examples of the components of a URL that are separated by dots.

For example, a URL like "https://www.exmaple.com/my-page" contains two dots, one after "https://" and another between "exmaple" and "com".

8.4.2.1.15   Ratio of Path Delimiter Count and URL Delimiter Count

The ratio of delimiter count in the URL and delimiter count in the path is a metric that can provide insights into the structure and complexity of a URL.

To calculate this ratio, you would count the number of delimiters (such as slashes, dots, question marks, and ampersands) in the URL and divide that by the number of delimiters in the path segment of the URL.

A high ratio may indicate that the URL is more complex or has a longer path, whereas a low ratio may indicate that the URL is simpler or has a shorter path.

For example, if a URL has a total of 10 delimiters (slashes, dots, question marks, and ampersands) and 6 of those delimiters are in the path segment, the ratio would be 10/6 or approximately 1.67.

TABLE 8.1
Results of feature selection algorithms with different classifiers using selected top features

Feature Selection Algorithm	Classifiers	Result (%)
Recursive Feature Elimination	LR	87.0
	SVM	92.3
	KNN	93.0
	GB	93.0
K best using Chi^2	LR	83.2
	SVM	93.1
	KNN	93.1
	GB	92.3
Principal Component Analysis (PCA)	LR	83.5
	SVM	93.5
	KNN	93.4
	GB	93.0

8.4.3 FEATURE SELECTION

After extracting the characteristics, we use feature selection techniques to determine which features are most important for classifying phishing URLs. Typical methods for feature selection include:

- KBest using Chi^2 feature selection: The "K" in KBest stands for the number of best features that will be selected from the data set. The "chi^2" part of the technique stands for the statistical test that was used to determine the significance of each feature.
- Recursive Feature Elimination: This approach involves repeatedly removing the least important features until the desired number of features is obtained.
- Principal Component Analysis: This technique reduces the dimensionality of the data set by grouping the features into an easier-to-manage set of uncorrelated variables.

Table 8.1 shows the results of various classifiers using different feature selection algorithms.

We identified the top ten features after examining all of the aforementioned feature selection algorithms. Table 8.2 presents a list of all the features we used for our model.

8.5 RESULTS AND DISCUSSION

8.5.1 EXPERIMENTAL SETUP

On a PC with a core i5 eighth generation processor, 8 GB RAM, and 4 GB graphics card, experiments have been carried out. The suggested methodology has been fully

TABLE 8.2
Distribution of features

S.No.	Feature	Min Val.	Max Val
1.	Domain token count	1	21
2.	Symbol count in URL	0	207
3.	Ratio of path token and URL token	0	24
4.	Ratio of path length and URL length	0	0.99
5.	Ratio of path token and symbol count in path	1	17
6.	Ratio of domain length and URL length	0.006	0.96
7.	Path symbol count	0	141
8.	Contain prefix or suffix("-")	False	True
9.	Contain tiny URL	False	True
10.	Count dots in URL	0	20

	Phishing	Legitimate
Phishing	True Positive (TP)	False Positive (FP)
Legitimate	False Negative (FN)	True Negative (TN)

FIGURE 8.6 Confusion matrix.

implemented using Python and the Jupyter Notebook IDE. We used a number of Python packages to put our anti-phishing strategy into practice.

8.5.2 PERFORMANCE EVALUATION PARAMETERS

The testing set is used to put the hybrid model into test. Accuracy, precision, recall, F1 score, and response time can all be used to evaluate the model. Accuracy is the fraction of correct predictions, precision is the proportion of true positives among all positive forecasts, and recall is the proportion of true positive instances. F1 score, which is the harmonic mean of precision and recall, balances model performance. Model prediction time is determined by reaction time. Many indicators may be used to analyze the model's performance and indicate areas for improvement [23]. A summary of properly and wrongly predicted instances of a class in a classification task is shown in Figure 8.6 as a confusion matrix. The results are evaluated using a variety of indicators, including the confusion matrix.

Mathematically, precision, recall, F1-score, and accuracy can be defined as follows:

$$Accuracy = \frac{TP + TN}{TP + TN + FP + FN} \tag{8.1}$$

$$Precision = \frac{TP}{TP + FP} \tag{8.2}$$

$$Recall = \frac{TP}{TP + FN} \qquad (8.3)$$

$$F1\text{-}score = 2 \cdot \frac{Precision \cdot Recall}{Precision + Recall} \qquad (8.4)$$

8.5.3 IMPLEMENTATION AND RESULTS

Initially, we used the same data set and features as in the suggested study by Gupta et al. [19], and the result is displayed in Table 8.3 for a comparison of our results and their results.

The results are displayed in Table 8.4. We realized that the proposed research [19] is not accurate on the newly formed data set after implementing our newly created data set as described in section 8.4.1 and using the same features and classifier provided by Gupta et al. [19]. The model is not reliable for the most recent phishing URL scams.

With the help of several ML models, we developed a hybrid model for our suggested method to increase prediction accuracy. To identify phishing attacks in this instance, we combined RF and MLP models. MLP is a form of artificial neural network that is frequently used for classification tasks, whereas RF is an ensemble learning technique that combines many decision trees to create predictions.

In order to maintain consistency and facilitate reproducibility and debugging, we have incorporated a mechanism to generate a consistent sequence of random numbers

TABLE 8.3

Results of the proposed method by Gupta et al. [19] in comparison to our results using the same data set and features proposed in [19]

Classifier	Accuracy (%) by Gupta et al. [19].	Our Results (%)
RF	99.57	98.0
KNN	99.04	97.8
SVM	97.64	96.7
LR	95.56	92.5

TABLE 8.4

Results after implementing the proposed method by Gupta et al. [19] on our newly created data set

Classifier	Accuracy (%)
RF	91.0
KNN	91.1
SVM	90.2
LR	87.0

TABLE 8.5
Performance of our proposed method

Hybrid Model	Accuracy (%)	Precision (%)	Recall (%)	F1-score (%)	Response Time (sec)
RF, MLP, and SVM	97.26	97.58	95.47	96.51	0.95
RF, MLP, and LR	97.20	97.57	95.47	96.50	0.61
RF, MLP, and KNN	97.17	97.50	95.46	96.48	0.29
RF, MLP, and GB	97.20	97.58	95.46	96.51	5.23

every time the code is run. This approach proves valuable for ensuring that the train-test split of the data set, as well as the ML classifier, produces identical results across different executions. To achieve this, we have set the random seed to 42 for both the train-test split and the ML classifier.

The training set trains an RF. Each RF decision tree is trained on a separate subset of the training data. Each RF model tree guesses the input class separately, and the final prediction is the majority vote. The RF classifier class takes parameters like the forest's tree count as 100 and the random state as 42. Fitting the RF classifier instance on training data trains the model.

The training set trains an MLP model. Each neuron in the MLP model is linked to all its predecessors. Backpropagation modifies neuron weights to decrease the error between predicted and actual outputs in the MLP model. Sequential instances represent linear layers. We add three dense layers to the model.

The input dimension of the first layer is 10, or the number of input features, and the activation function is ReLU. The first layer includes 64 neurons. Thirty-two ReLU-activated neurons are located in the second layer. The last layer has a single neuron and has a sigmoid activation function, which is frequently employed for binary classification problems. We assess the model using accuracy as the measure, Adam as the optimizer, and binary cross-entropy as the loss function.

We combine the predictions of the RF model and MLP model as features for a new model. We concatenate the predictions as columns to create a new feature matrix, which we use to train a new model. Logistic Regression (LR), KNN, SVM, and GB are used to train the new model.

The experiments give positive results, and Table 8.5 shows that the strategy is successful. In all the experiments performed, there was no evidence of overfitting to the trained and evaluated data. When compared to all the implemented models, the hybrid model of RF, MLP, and SVM was found to perform best overall, with an accuracy of 97.26%. Nevertheless, the other implemented model also achieved quite high accuracy. In addition, the RF, MLP, and GB models achieved the worst response time of all the other classifiers, with an accuracy of about 97.2%.

The deployment of web application or cloud service is constructed in the next step, using the saved model to generate predictions. The client initiates the process by submitting an HTTP request to the active web service. The URL is collected and processed from the HTTP request once it reaches the web service. The feature extraction units receive the parsed URL after that. Then advised extracting ten features directly from the URL. The data is neatly put into the trained model once all the features have been acquired. The trained hybrid ML classifier then outputs its prediction as either "phishing" or "benign", depending on the input. Following that, the client retrieves this response in order to block phishing URLs.

8.6 CONCLUSION

This study discusses the use of a lexical analysis approach to URLs and the development of hybrid ML models using RF and MLP to detect phishing attacks in a cloud-based environment. The predictions of the RF and MLP models were combined and used as features for a new model, which was trained using different classification algorithms. The proposed hybrid model achieved an accuracy of 97.26%. The study not only highlights the effectiveness of ML strategies in the field of information security but also suggests further research to be conducted to improve the accuracy of the model, including identifying new features and evaluating other algorithms such as deep learning and convolutional neural networks. The chapter concludes that the development of more sophisticated and effective methods for detecting phishing attacks is crucial in the fight against cyber threats.

REFERENCES

1. Domain Register Report Available at: www.deepinfo.com/blog/hello-deepinfo/ (Last accessed 27 March, 2023).
2. Anti Phishing Working Group Report Available at: apwg.org/trendsreports/ (Last accessed 27 March, 2023).
3. G. S. Puri, R. Tiwary and S. Shukla (2019). "A Review on Cloud Computing," *2019 9th International Conference on Cloud Computing, Data Science & Engineering (Confluence), Noida, India*, pp. 63–68. doi: 10.1109/CONFLUENCE.2019.8776907
4. S. Afroz and R. Greenstadt (2011), "PhishZoo: Detecting Phishing Websites by Looking at Them," *2011 IEEE Fifth International Conference on Semantic Computing, Palo Alto, CA, USA*, pp. 368–375. doi: 10.1109/ICSC.2011.52
5. Ankit Kumar Jain & B.B. Gupta (2022). A survey of phishing attack techniques, defence mechanisms and open research challenges, *Enterprise Information Systems*, 16(4), 527–565. DOI: 10.1080/17517575.2021.1896786
6. S. Haruta, F. Yamazaki, H. Asahina and I. Sasase (2019), "A novel visual similarity-based phishing detection scheme using hue information with auto updating database," *2019 25th Asia-Pacific Conference on Communications (APCC), Ho Chi Minh City, Vietnam*, pp. 280–285. doi: 10.1109/APCC47188.2019.9026498
7. A.K. Jain, B.B. Gupta (2018). PHISH-SAFE: URL features-based phishing detection system using machine learning. In: Bokhari, M., Agrawal, N., Saini, D. (eds). *Cyber Security. Advances in Intelligent Systems and Computing*, vol. 729. Springer, Singapore. https://doi.org/10.1007/978-981-10-8536-9_44
8. Mohammed Alwanain (2019). Effects of user-awareness on the detection of phishing emails: A case study. *International Journal of Innovative Technology and Exploring Engineering*, 8, 480–484.

9. Ankit Kumar Jain and B. B. Gupta (2017). Phishing detection: Analysis of visual similarity based approaches, Security and Communication Networks, vol. 2017, Article ID 5421046, 20 pages. https://doi.org/10.1155/2017/5421046

10. R.S. Rao, A.R. Pais (2019). Detection of phishing websites using an efficient feature-based ML framework. *Neural Computing and Applications* 31, 3851–3873 doi: https://doi.org/10.1007/s00521-017-3305-0.

11. M. Hara, A. Yamada and Y. Miyake (2019), Visual similarity-based phishing detection without victim site information, *2009 IEEE Symposium on Computational Intelligence in Cyber Security, Nashville, TN, USA*, pp. 30–36. doi: 10.1109/CICYBS.2009.4925087

12. Y. Zhou, Y. Zhang, J. Xiao, Y. Wang and W. Lin (2014). Visual similarity based anti-phishing with the combination of local and global features. *2014 IEEE 13th International Conference on Trust, Security and Privacy in Computing and Communications, Beijing, China*, pp. 189–196. doi: 10.1109/TrustCom.2014.28

13. Kang Leng Chiew, Ee Hung Chang, San Nah Sze, and Wei King Tiong (2015). Utilisation of website logo for phishing detection. *Computers & Security*, 54, 16–26. ISSN 0167-4048. https://doi.org/10.1016/j.cose.2015.07.006

14. Waleed Ali (2017). Phishing website detection based on supervised machine learning with wrapper features selection. *International Journal of Advanced Computer Science and Applications(IJACSA)*, 8(9). doi: http://dx.doi.org/10.14569/IJACSA.2017.080910

15. Rishikesh Mahajan and Irfan Siddavatam (2018). Phishing website detection using machine learning algorithms. *International Journal of Computer Applications*. 181, 45–47. 10.5120/ijca2018918026

16. Alexa Top Sites Dataset. Available at: www.alexa.com/topsites (Last accesed 27 March 2023).

17. Dataset of PhishTank. Available at: https://phishtank.org/developer_info.php (Last accesed 27 March 2023).

18. A.K. Jain, B.B. Gupta (2018). Towards detection of phishing websites on client-side using ML based approach. *Telecommunication Systems* 68, 687–700. doi: https://doi.org/10.1007/s11235-017-0414-0

19. Brij B. Gupta, Krishna Yadav, Imran Razzak, Konstantinos Psannis, Arcangelo Castiglione and Xiaojun Chang (2021), A novel approach for phishing URLs detection using lexical based ML in a real-time environment. *Computer Communications*, 175, 47–57, ISSN 0140-3664. https://doi.org/10.1016/j.comcom.2021.04.023

20. University of New Brunswick (ISCXURL-2016) Dataset, Available at: http://205.174.165.80/CICDataset/ISCX-URL-2016/ (Last accesed 27 March 2023).

21. Legitimate and Phishing URLs Dataset, Available at: www.kaggle.com/data~sets/siddharthkumar25/malicious-and-benign-URLs/ (Last Accesed 27 March 2023).

22. COVID-19 Themed Domain Dataset. Available at: https://spycloud.com/resource/covid19-domain-data set/ (Last accesed 27 March 2023).

23. What is performance evaluation parameter ? www.analyticsvidhya.com/blog/2020/10/quick-guide-to-evaluation-metrics-for-supervised-and-unsupervised-machine-learning/ (Last accesed 27 March 2023).

9 Toward the Sustainable Development of Smart Cities through Cloud of Things

Saurabh Singhal and Rakesh Kumar

9.1 INTRODUCTION

With the evolution of the Internet, various new computing paradigms have evolved. Cloud computing is a new type of Internet-based computing that is gaining popularity. It uses Information Technology (IT) resources as a service. The Internet of Things (IoT) may make smart devices more productive, enhance throughput, and make them more efficient when they work with cloud infrastructure. Smart towns are locations where people work hard to figure out where information and communication technologies are heading, to protect the environment, to enhance health, and to learn [1, 2, 3]. Different tools and services are shared by the service provider through a network. The IoT can perform better, handle more, and be more valuable using cloud technologies. As cloud computing has advanced, it has become easier to create, share, and bundle software and electronic enterprises. As a result, cloud computing and IoT are growing closer with new technologies that will be compatible with IoT systems.

With the IoT revolution, embedded devices that are network-connected can connect in new ways. As a result, it is becoming increasingly difficult to distinguish between the physical world and the Internet world. To handle this type of jumbled data, cloud computing and the IoT must collaborate [4, 5]. This all-in-one service is known as the Cloud of Things (CoT). The cloud's role in CoT architecture is similar to that of a central control unit that examines raw data in a methodical manner [6].

IoT is essentially concerned with problems that are dynamic and global. It includes a wide range of unusual and flexible devices that are resource constrained. These constraints halt and obstruct the development of IoT systems. They address difficult issues like availability, full functioning, efficiency, interoperability, and compatibility. Cloud computing is one of the most promising ways to leverage IoT to avoid these challenges. Because of the cloud, you can use shared resources such as networks, storage, computers, and software. These materials are easy to get, inexpensive, and appealing. Smart cities may use resources and services from the cloud for collecting, analyzing, processing, and storing data. All organizations, despite their nature and size, make use of the IoT. Cities are becoming increasingly connected as the

DOI: 10.1201/9781003390954-9

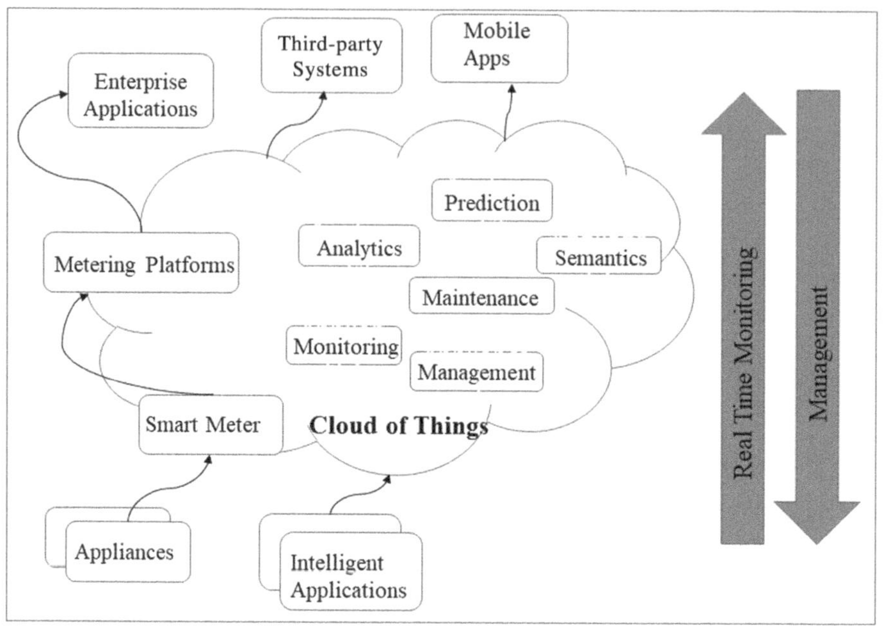

FIGURE 9.1 CoT in smart grid context.

IoT grows, allowing organizations to improve the quality and pace of infrastructure construction as well as the dependability and speed of emergency services.

Figure 9.1 depicts the Smart Grid's paradigm shift and transition to the CoT [7]. Peer-to-Peer (P2P) and cloud-based auxiliary services enable appliances with wireless networked embedded systems to integrate, interact, and cooperate [8]. This means that "edge" devices with limited resources can now employ "powerful" cloud-based ones to increase their capabilities. The applications can gain benefits from the interaction of physical resources with virtual resources by connecting them with global infrastructure and increasing bidirectional information exchange.

Improving IoT edges and system-level optimizations in a building, business, or smart grid city necessitates the ability to supply and consume data. Real-time measurements are provided via edge devices. Because of the great number of cloud resources available, energy and device operational metrics may be reviewed, and decisions for device-specific and global actions, such as massive consumption or generation system control, can be taken. This enables previously unfeasible or prohibitively expensive options, such as avoiding blackouts by drastically lowering energy use or rescheduling or altering certain categories of domestic appliances while prioritizing crucial sites such as hospitals.

As previously stated, leveraging the CoT enables resource-constrained devices to fully exploit cloud features such as virtualization and multi-tenancy. A manufacturer may use cloud-based services to monitor the deployed application, upgrade the application and firmware, and much more. Smart grid cities may monitor and adjust public

TABLE 9.1
Summary of previous studies

Reference	Description
[1]	Framework for security and privacy provided. Authentication verified using Scyther verification tool.
[14]	An architecture based on CoT to identify the existing and forthcoming challenges.
[39]	Describes the features and components of smart cities along with trends and challenges.
[40]	Achieved interoperability by using semantic web technologies for collecting and extracting data in a smart city.
[41]	Security concerns in smart cities are analyzed in the paper.
[42]	A model for street lights in a smart city using IoT, SOA, and SysML is given.
[43]	Load balancing considering fog environment for a smart city is given.

infrastructure to achieve objectives. Combining virtual and real data is required for innovation and efficiency. CoT makes information from widely dispersed locations in the real world visible at a low cost. The cloud can collect and analyze "big data" to help you make better decisions. As per the requirements of users, customized analytics are delivered to them. The latter may allow for more precise, near-real-time, and scalable services.

In CoT on-premise middleware and unique solutions will be limited. Real-time energy monitoring, customized information services, and other services will be provided by new service providers. The energy industry is shifting away from massive, monolithic projects and toward interactive, participatory efforts that leverage regional expertise. We foresee a new generation of personalized energy efficiency services at the smart grid city level by improving visibility through the gathering and analysis of energy-related data in near real time, providing analytics on it, and enabling selectable control [10].

Academics are eager to explore novel concepts for IoT-enabled smart communities in the years to come. Table 9.1 summarizes some of the previous studies undertaken by various researchers in the past.

9.2 IOT PLATFORMS

IoT aspires to create a world in which everything is online and capable of talking with one another on its own [11]. It looks to address the issues of a dynamic, heterogeneous environment. The authors in [12] show how these options enable a wide range of applications in a variety of areas. Various application areas of IoT are shown in Figure 9.2.

At the same time that the importance of IoT systems grew, so did the importance of middleware. Middleware has grown in popularity in recent years due to its ability to simplify the creation of new services and the integration of old and new technologies [13].

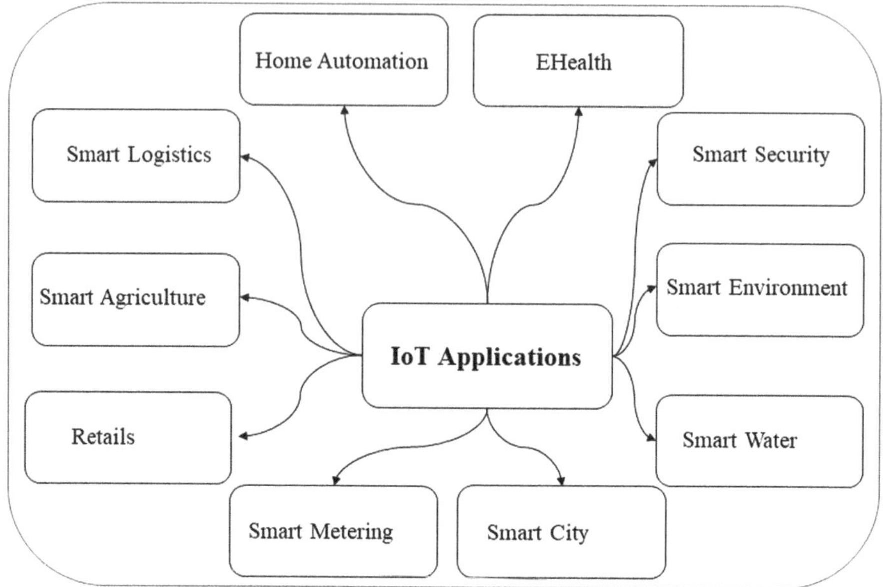

FIGURE 9.2   IoT application domain.

9.2.1   Internet Based Computing

Cloud computing represents the subsequent expansion of web-based computing. The provision of Information and Communication Technologies (ICT) services can be facilitated through transportation. Cloud computing facilitates communication between computers, networks, systems, business processes, and other essential tools [14]. The expansion of cloud computing facilitates the implementation of adaptable business models, which enable companies to utilize their resources in tandem with their business growth. Cloud computing is distinguished from conventional web-based services by its ability to provide immediate cloud delivery without the need for a protracted provisioning process. Cloud computing allows for the dynamic allocation and de-allocation of resources based on demand. Application Programming Interfaces facilitate the utilization of cloud services and enable communication between cloud-based applications and resource records. Invoicing and rating providers are used as means of remuneration. This facilitates assistance for providers to optimize rating aid and expedite payment processing, monitoring and evaluating performance. The cloud computing infrastructure, which provides a platform for managing and accessing performance in the carrier industry, complements the integration of physical computing and its methodologies. Cloud computing is founded on secure procedures designed to safeguard confidential information.

Cloud computing and similar services are increasingly being utilized for two primary business reasons.

- Commercial activity: Cloud computing provides organizations with the ability to obtain computer resources as needed, with flexibility, to achieve their objectives.
- Reducing Costs: Cloud computing reduces expenses by converting capital costs into operational costs. Cloud computing enhances resource organization and allocation by providing a transparent framework and prioritizing preexisting management protocols.

9.3 TOWARD CLOUD OF THINGS

Cloud computing is gaining popularity among academics and businesses worldwide due to its potential to revolutionize how services are delivered in the IT field as a whole. Expected speed, high availability, large fault tolerance, limitless scalability, and other requirements for service provisioning are specified [15].

According to [16], services can be classified into three categories. The users interact with the software via Software as a Service (SaaS). The software developers can develop and deploy applications using Platform as a Service (PaaS). Infrastructure as a Service (IaaS) provides computing resources such as processing or storage.

The authors explore the "Everything as a Service" model in [17]. Cloud computing operates on a "pay as you go" business model, allowing customers to utilize a service and only pay for what they consume. In the context of the IoT, this strategy gave rise to the so-called CoT, which integrates indexing and querying services for objects and makes them available as a service to end users, developers, and providers (Figure 9.3). In [20], the authors go into detail about an intriguing proposal for enabling CoT where the focus is on the Sensing as a Service paradigm, which is built on IoT infrastructure and comprises four conceptual layers:

- The Sensor and Sensor Owners Layer: It comprises sensors and their owners' management practices, such as whether or not they can be published.

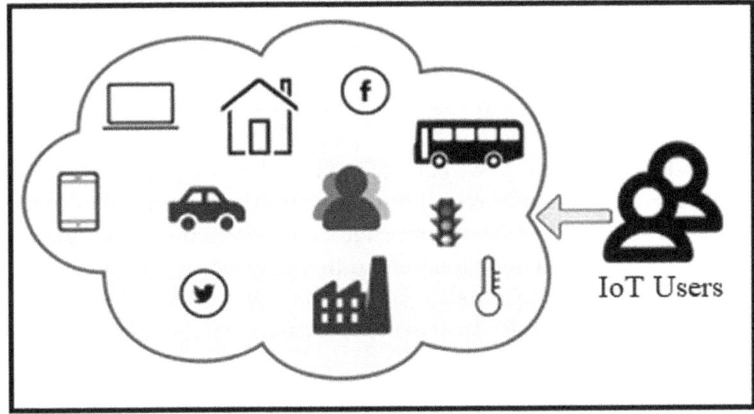

FIGURE 9.3 Cloud of Things.

- Sensor Publishers: It finds the available sensors and contacts their owners to gain permission to store the sensors in the cloud.
- Extended Service Providers: They choose sensors by working with multiple sensor publishers based on consumer preferences.
- Individuals: They must have an account to access monitoring data. Sharing and reusing sensor data is achieved via Sensing as a Service.

9.4 CLOUD CONVERGENCE AND IOT

IoT applications typically generate enormous amounts of data, and because they have numerous computational add-ons, integrating them with cloud computing infrastructure may result in cost savings. As their customer base grows and their product becomes more well known, small and medium-sized enterprises (SMEs) may collect and use more data. Cloud connectivity enables SMEs to manage and retain large amounts of data collected from numerous sources [21, 22]. Cloud-based building and system distribution can assist a smart city. Smart logistics, smart waste management, smart meters, and other IoT products are some of the traits of a smart city. They might also provide more data volume. The smart city manages these applications and the generated data effortlessly due cloud integration. The distribution of new packages and the extension of current ones, both of which previously generated severe problems over the availability of necessary computer resources, can also be accelerated using the cloud. By granting other parties access to its infrastructure and enabling them to incorporate IoT data and processing, a public cloud computing provider can expand the market.

The company can offer IoT data for assistance and access. This underlines the importance of keeping the IoT and cloud infrastructure up to date. Because of differences in IoT and cloud designs, integration has always been difficult. IoT devices usually have poor performance, shoddy support, and exorbitant prices. They usually concentrate on their immediate surroundings. Cloud computing resources, on the other hand, are typically located in a remote location, offering speed and flexibility. Sensors and devices are uploaded to the cloud, allowing them to spread across all resources and reduce discrepancies.

By putting services and sensors in the cloud, those services and sensors may be accessed in real time. Wireless Sensor Network (WSN) and sensory data may be uploaded to the cloud as a result of IoT and cloud connectivity. This massive infrastructure, which was widely used in Japan following earthquakes for radiation monitoring and radiation maps, was one of the initial developments. Customers can pay-per-use these public cloud service providers to store IoT packages in the cloud. The majority of providers offer cutting-edge developer tools for improving cloud systems and changing them into cloud-based IoT services.

IaaS: Users can connect to cloud-based sensors and actuators due to the IoT and clouds. IaaS enables IoT management to monitor assets and provide required services [23, 24].

The high-performance PaaS paradigm of IoT and cloud services is used to define the public IoT and cloud infrastructure. The PaaS architecture, which is a comprehensive framework for cloud-based development and deployment, enables the provision of cloud-based services ranging from the most basic to complex, cloud-capable organizations.

Customers can now access whole suite of softwares designed exclusively for the cloud and IoT due to services such as SaaS. SaaS packages, which are identical to traditional cloud-based packages, are used to access IoT sensors and devices. SaaS packages are frequently deployed on top of PaaS infrastructure to support business models that rely on IoT products and services.

9.5 SMART CITY REQUIREMENTS

A smart city employs ICT to improve its residents' quality of life, operational efficiency, and infrastructural sustainability. Because the concept of a smart city is continuously growing, there is no singular definition. Smart cities rely on a wide range of technologies, including the IoT, big data, and Artificial Intelligence (AI). These technologies are used for data collection, system monitoring, and process automation. However, some of the essential features of smart cities are as follows:

- Use of ICT: Smart cities use ICT to increase efficiency in a range of areas, including transportation, electricity, water, waste management, and public safety. Smart cities, for example, may utilize sensors to monitor traffic flow and alter traffic signals accordingly or smart meters to measure energy usage and provide feedback to inhabitants on how to conserve energy.
- Data-driven decision-making: Smart cities use data to make better decisions about resource allocation and service improvement. A smart city, for example, could use sensor data to identify crime hotspots and then send extra police personnel to those locations. Citizens are involved in the design and development of smart city initiatives in smart cities. This ensures that activities are matched with community needs and that residents feel ownership of the projects.
- Sustainability: Smart cities are intended to be both economically and environmentally sustainable. This means that they use resources wisely and have a low environmental impact.

Some of the advantages of smart cities are as follows:

- Improved quality of life: Smart cities can improve their inhabitants' quality of life by enhancing access to services, reducing transportation congestion, and improving air quality.
- Increased economic growth: By creating a more efficient and appealing environment, smart cities can attract firms and investment.
- Reduced environmental impact: Smart cities can lower their environmental impact by making better use of resources and reducing pollution.

Smart city development is a complex and difficult process. The potential benefits of smart cities, on the other hand, are substantial, and many cities throughout the world are attempting to become smarter. Smart city development is still in its early phases, but it is a rapidly expanding subject. As technology advances, we may anticipate even more inventive and sustainable smart cities in the future. Before combining the IoT and CoT paradigms in a smart city, it is vital to understand the city's core services and the solution requirements. To do so, in [25], the authors distinguish between two types of needs:

- Service and application needs as perceived by individuals
- Operational needs as perceived by city officials and network managers

End-user devices outfitted with several wireless technologies, as well as various sensors and actuators strategically positioned across cities, allow for the differentiation of specialized services and apps for individuals in terms of service and app elements. These services will be distinguished by the following characteristics:

- User-centric: The specific context and preferences according to the users
- Ubiquitousness: Which makes them accessible from any location and device
- High integration: Various applications and platforms are integrated so that the data can be made available to different users.

Key elements of these new services that increase the sustainability of the city will surely benefit the different stakeholders of the city along with its citizens. From the viewpoint of administrations and network providers, the network architecture of a smart city can be:

- Highly interconnected, allowing ubiquitous connectivity for heterogeneous devices
- Energy-efficient, meeting the requirements of green applications by effectively utilizing resources
- Cost-efficienting, installating such a network architecture is relatively simple.

9.6 SMART CITY – IOT AND COT BUILDING BLOCKS

Even though there is no formal definition of the term that is agreed upon by all parties, Hancke et al. [26] defines a smart city as one that employs sophisticated technologies for monitoring and control to assure long-term sustainability and efficiency. It accomplishes this by consolidating all of its infrastructure and services into a single entity. According to this interpretation, the smart city concept asks for interoperability between multiple IoT implementations, which are now predominantly locked and vertically integrated into specialist application areas [27].

The IoT landscape is tremendously fragmented because of divergent architectures, protocols, and platforms, making integration difficult. According to [28], the IoT framework is categorized into five layers, namely:

- Device Layer: This uses a sensing device to identify and learn more about specific items.
- Network Layer: It transmits information from the Device Layer to the information processing system via the Network Layer.

TABLE 9.2
Requirements of smart cities

Requirements	Descriptions	Variance
Services	The helpful information offered to the recipients.	Noise, air waste, energy
Platforms	Smart city components and services are developed, organized, and maintained by the system.	Government, enterprises, business
IoT-Infrastructure	The method used to handle the physical elements and transferred data.	Data and network
Devices and enabling technologies	The infrastructure equipment involved in the smart city infrastructure, including their design.	Sensors, mobiles, data portals
Connectivity models and features	The mechanism by which the tools and technology are interconnected.	Ubiquitous, autonomic

- Middleware Layer: This analyzes data and makes a decision based on the results.
- Application Layer: This enables global application administration based on the data acquired from the middleware.
- Business Layer: This controls the complete IoT system.

Table 9.2 shows a summary of the requirements of a smart city.

In the context of the smart city, to solve large-scale processing problems, the CoT uses distributed resources more efficiently and achieves better throughput [29]. As a result, the smart city idea and horizontal integration of diverse IoT platforms are made viable. Furthermore, the users are allowed to specify the service they require by quickly returning the relevant data to users without requiring to select the sensors manually. The variability and diversity of sensor types, as well as communications and cloud compatibility, are undoubtedly the most critical research challenges that CoT must face. An abstraction level must be built to bridge the gap between the various technologies.

Data abstraction, comprising of methods to organize, exchange, and make sense of IoT data and make it easier to transform it into meaningful information and intelligence across a wide range of application domains, provides a remedy for the problem. Ontologies, semantic web services, and linked data are just a few of the semantic web technologies that have gained attraction in the field of diverse sensor types recently [30].

9.6.1 Importance of CoT in Smart Cities

There are numerous reasons for the adoption of IoT technologies by smart cities. Some of them are as follows:

- IoT systems make it possible for sensors to gather data, potentially saving money.

- Cost is an important consideration when selecting whether to go online or offline because IoT initiatives are easier to build and run. Another significant variable is efficiency, and costs are falling while communications are stronger and more successful than ever.
- For the most complex systems, wireless communication enables the administration and monitoring of IoT broadcasts, reducing the need for a physical visit from the service provider.
- Administrators can update firmware, apply security updates, and can get alerts for problems automatically.
- In operational contexts, declining assistance is typically to blame when smart road illumination and tracking equipment are maintained.

9.7 APPLICATIONS OF COT IN SMART CITIES

This section examines a few of the CoT applications in smart cities. The use of cloud computing has hastened the growth of machine-to-machine communication, which began with the introduction of the IoT [31, 32]. Some of the fields where CoT is used in smart cities are shown in Figure 9.4 and are described next.

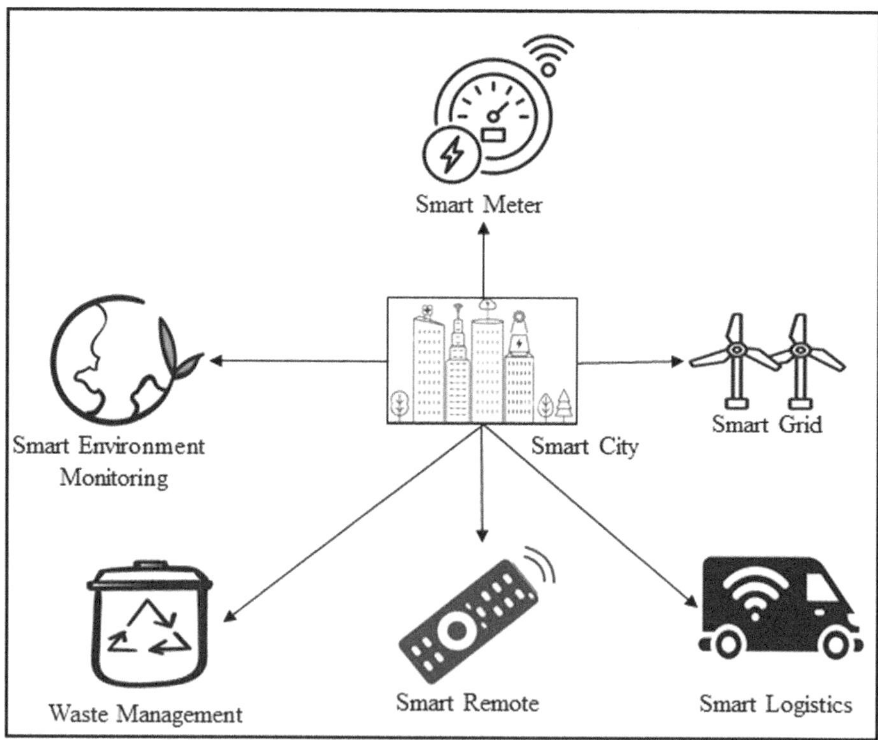

FIGURE 9.4 Application of CoT in smart cities.

9.7.1 SMART GRID FOR SMART CITIES

Because IoT can only process data locally, its use in regulating energy usage and distribution is limited. The cloud for big-data computing is now available. Cloud computing resources provide a dependable, secure, and independent networking infrastructure. Cloud computing technology efficiently manages massive amounts of data and information received from various sources [33].

Smart cities integrate technology into urban life, including transportation, energy, and communication systems. Common components include smart sensors, data analytics platforms, and IoT devices [40]. However, challenges like cybersecurity threats and privacy concerns must be addressed for the successful implementation and sustainability of these initiatives [42].

The integration of semantic web technologies in smart cities has achieved interoperability, necessitating partnerships with private sector companies and government agencies for funding and implementing projects [41].

The text presents a smart city street light model using IoT, SOA, and SysML for optimizing resource usage and improving quality of life [43]. The goal of managing a smart city in a fog environment is to balance the load among nodes. Load balancing is achieved in a smart city using fog computing to distribute resources efficiently, minimize energy consumption, and reduce latency in data processing [44].

9.7.2 SMART ENVIRONMENT MONITORING

Monitoring the environment is critical to reaching the objectives of a cleaner environment. In circumstances where direct monitoring is difficult, cloud technology is used. Monitoring can be difficult where there are non-point sources of contamination, such as gas concentrations in mines, laboratories, and automobiles. Long-term sensor data gathering demands efficient administration and processing techniques, as well as sufficient storage, which CoT may provide [34].

9.7.3 SMART LOGISTICS

A well-balanced diet is one of life's most basic requirements. Food is becoming increasingly vital as the world population grows. To meet this need, increased food production must be linked with loss avoidance. In this case, advanced technologies prove to be extremely beneficial. Because of sophisticated logistics techniques such as packing facilities and refrigerated vehicles, food waste is drastically reduced, and utilization is maximized. The establishment of the supply chain is also highly useful in the modern sector. It facilitated effective food marketing and enabled farmers to interact with customers [35].

9.7.4 WASTE MANAGEMENT

Waste management is critical to maintaining a clean city. A variety of processes are required for efficient and organized waste management. This chain involves waste collection regularly, transportation to the right location, and, depending on the type of waste, recycling or degradation, as well as waste processing, management, and monitoring. This archaic method comes at a significant time, effort, and financial cost. As a result, improved waste management lowers costs and frees up funds for other development goals [36].

To improve trash management, competent authorities must take strategic initiatives. When developing a waste collection strategy, sensor data could be a valuable resource. If this happens, garbage vehicles' fuel expenses may be decreased. These data can be used by various recycling companies to predict and monitor the waste sent to their facilities, allowing them to adjust their processing schedules accordingly. Automatic monitoring will help achieve even larger cost savings by simplifying human monitoring and saving time. Cloud-based technologies can support each of these. Savings from employing CoT in trash management can thus be utilized for other aspects of smart cities.

9.7.5 Smart Remote

Citizens may be able to get application administration services through IoT-based smart city technologies. Citizens will be able to use remote controls on smart remotes for products [37]. A homeowner, for example, could use his mobile phone to turn off the heating in his home. Utility companies can also notify residents of crises and dispatch personnel to address the issues.

9.7.6 Smart Meters

Cities with the deployment of smart meters may provide citizens with significant connections to service delivery systems. Smart meters now deliver precise meter readings by communicating data to other apps across the network in real time [38]. The service provider can provide a more precise bill to the subscriber for different utilities with the cooperation of the entire group.

9.8 SECURITY AND PRIVACY CONCERNS RELATED TO THE CLOUD OF THINGS IN SMART CITIES

The Cloud of Things has emerged as a critical component as smart cities continue to expand and integrate new technologies. CoT facilitates the storage, processing, and analysis of massive amounts of data generated by smart cities' interconnected devices, sensors, and infrastructure. While CoT provides tremendous advantages in terms of efficiency, sustainability, and improved services, it also raises important security and privacy problems that must be addressed properly. The key security and privacy concerns connected with CoT in smart cities, emphasizing the significance of reducing these risks to ensure residents' confidence, security, and privacy, are discussed in the following:

- Data Security Risks: Data security is a major worry in CoT-enabled smart cities. The gathering, transmission, and storage of huge amounts of data pose several problems that must be addressed. Hackers and bad actors may attempt to acquire unauthorized access to or modify data by exploiting weaknesses in data collection devices, communication networks, or the cloud infrastructure itself. Data breaches can have serious ramifications, such as identity theft, financial loss, and harm to key infrastructure. To mitigate these concerns, it is important to deploy strong security measures such as encryption, access controls, and regular security audits to safeguard sensitive data in transit and at rest.

- Network Security Risks: The interconnected network of CoT-enabled smart cities' gadgets and sensors brings major network security vulnerabilities. Cyber-criminals may attempt to exploit device vulnerabilities or penetrate network infrastructure to interrupt vital services or acquire unauthorized data access. Denial of Service (DoS) and botnet assaults are particularly problematic because they can overwhelm the network and render it inoperable. Furthermore, the penetration of a single device inside the network can undermine the security and integrity of the entire ecosystem. Robust network architecture, intrusion detection systems, and frequent security updates are required to alleviate network security concerns. Strong authentication systems and encryption protocols can improve the security of data transfers and communication channels even further.

- Privacy Concerns: CoT-enabled smart cities rely on the collection and analysis of massive amounts of data, such as personal information, behavioral patterns, and location data. This massive data collection poses serious privacy concerns. Citizens may be concerned about continuous monitoring and tracking of their activities, which could lead to surveillance, profiling, and exploitation of personal information. It is critical to strike a balance between data collection for service improvement and privacy rights protection. To create confidence and address privacy concerns in CoT-enabled smart cities, transparency in data governance, clear privacy regulations, and procedures for getting informed consent from individuals are required.

- Transparency and Trust: Trust is essential for the successful adoption of CoT in smart cities. Citizens must have confidence that their data will be handled securely and that their privacy rights will be honored. Transparent data governance practices, in which individuals have full visibility into how their data is acquired, utilized, and shared, are critical for building trust. Building trust between stakeholders can be aided by establishing good communication channels and giving citizens options to express concerns and provide feedback. Collaboration among government agencies, private-sector companies, and individuals is essential for ensuring transparency in data practices and promoting accountability in data use.

- Legal and Regulatory Issues: CoT-enabled smart cities operate in a complex legal and regulatory environment. Data protection legislation, cybersecurity laws, and privacy frameworks differ between jurisdictions, making the development and implementation of CoT difficult. It is critical to harmonize regulations and ensure compliance with current legislation to guarantee the security and privacy of people's data. Governments and legislators must create broad legal frameworks that could handle the special difficulties of CoT in smart cities, such as cross-border data flows, data retention, and consent methods. Establishing regulatory authorities and supervision systems can also aid in enforcing compliance and ensuring responsible data practices.

9.8.1 MITIGATION STRATEGIES AND BEST PRACTICES

Several mitigation measures and recommendations can be applied to address the security and privacy risks associated with CoT in smart cities:

- Comprehensive Security Framework: It is critical to implement a comprehensive security framework that includes end-to-end security measures, such as device security, network security, and cloud security.
- Risk Assessment: Regular risk assessments can help successfully mitigate risks by identifying vulnerabilities, evaluating potential threats, and implementing appropriate security controls.
- Encryption and Authentication: By using strong encryption technologies to safeguard data in transit and at rest, as well as strong authentication processes, CoT-enabled smart cities can be more secure.
- Access Control and Authorization: It is critical to implement access control methods to guarantee that only authorized organizations have access to sensitive data and resources.
- Security Audits and Testing: Regular security audits and penetration testing of CoT infrastructure can assist in identifying and addressing issues.
- Privacy by Design: Integrating privacy considerations from the beginning of CoT development helps ensure that privacy is a basic component of system design and architecture.
- User Awareness and Education: Raising citizen awareness of the security and privacy issues connected with CoT, as well as giving education on best practices for data protection, can enable people to take the appropriate safeguards.
- Public–Private Partnerships: Collaboration between government agencies, private-sector entities, and academics can help establish robust security frameworks, share threat intelligence, and stimulate security solution innovation.
- Ethical Rules: Creating and following ethical rules that address the responsible and ethical use of data can assist in ensuring that CoT is implemented in a way that respects privacy, human rights, and societal values.

9.9 FUTURE TRENDS AND OPPORTUNITIES

Smart cities powered by CoT have the potential to transform urban living by harnessing technology, data, and connections. As technology advances, new trends and possibilities emerge, which can improve the capabilities and effects of smart cities. This section examines some of the potential implications of future trends and possibilities in CoT-enabled smart cities.

9.9.1 5G and Edge Computing

The deployment of 5G networks, as well as the integration of edge computing technologies, will greatly improve the capabilities of CoT-enabled smart cities. The high-speed, low-latency connectivity of 5G will permit real-time data transmission and a larger network of networked devices. With its decentralized processing capabilities, edge computing can enable faster data analysis and decision-making at the network's edge, lowering latency and improving the responsiveness of smart city applications.

9.9.2 Artificial Intelligence and Machine Learning

AI and machine learning (ML) technologies will create new opportunities for data analysis, predictive modeling, and wise decision-making in CoT-enabled smart cities. Massive amounts of data gathered from many sources can be analyzed by AI and ML algorithms, leading to more precise forecasts, improved resource allocation, and individualized services. AI-powered technology can be used in smart city applications to automate processes, boost productivity, and enhance inhabitants' quality of life in general.

9.9.3 Data Analytics and Insights

As CoT-enabled smart cities generate more data, the emphasis will shift to extracting valuable insights from it. Big data analytics, real-time analytics, and predictive analytics are examples of advanced data analytics methodologies that will allow important insights to be extracted for decision-making, service optimization, and anomaly identification. The integration of data from various sources, including social media, sensors, and public records, will provide a more comprehensive view of urban environments, allowing for more informed and data-driven decision-making.

9.9.4 Sustainable and Resilient Infrastructure

The next generation of smart city infrastructure will place a greater emphasis on sustainability and resilience. CoT-enabled systems can aid in better resource management, reduced energy use, and improved environmental monitoring. Integrating renewable energy sources, smart grids, and intelligent transportation systems can result in a more ecologically friendly and sustainable urban environment. Resilience measures can also be included in smart city infrastructure to mitigate the effects of natural disasters and improve the city's capacity for recovery and adaptation.

9.9.5 Digital Inclusion and Equity

Addressing digital inclusion and equality issues is critical for the creation and long-term viability of CoT-enabled smart cities. Future trends will focus on narrowing the digital divide by ensuring low-cost access to technology, digital literacy programs, and inclusive design and implementation of smart city solutions. Efforts will be made to ensure that benefits are delivered to all sectors of society, particularly marginalized populations, to reduce imbalances and build a more equitable and inclusive urban environment.

Future trends and opportunities will expand the capabilities of CoT-enabled smart cities, which have the potential to drastically alter urban living. Smart cities may become more successful, sustainable, and citizen-focused as connectivity, AI, IoT, and data analytics increase. However, substantial care must be given to ensuring privacy, security, and equal access to technology. By embracing these potential future trends and opportunities, CoT-enabled smart cities may pave the way for a more linked, resilient, and livable urban future.

9.10 CONCLUSION

Experts, businesses, and individuals may be able to access, analyze, and manage accurate data with the use of cloud-based enabling technologies, allowing them to make better decisions that increase people's standard of living. Citizens of smart cities use mobile devices to communicate with one another via connected cars and smart houses. To minimize costs and increase efficiency, devices and information can be connected to the physical infrastructure of a smart city. Smart cities may benefit from IoT by having more efficient garbage collection, fewer accidents, and better resource distribution. Cloud-based IoT applications are researched and analyzed in this chapter concerning their functions in smart cities.

REFERENCES

1. Khan, Z., Pervez, Z., & Ghafoor, A. (2014, December). Towards cloud based smart cities data security and privacy management. In *2014 IEEE/ACM 7th International Conference on Utility and Cloud Computing* (pp. 806–811). IEEE.
2. Khan, Z., & Kiani, S. L. (2012, November). A cloud-based architecture for citizen services in smart cities. In *2012 IEEE Fifth International Conference on Utility and Cloud Computing* (pp. 315–320). IEEE.
3. Suciu, G., Vulpe, A., Halunga, S., Fratu, O., Todoran, G., & Suciu, V. (2013, May). Smart cities built on resilient cloud computing and secure internet of things. In *2013 19th International Conference on Control Systems and Computer Science* (pp. 513–518). IEEE.
4. Gomes, M., da Rosa Righi, R., & da Costa, C. A. (2014, October). Internet of things scalability: Analyzing the bottlenecks and proposing alternatives. In *2014 6th International Congress on Ultra Modern Telecommunications and Control Systems and Workshops (ICUMT)* (pp. 269–276). IEEE.
5. Botta, D. D., Botta A., de Donato W., Persico V., & Pescapè A. (2016). Integration of cloud computing and internet of things: A survey, *Future Generation Computer Systems* 56, 684–700.
6. Khanna, A. (2015). An architectural design for cloud of things. *Facta Universitatis, Series: Electronics and Energetics*, 29(3), 357–365.
7. Karnouskos, S. (2013). Smart houses in the smart grid and the search for value-added services in the cloud of things era. In: *IEEE International Conference on Industrial Technology (ICIT 2013)*, Cape Town, South Africa. (pp. 2016–2021). IEEE.

8. Karnouskos, S. (2010, June). The cooperative internet of things enabled smart grid. In *Proceedings of the 14th IEEE International Symposium on Consumer Electronics (ISCE 2010)*, June (pp. 07–10).

9. Ilic, D., Da Silva, P. G., Karnouskos, S., & Griesemer, M. (2012, June). An energy market for trading electricity in smart grid neighbourhoods. In *2012 6th IEEE International Conference on Digital Ecosystems and Technologies (DEST)* (pp. 1–6). IEEE.

10. Valocchi, M., Juliano, J., & Schurr, A. (2012). Knowledge is power–driving smarter energy usage through consumer education. IBM Institute for Business Value, Tech. Rep.

11. Perera, C., Zaslavsky, A., Christen, P., & Georgakopoulos, D. (2013). Context aware computing for the internet of things: A survey. *IEEE Communications Surveys & Tutorials*, 16(1), 414–454.

12. Liu, H., Ning, H., Mu, Q., Zheng, Y., Zeng, J., Yang, L. T., & Ma, J. (2019). A review of the smart world. *Future Generation Computer Systems*, 96, 678–691.

13. Atzori, L., Iera, A., & Morabito, G. (2010). The internet of things: A survey. *Computer Networks*, 54(15), 2787–2805.

14. Roy, S., & Sarddar, D. (2017). The role of cloud of things in smart cities. arXiv preprint arXiv:1704.07905.

15. Alam, T. (2021). Cloud-based IoT applications and their roles in smart cities. *Smart Cities*, 4(3), 1196–1219.

16. Modi, P. P., Sunny, M. S. H., Khan, M. M. R., Ahmed, H. U., & Rahman, M. H. (2022). Interactive IIoT-based 5DOF robotic arm for upper limb telerehabilitation. *IEEE Access*, 10, 114919–114928.

17. Duan, Y., Fu, G., Zhou, N., Sun, X., Narendra, N. C., & Hu, B. (2015, June). Everything as a service (XaaS) on the cloud: Origins, current and future trends. In *2015 IEEE 8th International Conference on Cloud Computing* (pp. 621–628). IEEE.

18. Saha, H. N., Auddy, S., Chatterjee, A., Pal, S., Sarkar, S., Singh, R., & Maity, A. (2017, August). IoT solutions for smart cities. In *2017 8th Annual Industrial Automation and Electromechanical Engineering Conference (IEMECON)* (pp. 74–80). IEEE.

19. Alam, T. (2021). Cloud-based IoT applications and their roles in smart cities. *Smart Cities*, 4(3), 1196–1219.

20. Perera, C., Zaslavsky, A., Christen, P., & Georgakopoulos, D. (2014). Sensing as a service model for smart cities supported by internet of things. *Transactions on Emerging Telecommunications Technologies*, 25(1), 81–93.

21. Visvizi, A., & Lytras, M. D. (2019). Sustainable smart cities and smart villages research: Rethinking security, safety, well-being, and happiness. *Sustainability*, 12(1), 215.

22. Talari, S., Shafie-Khah, M., Siano, P., Loia, V., Tommasetti, A., & Catalão, J. P. (2017). A review of smart cities based on the internet of things concept. *Energies*, 10(4), 421.

23. Syed, A. S., Sierra-Sosa, D., Kumar, A., & Elmaghraby, A. (2021). IoT in smart cities: A survey of technologies, practices and challenges. *Smart Cities*, 4(2), 429–475.

24. Almalki, F. A., Alsamhi, S. H., Sahal, R., Hassan, J., Hawbani, A., Rajput, N. S., & Breslin, J. (2021). Green IoT for eco-friendly and sustainable smart cities: Future directions and opportunities. *Mobile Networks and Applications*, 1–25.

25. Aloi, G., Bedogni, L., Felice, M. D., Loscri, V., Molinaro, A., Natalizio, E., & Zema, N. R. (2014). STEM-Net: An evolutionary network architecture for smart and sustainable cities. *Transactions on Emerging Telecommunications Technologies*, 25(1), 21–40.

26. Hancke, G. P., de Carvalho e Silva, B., & Hancke Jr, G. P. (2012). The role of advanced sensing in smart cities. *Sensors*, 13(1), 393–425.

27. Sánchez, L., Gutiérrez, V., Galache, J. A., Sotres, P., Santana, J. R., Casanueva, J., & Muñoz, L. (2013, June). SmartSantander: Experimentation and service provision in the smart city. In *2013 16th International Symposium on Wireless Personal Multimedia Communications (WPMC)* (pp. 1–6). IEEE.

28. Khan, R., Khan, S. U., Zaheer, R., & Khan, S. (2012, December). Future internet: The internet of things architecture, possible applications and key challenges. In *2012 10th International Conference on Frontiers of Information Technology* (pp. 257–260). IEEE.

29. Caragliu, A., Del Bo, C., & Nijkamp, P. (2009, October). Smart cities in Europe. In *Proceedings of the 3rd Central European Conference in Regional Science. Košice, Slovak Republic* (pp. 7–9). International Journal on Semantic Web and Information Systems.

30. Berners-Lee, T. (2006). Linked data. W3C design issues. *International Journal on Semantic Web and Information Systems*, 4, W3C.

31. Khanna, A. (2015). An architectural design for cloud of things. *Facta Universitatis, Series: Electronics and Energetics*, 29(3), 357–365.

32. Zhou, J., Leppanen, T., Harjula, E., Ylianttila, M., Ojala, T., Yu, C., & Yang, L. T. (2013, June). Cloudthings: A common architecture for integrating the internet of things with cloud computing. In *Proceedings of the 2013 IEEE 17th International Conference on Computer supported Cooperative Work in Design (CSCWD)* (pp. 651–657). IEEE.

33. Yun, M., & Yuxin, B. (2010, June). Research on the architecture and key technology of Internet of Things (IoT) applied on smart grid. In *2010 International Conference on Advances in Energy Engineering* (pp. 69–72). IEEE.

34. Gubbi, J., Buyya, R., Marusic, S., & Palaniswami, M. (2013). Internet of Things (IoT): A vision, architectural elements, and future directions. *Future Generation Computer Systems*, 29(7), 1645–1660.

35. Li, W., Zhong, Y., Wang, X., & Cao, Y. (2013). Resource virtualization and service selection in cloud logistics. *Journal of Network and Computer Applications*, 36(6), 1696–1704.

36. Perera, C., Zaslavsky, A., Christen, P., & Georgakopoulos, D. (2014). Sensing as a service model for smart cities supported by internet of things. *Transactions on Emerging Telecommunications Technologies*, 25(1), 81–93.

37. Khalifeh, A., Darabkh, K. A., Khasawneh, A. M., Alqaisieh, I., Salameh, M., AlAbdala, A., & Rajendiran, K. (2021). Wireless sensor networks for smart cities: Network design, implementation and performance evaluation. *Electronics*, 10(2), 218.

38. Miyasawa, A., Akira, S., Fujimoto, Y., & Hayashi, Y. (2021). Spatial demand forecasting based on smart meter data for improving local energy self-sufficiency in smart cities. *IET Smart Cities*, 3(2), 107–120.

39. Silva, B. N., Khan, M., & Han, K. (2018). Towards sustainable smart cities: A review of trends, architectures, components, and open challenges in smart cities. *Sustainable Cities and Society*, 38, 697–713.

40. Rubí, J. N. S., & de Lira Gondim, P. R. (2021). IoT-based platform for environment data sharing in smart cities. *International Journal of Communication Systems*, 34(2), e4515.

41. Hyman, B. T., Alisha, Z., & Gordon, S. (2019). Secure controls for smart cities; applications in intelligent transportation systems and smart buildings. *International Journal of Science and Engineering Applications*, 8(6), 167–171.

42. Delsing, J. (2021). Smart city solution engineering. *Smart Cities*, 4(2), 643–661.

43. Singhal, S., Athithan, S., Alomar, M. A., Kumar, R., Sharma, B., Srivastava, G., & Lin, J. C. W. (2023). Energy aware load balancing framework for smart grid using cloud and fog computing. *Sensors*, 23(7), 3488.

10 Data-Level Cyber Deception in Cloud of Things

Prospects, Issues, and Challenges

Nilin Prabhaker, Ghanshyam S. Bopche, and Michael Arock

10.1 INTRODUCTION

Internet of Things (IoT) and cloud-based technologies offer a new way of delivering traditional information and communications technology (ICT) services to organizations and governments by combining platforms, operating systems, storage elements, databases, and other ICT equipment. Such Cloud of Things (CoT) are increasingly used in homes and workplaces to improve service delivery and productivity. Early adopters of IoT and cloud technologies include various sectors but not limited to fitness [43], healthcare [34], telecommunication industry [64], retail[23], manufacturing industries [63], real estate [52], transport [7], governments [4], life-sustaining critical infrastructures such as power plant [36], water treatment plants [40], defense (e.g., Internet of Battle things [37]), etc. While CoT infrastructure offers numerous benefits to organizations, governments, and end users, it presents unique security and privacy challenges.

Figure 10.1 shows the typical CoT infrastructure for providing affordable healthcare services to remote home care patients. The Internet of Medical Things (IoMT) [57] collect the patient's health-related parameters and transmit them to the cloud via the intermediate fog layer. The predictive analytics about the patient's health is performed over the cloud (based on past medical history and current data) using various machine learning or deep learning algorithms. Based on the generated results, a team of medical experts will advise medication to the patient. Suppose the collected health parameters of the patient are not within the normal range, and the predictive analytics outcome forecasts a severe medical event soon, the air ambulance [30] can be deployed immediately and the closest hospitals will get alerted about the patient's health.

Patient health parameters and past medical history are considered sensitive information. To offer affordable medical services, healthcare providers need to collect patient data, transmit it over the public Internet, and store it on the cloud. However, multiple avenues of potential cyberattacks exist to steal and compromise the patient's

DOI: 10.1201/9781003390954-10

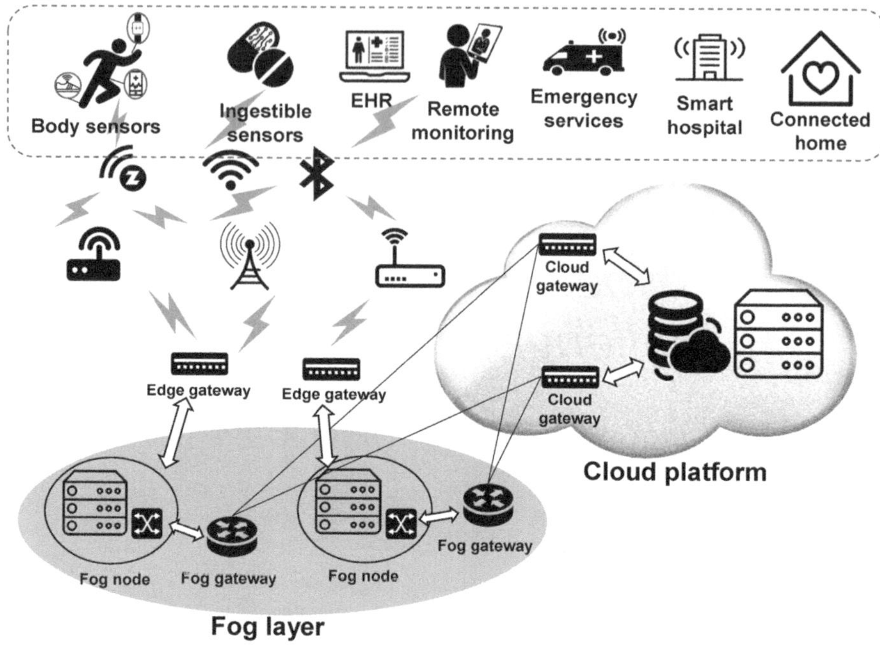

FIGURE 10.1   Typical Cloud of Things infrastructure for healthcare services [21].

medical data. Potential adversaries may compromise the deployed medical devices (IoMT), breach the intermediate fog layer's security, steal or poison the stored data on the cloud servers, or in the worst case, create a hostage situation by encrypting the data. Data poisoning may lead to the wrong prediction, prescription, and loss of human life. Ransomware attacks lead to the unavailability of healthcare services [56]. As CoT technology is increasingly being used in critical infrastructures and business organizations, it is essential to secure the data during its entire life cycle. Due to the lack of necessary built-in security controls in the IoT devices, the use of various cloud deployment models, the lack of security compliance, and the increasing sophistication of cyberattacks, cybersecurity risks have become significant in the CoT environment. In other words, the increased complexity of data access due to the diversity of platforms leads to multiple leakage scenarios while data is being created, accessed, and utilized.

The cloud offers cheap storage and data processing capabilities. If an adversary likes to steal or compromise the data in the CoT environment, they may prefer to do it on the cloud storage servers. It is because of the volume of sensitive data available in the cloud, its accessibility worldwide over the Internet, and the enormous cloud threat landscape. Many security solutions are available to protect cloud storage at different levels of abstraction. To protect the outsourced data, security administrators must first classify the data, enforce robust encryption, and perform logging and reporting to ensure the authorized use of data. Next, the administrator must implement strong access control policies and authentication mechanisms and perform behavior analysis at the user level to prevent potential data leakage. To protect the cloud at the infrastructure level, techniques such as network segmentation [5], configuration hardening [10], and key management [15] need to be performed.

Despite having security at different levels and the agreement with the cloud service provider (CSP), maintaining privacy, confidentiality, integrity, and availability of outsourced data remains challenging for the data owners. As per the Attivo Networks report [6], data loss and leakage, threats to data privacy, and breaches of confidentiality are the top three cloud security challenges. Securing outsourced data in the cloud environment is one of the most significant security challenges for businesses and governments due to the limited level of control over the security of their data once it is outsourced. Adversaries may gain unauthorized access to the outsourced data by doing social engineering, using harvested credentials, exploiting technical software vulnerabilities (zero-day or well-known vulnerabilities), and bypassing existing security controls.

Cyber deception is one of the cyberdefense techniques proposed in the literature to deceive the attacker by creating deceptive artifacts such as honeypots [49], honeynet [50], honeytoken [16], honeyservers [35], honeyaccounts [1], honeyfiles [62], believable decoy documents [17, 33], etc. Essentially, the objective of using deceptive artifacts is to mislead the attacker from the actual target and keep track of an adversary's actions during the real-time network intrusion. Even post-data breach, the deceptive artifacts can confuse the attacker, slow him down, and waste their resources in identifying the actual documents, thereby increasing the overall attack cost. However, most data-level cyber deception solutions proposed in the literature protect the resources in the local environment (i.e., on-premise company infrastructure). Since the cloud refers to virtualized resources, can the existing data-level deception techniques be used as it is in the cloud infrastructure? It is evident from the literature that the use of deception [8, 59, 53] in the cloud and IoT environment is possible. However, very few of them focused on data-level cyber deception. This chapter aims to study the feasibility of using data-level cyber deception in CoT environments. Our primary focus is on protecting the CoT infrastructure from potential data-exfiltration attacks.

The rest of the chapter is organized as follows: Section 10.2 discusses the array of potential cyber risks in the IoT and cloud infrastructure. Section 10.3 explores existing security controls for securing data in the CoT environment. The use of deceptive technologies as a final line of defense against the ongoing cyberattack and post-attack is covered in Section 10.4. Section 10.5 discusses the open issues and challenges in using data-level cyber deception in the CoT environment. Section 10.6 focuses on the future research direction to safeguard sensitive and critical data in the CoT environment. Finally, Section 10.7 concludes the chapter.

10.2   CLOUD AND IOT THREAT LANDSCAPE

With the advancement in computing technologies, adversaries always find new ways of exfiltrating business or mission-critical data. Although the cloud infrastructure provides tremendous benefits over traditional services, it is not secure by default. For data owners, the security of outsourced data is crucial. Any violation of security

TABLE 10.1
OWASP top ten risks for IoT and cloud infrastructure [46]

	IoT Top Ten	Cloud Top Ten
1	Weak, guessable, or hardcoded unchangeable passwords	Accountability and data risk
2	Insecure network services	User identity federation
3	Insecure ecosystem interfaces	Legal and regulatory compliance
4	Lack of secure update mechanism	Business continuity and resiliency
5	Use of insecure or outdated components	User privacy and secondary usage of data
6	Insufficient privacy protection	Service and data integration
7	Insecure data transfer and storage	Multi-tenancy and physical security
8	Lack of device management	Incidence analysis and forensic
9	Insecure default setting	Infrastructure security
10	Lack of physical hardening	Non-production environment exposure

policies leads to a successful data breach, which may lead to severe consequences, such as identity theft, financial loss, or reputation damage. As per the Cloud Security Alliance report [19], data-exfiltration attack on the cloud is a leading concern, which underpins and breeds other security issues. For example, an undesirable data flow violates data privacy, leading to regulatory compliance issues. Compromise of credentials leads to unauthorized access to the system and, finally, access to data. Data exfiltration may also lead to a downgrade in the performance of an organization or even discontinuity of the business.

Table 10.1 shows the top ten security risks in the IoT and cloud environments compiled by Open Web Application Security Project (OWASP) [46]. The table shows that data privacy is a significant security concern in the CoT infrastructure. As per the Cloud Security Alliance report [19], the most crucial consideration is the privacy of the outsourced data in the CoT environment. Figure 10.2 shows the typical threat landscape for the CoT infrastructure. Each security concern may lead to a successful data breach by unauthorized users. The potential adversary may attempt to bypass authentication mechanisms to gain access to the CoT, listen to ongoing communication between CoT devices, manipulate the in-transit data, modify device settings physically to make it more vulnerable, exploit vulnerabilities in IoT devices, and introduce malware into the CoT through infected devices. An adversary can be an external malicious entity or an insider disgruntled employee. The net effect of such attacks is unauthorized access or disclosure of the data during transit or at rest in the cloud. Furthermore, due to the adoption of the shared responsibility model in the cloud platform (Figure 10.3), the customer in the Software-as-a-Service (SaaS) environment has minimal access to the native cloud resources. Usually, it is very challenging for users to protect their data when they need help to visualize and analyze cloud services. Last but not least, the abusive and nefarious use of cloud services leads to successful data breaches.

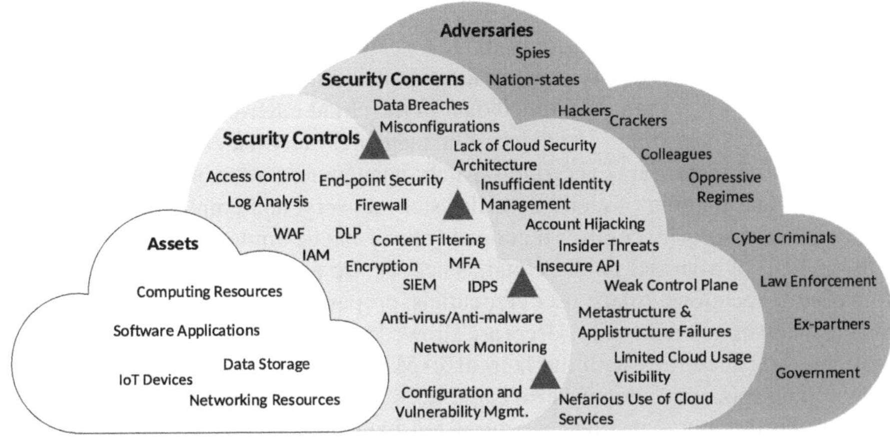

FIGURE 10.2 Threat landscape for Cloud of Things (CoT) environment.

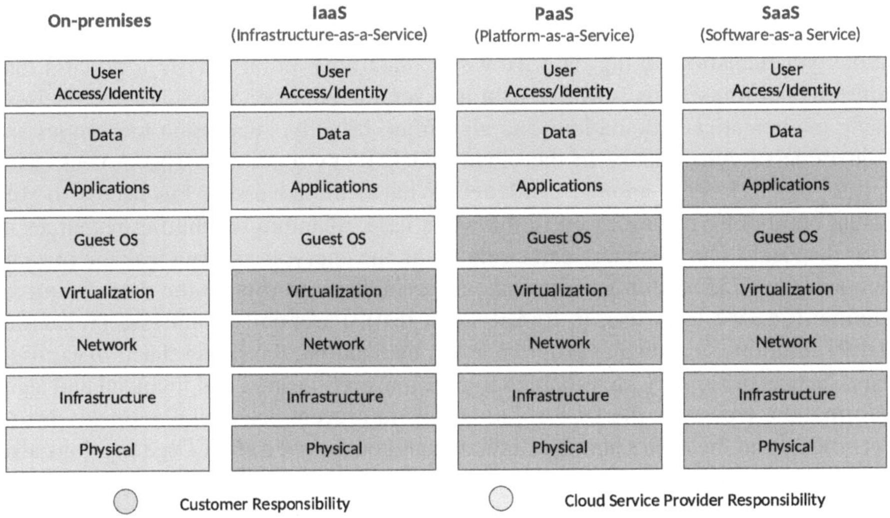

On-premises	IaaS (Infrastructure-as-a-Service)	PaaS (Platform-as-a-Service)	SaaS (Software-as-a Service)
User Access/Identity	User Access/Identity	User Access/Identity	User Access/Identity
Data	Data	Data	Data
Applications	Applications	Applications	Applications
Guest OS	Guest OS	Guest OS	Guest OS
Virtualization	Virtualization	Virtualization	Virtualization
Network	Network	Network	Network
Infrastructure	Infrastructure	Infrastructure	Infrastructure
Physical	Physical	Physical	Physical

○ Customer Responsibility ○ Cloud Service Provider Responsibility

FIGURE 10.3 Shared responsibility model in cloud environment [45].

10.3 EXISTING SECURITY CONTROLS IN THE IOT AND CLOUD

The CoT presents unique security challenges due to the large number of connected devices and the vast amounts of data generated, transmitted, stored, and processed over the distributed infrastructure. Since it integrates two technologies, the attack surface has a multiplicative increase. Therefore, there is a need to protect both the IoT environment and cloud infrastructure equally. Organizations can implement various security controls to mitigate security risks and protect data and systems. The most prominent security controls to protect data in an IoT environment are strong authentication and access control policies [20], robust encryption algorithms [61], regular firmware updates, network segmentation [5], continuous network monitoring and logging, physical security of the IoT devices, secure software development, hardware and software supply chain management, and user awareness.

However, to protect the outsourced data in the cloud environment, the significant security controls are identity and access management (IAM), compliance and regulatory requirements, data encryption, patch management, network security, secure monitoring and logging of cloud data, disaster recovery, and employee training and awareness. The security controls mentioned above can be implemented and deployed with the help of a virtual public cloud. Since we are accessing the cloud resources from remote locations, robust user credentials, proper authentication, and authorization mechanisms are necessary. Users are allowed to access cloud resources only after verifying their identities with the help of IAM services [3]. IAM provides authentication, authorization, identity provision, access control, monitoring, and auditing services. Furthermore, it offers a centralized management system where business users can manage all services and programs in one place with cloud-based services. Data masking and data encryption [61] is another set of security controls that provides end-to-end data protection in the cloud environment.

Data masking is the process of permanently replacing sensitive data with fictitious yet realistic-looking data. However, data encryption [61] is a process that converts and transforms sensitive data into scrambled, often unreadable, ciphertext using mathematical calculations and algorithm. Usually, encryption techniques are vulnerable to cyberattacks in the cloud and IoT environments. The protection of cryptographic keys is another challenge. What if the adversary has access to sufficient computing resources or, in the worst case, quantum computing resources at their disposal? The adversary can easily break the encryption. Data leakage prevention solutions [55] are in use throughout the industry to protect the data during its lifetime (i.e., while in use, in transit, or at rest). Essentially, data loss prevention (DLP) solutions save data at different levels by handling the risks related to exposed data, such as personally identifiable information, credit cards, and financial and legal information. However, the DLP solution fails to prevent data leaks in transit due to encryption and the high volume of inbound and outbound traffic. Organizations also need help implementing DLP solutions due to their complexity in configuration and management.

Network segmentation [5] is another security control for the CoT environments that allows network administrators to control network traffic flow between sub-nets based on fine-granular policies. By isolating the network traffic, the administrator may reduce the attack surface and obstruct the possibility of lateral movement. However, network segmentation may increase the management and routing overhead. Network access control lists [32] govern access to resources within the network segments. To allow or deny service requests to access outsourced data and the CoT resources, fine-grained access controls [2] can be used. It maintains the principle of least privilege, privacy and confidentiality, and centralized data control. Secure configuration of the hardware and software components within the CoT environment is of utmost priority to deter potential cyberattacks, if any. Cloud configuration management [12] allows CSPs to automate various configuration states to help

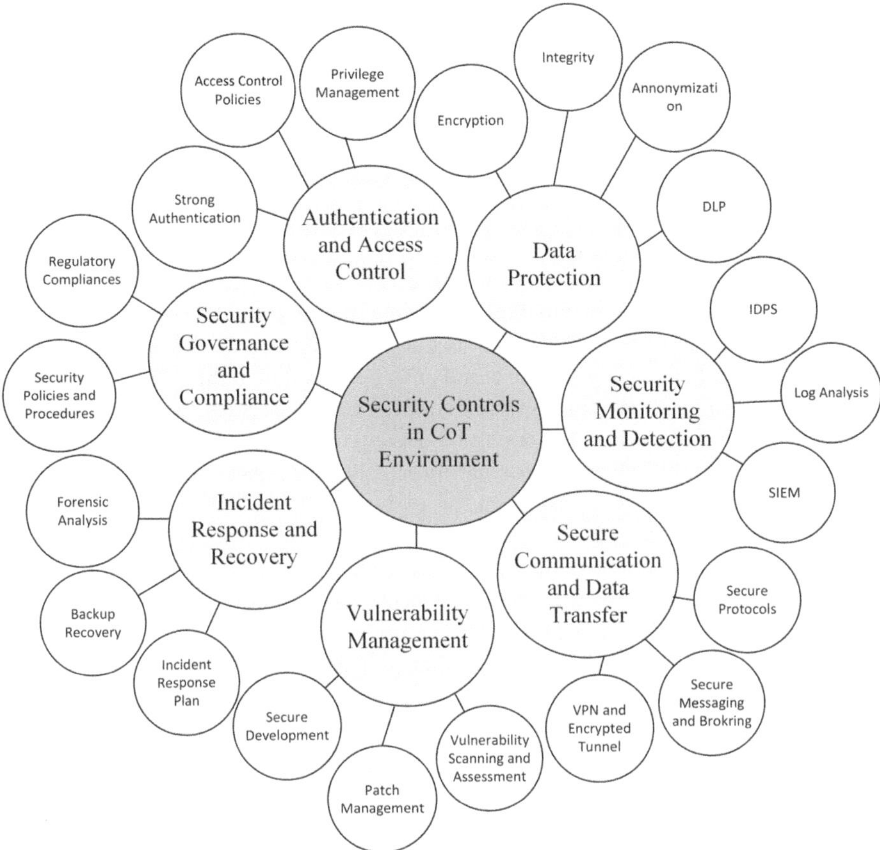

FIGURE 10.4 Existing security controls in CoT environment.

ensure deployed devices meet proper configuration requirement. However, achieving a secure CoT environment is challenging due to the environment's heterogeneity. Security Information and Event Management (SIEM) solutions [11] are widely used throughout the industry to identify data breaches. SIEM collects data from different sources (e.g., network devices, domain controllers, and routers), aggregates it, analyzes it to detect ongoing attacks, and notifies the security administrator.

The predominant security controls used to address the attacks and threats in CoT environment are presented in Figure 10.4, which provides a high-level classification and many solutions across different categories. Despite having numerous security solutions to protect the data at various stages in the CoT environment, cyberattacks on such infrastructure keep increasing in volume and sophistication. The consequences of such an attack may lead to financial loss, damage to reputation, loss of intellectual property, operation discontinuity, and regulatory compliance. There is a need for security defense techniques to detect the attack proactively and prevent the CoT environment. Cyber deception is a defense technique that can protect the data even after an attack. Essentially, the security administrator can deploy the deceptive artifacts in a real-time environment and mislead the attacker, increasing the attack's cost and thereby slowing down the attacker. In the next section, we discuss deception in IoT and cloud to proactively detect attacks and prevent data breaches.

10.4 USE OF CYBER DECEPTION IN THE IOT AND CLOUD ENVIRONMENT

Deception technology is invaluable in solving the asymmetry problem between adversaries and defenders. Organizations may generate and deploy genuinely looking deceptive artifacts to convert the cloud or IoT environment into a trap. The benefit of planning such an active trap is that whenever adversaries try to search and exfiltrate the sensitive resources, the deployed deceptive artifacts may trigger a security alert that the security administrator can precisely investigate. To conclude, active cyberdeception platforms are essential for analyzing the real-time alert, tracking the adversary, and deploying countermeasures for attack remediation. On the other hand, the passive deceptive artifacts may complicate the job of adversaries even after the successful data breach. Essentially, adversaries must identify the valuable, sensitive data hidden among the misleading, deceptive artifacts. The deceptive data may increase the attacker's confusion, affect their decision-making capability, and slow them down.

Numerous deception techniques are proposed in the literature to protect the IoT infrastructure. Honeypot [51] is one of the deception techniques which uses decoy computers to attract cyberattackers, attention. Pa et al. [47] implemented the first tailored honeypot, IoTPOT, which can mimic the IoT device to lure the attacker, and IoTBOX, the first malware analysis environment for IoT devices. IoTBOX can analyze malware at different CPU architectures and consider only the telnet protocol in their honeypot environment. Jicha et al. [31] proposed the Supervisory Control and Data Acquisition (SCADA) honeypot, Conpot, to detect tampering of devices within the SCADA network. Guarnizo et al. [28] proposed SIPHON, a high-interaction honeypot for the IoT environment. SIPHON analyzes the Secure Shell (SSH) and HTTP traffic to detect the presence of adversaries. To capture the attack traffic for retrospective analysis, Dowling et al. [22] proposed the ZigBee honeypot, which simulates a gateway to assess the presence of a ZigBee attack. T. Luo et al. [38] proposed the IoTCandyJar honeypot, which can dynamically redirect the IoT environment from low interaction to high interaction. IoTCandyJar focuses on IoT communication protocols and uses machine learning technology to learn behavioral knowledge of IoT devices. Wang et al. [58] proposed ThingPot, a honeypot IoT platform to analyze the attacker activities in a household environment. ThingPot focuses only on the HTTP and XAMP protocols. To capture the IoT attacks on the IoT environment's universal plug-and-play protocols, Hakim et al. [29] proposed a novel framework U-Pot, which creates a honeypot from the device description documents automatically. X. Luo et al. [39] proposed software-defined network (SDN)-based honeypots, which mimic IoT devices to protect against distributed denial-of-service attacks. To understand the attacks against IoT cameras, Tabari et al. [54] proposed the multi-phased multi-faceted honeypot ecosystem. The authors proposed a recorded video-based low-interaction IoT camera honeypot to detect and prevent video tampering-based attacks. Guan et al. [27] presented scalable high-interaction IoT camera honeypots, which reduce the cost of deployment of a real IoT camera as a honeypot. To automatically generate the intelligent-interaction honeypot, Yamamoto et al. [60] proposed a framework called FirmPot. The FirmPot uses a firmware emulator to create a honeypot capable of learning the behavior of emulated applications. However, the FirmPot does not cover misconfiguration attacks. Mfogo et al. [42] proposed a honeypot, which automatically learns and interacts with attacker activities. The authors claimed that the proposed honeypot improves the attacker's session length and can capture numerous attacks.

Cloud infrastructure also uses honeypot technology to lure attackers and distract them from reaching valuable resources. Balamurugan and Poornima [8] proposed honeypots-as-a-service to track attackers and secure the actual instances of resources. The service can detect adversarial actions in the cloud environment but does not recommend remedial action. Honeypots may be low-interaction or high-interaction, depending on their deployment objective. Nithin Chandra and Madhuri [44] analyzed the use of honeypots in the cloud environment with their advantage and limitation. Brown et al. [14] have used honeypots such as Dionaea, Kippo, and Amun on various cloud instances and gathered data about the source of attacks, their types and variation to cloud instances. The study found that Dionaea and Kippo are the most effective in the cloud environment. To summarize, the honeypot monitors the native cloud infrastructure and the incoming and outgoing traffic. It also provides complete isolation from any production network in a public cloud environment. The ultimate objective of deploying honeypots is to distract the adversary from the actual targets (here, the production server), monitor the attacker's behavior to understand their tools, techniques, and practices and gather the threat intelligence. However, honeypots require regular maintenance to deceive the attackers successfully. An adversary may also compromise the honeypots and use them for lateral movement if connected to the production network.

Using genuinely looking decoy objects such as honeyfiles, honeytokens, honeyservices, etc., is advocated throughout the literature to lure attackers away from essential data stored on the cloud and learn about the attackers' intention (potential target) and techniques. Despite many solutions for generating and deploying honey objects, only some are proposed to protect outsourced data in the cloud environment. Stolfo et al. [53] proposed an offensive decoy technology that monitors and detects abnormal data access patterns. The authors proposed launching a disinformation attack by returning a large amount of fake information to adversaries against the misuse of users' data. Wilson and Avery [59] proposed to generate the deceptive object of each file uploaded on the SaaS platform. The proposed system consists of a generation engine to produce decoy objects, a detection engine to monitor access requests, a prevention engine for control, and a threat engine to adjust protection levels based on threat types and severity. Bourke and Grzelak [13] generated the honeytoken of Amazon Web Services (AWS) credentials to detect the adversaries and create an alert for the access of each honeytoken. The authors have developed a deployable product

SPACECRAB which provides an application programming interface (API) to create, update, and dispose of AWS credentials. Bao et al. [9] proposed a deep learning approach to generate realistic network flow data based on user and IoT device interaction. The authors resolved the problem of scarcity of network signal in the IoT honeypot environment to analyze the attacker activity.

To conclude, deception technology offers numerous benefits, such as detecting attack attempts just in time (with the minimum number of false positives), affecting the attacker's decision-making capabilities (thereby slowing them down), and wasting the attacker's time and resources identifying the real assets. Organizations can use deception technology in isolation or integration with the existing security controls. It offers active cyber deception during the real-time attack attempt and affects attackers' decision-making capabilities while identifying the legitimate sensitive data hidden among the believable fakes. However, deploying deceptive artifacts is tricky and challenging in the CoT environment as IoT devices are resource-constrained, and the cloud is resourceful. Furthermore, the generated deceptive artifacts must be created carefully, deployed consciously (at vantage points), and maintained periodically to increase attackers, uncertainty. However, there are issues and challenges in using deception in the CoT environment. Section 10.5 discusses such challenges.

10.5 OPEN ISSUES AND CHALLENGES IN IMPLEMENTING DATA-LEVEL CYBER DECEPTION IN COT ENVIRONMENT

Data-level cyber deception uses believable fake artifacts to lure the attacker. The decoy data appears real and misleads the attackers from reaching the sensitive and valuable information. Nowadays, CSPs have started offering a new type of security service named Deception-as-a-service [18, 24] that creates and maintains a wide range of deceptive solutions such as honeypot, honeyserver, honeytoken, honeyfiles, etc., to improve the organization's cybersecurity posture on demand. However, there are inherent challenges in implementing the deception-as-a-service model in the CoT environment.

CoT infrastructure generates, transmits, stores, and processes various types of data such as structured, unstructured, and semi-structured (shown in Figure 10.5), which may be classified further into various subcategories such as signal, video, audio, image, text, chart, graph, table, and schema. Such diversity of data in organizations poses a unique challenge while generating deceptive artifacts. Developing the deceptive artifact for each data type has its own complexity, requires different strategies, and has its own resource requirements. Figure 10.6 depicts the complexity of deceptive artifact creation and detection for signal, video, audio, and textual data.

We live in an age of misinformation, where the deepfake technology [41] has the potential to create believable fake videos, audio, image, and text. The first challenge while adopting the data-level cyber deception comes during the generation of deceptive artifacts. Developing conceivable misleading signals that mimic real-world signals is challenging due to insufficient high-quality data, the cost of model training, and the model's generalization to signals of different types. Furthermore, human evaluation of the generated believable signal is time-consuming. However,

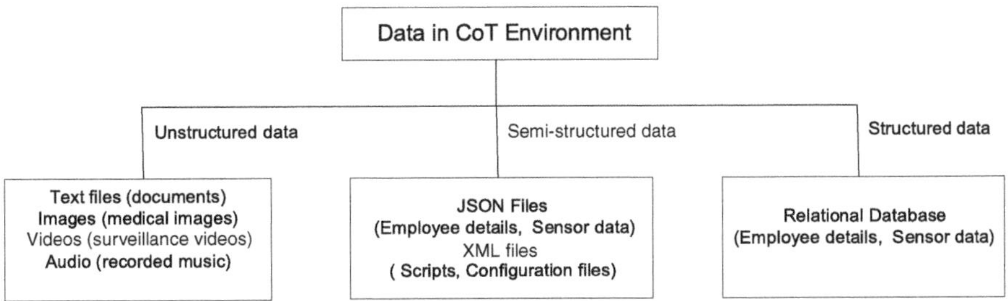

FIGURE 10.5   Types of data in CoT environment.

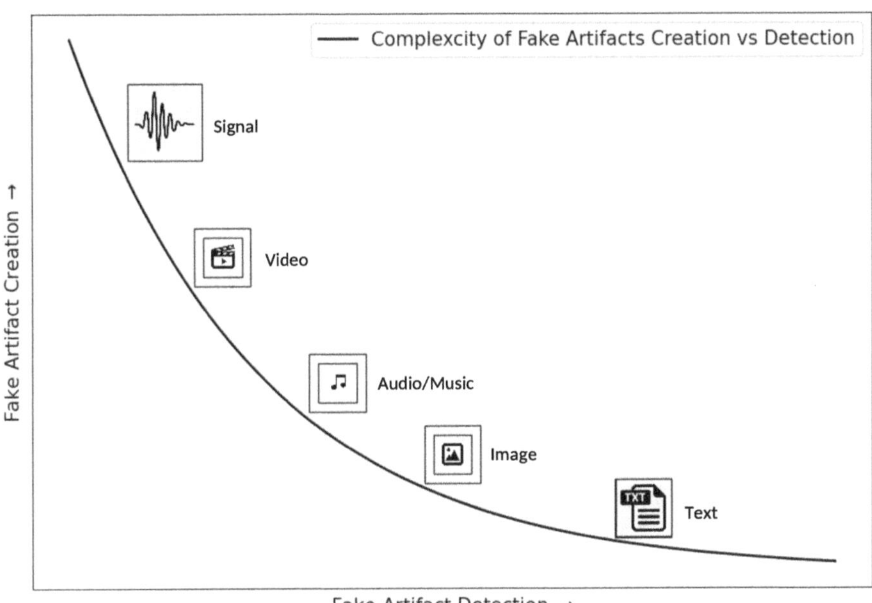

FIGURE 10.6   Complexity of believable fake artifacts creation and detection: Defender and attacker perspective.

on the detection side, even a tiny change in a signal can be easily detectable. Developing a credible fake video also takes an enormous amount of videos as input and computing resources to train the model, which increases the overall cost. Like video, generating fake audio requires massive memory space and computing power. However, creating deceptive audio is more straightforward than videos. On the other hand, detecting fake audio is moderately complex compared to deceptive video and signal. Due to the limited availability of resources (memory and computing power) in the IoT environment, deploying such artifacts is not feasible. However, it is feasible only in the cloud environment. Believable image generation using generative adversarial networks (GAN) [25, 48] also needs enormous computing power and memory. Detecting the fake image is challenging compared to audio, video, and signal data.

Due to the structural invariant property of the text, creating multiple copies of believable fake text is easy. However, comparatively, it is tough to detect fake textual content. Other challenges with generating diverse deceptive artifacts are real-time generation, model generalization, and human evaluation [26].

Usually, most organizations take the help of external penetration testers or red teams to test the resiliency of on-premise or cloud-based production environments. Such authorized penetration tests pose a significant danger to resource security, for example, data privacy. The red team may steal real-time data while conducting penetration tests or may not disclose all the uncovered vulnerabilities with the organization. To solve the problem of data privacy, organizations may mix the decoy data with the actual data during real-time penetration tests. The decoy artifacts could be honeypassword, honeytoken, honeyaccount, honey file, honeyserver, fake certificates, fake serverless functions, fake storage buckets, fake APIs, or even fake virtual machines. To solve the problem of nondisclosure of identified vulnerabilities, organizations may run some honeyservices. As the red team imitates real-time attacks, another advantage of using deceptive artifacts is to test the artifact's effectiveness in detecting real-time attacks and generating the alert. However, there is another risk of a deceptive environment or artifact getting identified by the adversary during real-time network intrusion. If so, the adversary will try to avoid decoys or honeypots, making it more challenging for the organizations to detect and prevent attacks in real time.

The generation and deployment of believable fake artifacts have some legal and ethical concerns. Hence, ensuring their responsible use in organizations' sensitive and regulated environment is a significant challenge. The use of cyber deceptions demands continuous maintenance and updates to stay ahead of potential adversary's constantly changing attack tactics and techniques. Regular updating and maintenance of deception environments can be time-consuming and resource-intensive. The limitation of using cyber deception in organizations is the risk of false positives wherein the legitimate user is mistakenly flagged as an attacker, leading to unwanted security alerts and disruptions. Investigation of such false positives leads to wasted time and resources and finally creates distrust among the security team about the effectiveness of cyber deception. Figure 10.7 depicts the effect of security controls on the usability and resource requirements. In any environment, the usability of resources such as data and services typically decreases with an increased security controls. In contrast, the requirement of resources (memory and computing power) goes up with increase in security controls. Deploying cyber deception could be resource-intensive. Therefore, it has to be deployed carefully in the CoT infrastructure to keep usability and resource requirements optimum.

Using data-level cyber deception in the CoT environment poses some more inherent challenges. It is due to the heterogeneity of IoT devices, operating systems, cloud platforms, and available services in the CoT infrastructure. Furthermore, the storage and processing of large volume of data also pose a significant challenge. The deceptive artifacts may consume a large amount of storage space. The increased amount of data on the cloud may also affect the monitoring process, which is essential to ensure security, performance, capacity planning, managing resources, and troubleshooting.

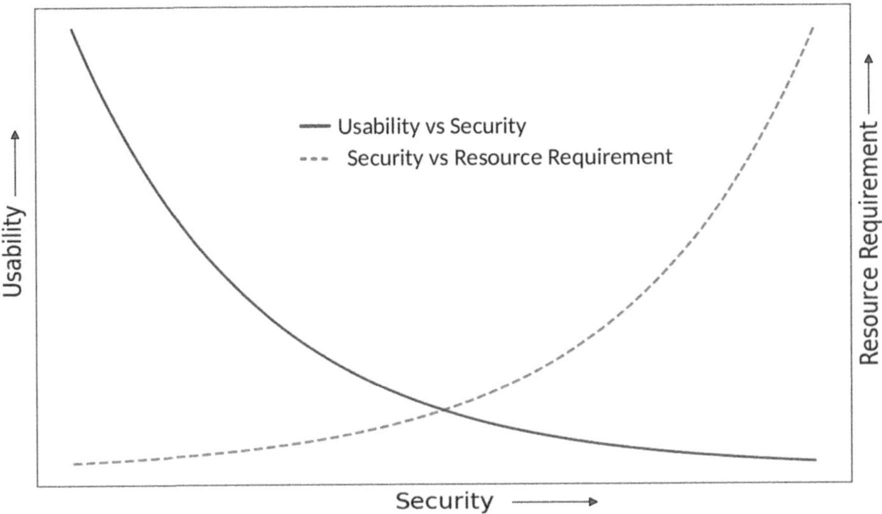

FIGURE 10.7   Security versus usability versus resource requirement.

Deceptive objects (e.g., deceptive traffic) may consume a significant amount of network bandwidth as well. Creating fake copies of all the potentially sensitive CoT resources can increase the organizations security budget. Organizations can lower the cost of cyber deception by classifying digital resources according to their sensitivity and creating deceptive copies for only sensitive components. Standardization is crucial for the wide acceptance of cyber deception by customers. Due to variations in sensitive resources and data, the mechanism to create deceptive artifacts still needs to be standardized. Furthermore, there is no metric support to evaluate the performance or efficiency of the employed cyber deception.

Integration of a deception technology adds to cloud visibility and cloud security. For example, the organization typically uses IAM as a primary defense line to protect the cloud environment from unauthorized access. However, detecting the attacker's lateral movement within the network is difficult using IAM security control alone. A security administrator may use deception techniques to create and deploy decoy objects to identify and monitor privileged accounts. However, before using deception techniques, organizations must be ready with a few things, such as the list of resources that need to be protected, information about the potential adversaries, thoroughly tested deception platform or artifacts, alert generation and analysis module, and countermeasure deployment module to contain or stop the ongoing attack.

10.6 RESEARCH DIRECTIONS

Nowadays, generative artificial intelligence (AI) can generate all types of synthetic data. Organizations increasingly use generative AI tools and techniques to synthesize diverse data for business requirements. Organizations and security researchers must investigate the possibility of using generative AI techniques to develop deceptive artifacts to protect business or mission-critical resources. Researchers must explore the ethical use of generative AI for believable deceptive content creation for active and passive cyber deception. Despite generating credible content, several significant directions, such as effective content generation, rigorous human evaluation, effective deployment of deceptive artifacts, ease of continuous maintenance, integration with existing solutions, the burden on genuine users, etc., must be explored.

When discussing effective content generation, the generated content must be similar enough to legitimate content to make the adversary believable yet dissimilar enough to make the adversary wrong. It must lure the adversary and affect his/her decision-making. Usually, the AI-generated content is similar to the original content, so it must be ensured that the generated content has some dissimilarity. The effectiveness of generated content can be evaluated using rigorous human evaluation techniques, which must be explored. Moreover, effective deployment of the generated content is also essential, which can be solved by categorizing the content based on the sensitivity of artifacts and deploying the deceptive content according to the level of protection it requires must be explored. The sensitivity-based classification and deployment also optimize the memory requirements. Since manual effort to measure the effectiveness of deployed artifacts is time-consuming and error-prone, metrics must automatically evaluate deployed contents' performance. With the increased sophistication of the adversary, there is a possibility of using deception against the deceptive environment, so there is a need to continuously maintain the artifacts to retain the effectiveness of misleading artifacts.

The use of cyber deception in the CoT environment for defense is still in its early stage, and robust tools and platforms still need to be improved. Integrating deception with existing security solutions will become a powerful security solution. Furthermore, the isolated solution or the integration with existing security may burden real users in detecting the actual object in a deceptive environment, which may affect usability. There are some ways to reduce the burden on real users, such as using a vault or by generating a hash with the MAC address of the system on which the content is created. However, there is a possibility to explore a new way by which the burden on real users can be reduced. With the advent of new technologies, the future of CoT is promisings. The advancement will enable more data to be gathered and processed for decision-making and efficient operations. Hence, there is a need for a dedicated framework that generates, deploys, and maintains the deceptive artifacts at every layer in the CoT environment.

10.7 CONCLUSION

Cyber deception is a promising defense strategy to enhance the security of the CoT environment. Deploying deceptive artifacts makes it possible to analyze the attackers' activity, detect the adversaries, and prevent attacks on outsourced data in the CoT environment. This chapter has discussed different threats and security controls to protect the outsourced data in the CoT environment. Furthermore, we discussed the prospects of using data-level cyber deception and highlighted the implementation issues and challenges for the CoT infrastructure. We have also shed light on integrating deception technologies with existing security controls and their effect on usability and performance.

REFERENCES

1. Mitsuaki Akiyama, Takeshi Yagi, Kazufumi Aoki, Takeo Hariu, and Youki Kadobayashi. Active credential leakage for observing web-based attack cycle. In *International Workshop on Recent Advances in Intrusion Detection*, pages 223–243, Berlin, Heidelberg. Springer, 2013.

2. Asma Alshehri and Ravi Sandhu. Access control models for virtual object communication in cloud-enabled IoT. In *IEEE International Conference on Information Reuse and Integration (IRI)*, pages 16–25. IEEE, 2017.

3. Amjad Alsirhani, Mohamed M. Ezz, and Ayman Mohamed Mostafa. Advanced authentication mechanisms for identity and access management in cloud computing. *Computer Systems Science and Engineering*, 43(3):967–984, 2022.

4. Hamidreza Arasteh, Vahid Hosseinnezhad, Vincenzo Loia, Aurelio Tommasetti, Orlando Troisi, Miadreza Shafie-khah, and Pierluigi Siano. IoT-based smart cities: A survey. In *16th International Conference on Environment and Electrical Engineering (EEEIC)*, pages 1–6. IEEE, 2016.

5. Murshedul Arifeen, Andrei Petrovski, and Sergey Petrovski. Automated microsegmentation for lateral movement prevention in industrial internet of things (IIoT). In Naghmeh Moradpoor, Atilla Elçi, and Andrei Petrovski, editors, *14th International Conference on Security of Information and Networks, SIN 2021, Edinburgh, United Kingdom, December 15–17, 2021*, pages 1–6. IEEE, 2021.

6. Attivo Networks. Enhancing cloud security with deception technology. www.attivonet works.com/wp-content/uploads/sites/13/documentation/ Attivo_Networks-Enhanced_ Cloud_Security.pdf, 2020. Online, Accessed on May 09, 2023.

7. Claudine Badue, Rânik Guidolini, Raphael Vivacqua Carneiro, Pedro Azevedo, Vinicius B Cardoso, Avelino Forechi, Luan Jesus, Rodrigo Berriel, Thiago M Paixao, Filipe Mutz, et al. Self-driving cars: A survey. *Expert Systems with Applications*, 165:113816, 2021.

8. M Balamurugan and B Sri Chitra Poornima. Article: Honeypot as a service in cloud. *IJCA Proceedings on International Conference on Web Services Computing (ICWSC)*, 1:39–43, November 2011.

9. Joseph Bao, Murat Kantarcioglu, Yevgeniy Vorobeychik, and Charles Kamhoua. IoT-FlowGenerator: Crafting synthetic IoT device traffic flows for cyber deception. *arXiv preprint arXiv:2305.00925*, 2023.

10. Robin Singh Bhadoria. Security architecture for cloud computing. In *Cyber Security and Threats: Concepts, Methodologies, Tools, and Applications*, pages 729–755. IGI Global, 2018.

11. Sandeep Bhatt, Pratyusa K Manadhata, and Loai Zomlot. The operational role of security information and event management systems. *IEEE Security & Privacy*, 12(5):35–41, 2014.

12. Muhammad Bilal. *Automatic configuration for cloud workloads*. PhD thesis, Catholic University of Louvain, Louvain-la-Neuve, Belgium, 2022.

13. Daniel Bourke and Daniel Grzelak. Breach detection at scale with AWS honey tokens. *Blackhat Asia*, www.blackhat.com/asia-18/briefings.htmlbreach-detection-at-scale-withaws-honey-tokens, pages 20–23, 2018.

14. Stephen Brown, Rebecca Lam, Shishir Prasad, Sivasubramanian Ramasubramanian, and Josh Slauson. Honeypots in the cloud. *University of Wisconsin-Madison*, 11, 2012. https://pages.cs.wisc.edu/ sbrown/academic-projects.html

15. Amar Ramesh Buchade and Rajesh Ingle. Key management for cloud data storage: Methods and comparisons. In *4th International Conference on Advanced Computing & Communication Technologies*, pages 263–270. 2014. IEEE.

16. Antanas Cenys, Darius Rainys, L Radvilavius, and N Gotanin. Implementation of honeytoken module in dbms oracle 9ir2 enterprise edition for internal malicious activity detection. *IEEE Computer Society's TC on Security and Privacy*, pages 1–13, 2005.

17. Tanmoy Chakraborty, Sushil Jajodia, Jonathan Katz, Antonio Picariello, Giancarlo Sperli, and V. S. Subrahmanian. A fake online repository generation engine for cyber deception. *IEEE Transactions on Dependable and Secure Computing*, 18(2):518–533, 2021.

18. N Choucri, S Madnick, and P Koepke. Cybercrime-as-a-service: Identifying control points to disrupt. Technical report, Working Paper CISL# 2017-06, Cybersecurity Interdisciplinary, Systems Laboratory, MIT, 2017.

19. Cloud Security Alliance. Top threats to cloud computing pandemic eleven. https://cloudsecurityalliance.org/artifacts/top-threats-to-cloud-computing- pandemic-eleven/, 2022. Online, Accessed on May 09, 2023.

20. Hakan Dalkılıç and Mehmet Hilal Özcanhan. Strong authentication protocol for identity verification in internet of things (IoT). In *6th International Conference on Computer Science and Engineering (UBMK)*, pages 199–203, 2021. IEEE.

21. L Minh Dang, Md Jalil Piran, Dongil Han, Kyungbok Min, and Hyeonjoon Moon. A survey on internet of things and cloud computing for healthcare. *Electronics*, 8(7):768, 2019.

22. Seamus Dowling, Michael Schukat, and Hugh Melvin. A zigbee honeypot to assess iot cyberattack behaviour. In *28th Irish Signals and Systems Conference (ISSC)*, pages 1–6. IEEE, 2017.

23. Nicolaie L Fantana, Till Riedel, Jochen Schlick, Stefan Ferber, Jürgen Hupp, Stephen Miles, Florian Michahelles, and Stefan Svensson. IoT applications—value creation for industry. In Ovidiu Vermesan and Peter Friess, Eds., *Internet of Things*, pages 153–206. River Publishers, 2022.

24. Daniel Fraunholz, Daniel Reti, Simon Duque Anton, and Hans Dieter Schotten. Cloxy: A context-aware deception-as-a-service reverse proxy for web services. In *Proceedings of the 5th ACM Workshop on Moving Target Defense*, pages 40–47. Association for Computing Machinery, New York, NY, United States, 2018.

25. Ian Goodfellow, Jean Pouget-Abadie, Mehdi Mirza, Bing Xu, David Warde-Farley, Sherjil Ozair, Aaron Courville, and Yoshua Bengio. Generative adversarial networks. *Communications of the ACM*, 63(11):139–144, 2020.

26. Matthew Groh, Aruna Sankaranarayanan, Andrew Lippman, and Rosalind Picard. Human detection of political deepfakes across transcripts, audio, and video. *arXiv preprint arXiv:2202.12883*, 2022.

27. Chongqi Guan, Xianda Chen, Guohong Cao, Sencun Zhu, and Thomas La Porta. HoneyCam: Scalable high-interaction honeypot for IoT cameras based on 360-degree video. In *IEEE Conference on Communications and Network Security (CNS)*, pages 82–90, 2022. IEEE.

28. Juan David Guarnizo, Amit Tambe, Suman Sankar Bhunia, Martín Ochoa, Nils Ole Tippenhauer, Asaf Shabtai, and Yuval Elovici. Siphon: Towards scalable high-interaction physical honeypots. In *Proceedings of the 3rd ACM Workshop on Cyber-Physical System Security*, pages 57–68, 2017. ACM.

29. Muhammad A Hakim, Hidayet Aksu, A Selcuk Uluagac, and Kemal Akkaya. U-pot: A honeypot framework for UpnP-based IoT devices. In *37th International Performance Computing and Communications Conference (IPCCC)*, pages 1–8. IEEE, 2018.

30. J Bethanney Janney, J Premkumar, S Krishnakumar, S Aishvariya Shivani, E Atchaya, and P Grace Kanmani. Air ambulance drone for medical surveillance. *Journal of Physics: Conference Series*, 2318(1):012023, Aug 2022.

31. Arthur Jicha, Mark Patton, and Hsinchun Chen. SCADA honeypots: An in-depth analysis of conpot. In *IEEE Conference on Intelligence and Security Informatics (ISI)*, pages 196–198. IEEE, 2016.

32. Neha Kaliya and Muzzammil Hussain. Framework for privacy preservation in IoT through classification and access control mechanisms. In *2nd International Conference for Convergence in Technology (I2CT)*, pages 430–434, 2017. IEEE.

33. Prakruthi Karuna, Hemant Purohit, Rajesh Ganesan, and Sushil Jajodia. Generating hard to comprehend fake documents for defensive cyber deception. *IEEE Intelligent Systems*, 33(5):16–25, 2018.

34. Mostafa Haghi Kashani, Mona Madanipour, Mohammad Nikravan, Parvaneh Asghari, and Ebrahim Mahdipour. A systematic review of IoT in healthcare: Applications, techniques, and trends. *Journal of Network and Computer Applications*, 192:103164, 2021.

35. Constantine Katsinis and Brijesh Kumar. A framework for intrusion deception on web servers. In *International Conference on Internet Computing (ICOMP'13)*, page P7. Worldcomp Proceedings, 2013.

36. Sheeraz Kirmani, Abdul Mazid, Irfan Ahmad Khan, and Manaullah Abid. A survey on IoT-enabled smart grids: Technologies, architectures, applications, and challenges. *Sustainability*, 15(1):717, 2023.

37. Alexander Kott. Challenges and characteristics of intelligent autonomy for internet of battle things in highly adversarial environments. *arXiv preprint arXiv:1803.11256*, 2018.

38. Tongbo Luo, Zhaoyan Xu, Xing Jin, Yanhui Jia, and Xin Ouyang. IoTcandyjar: Towards an intelligent-interaction honeypot for IoT devices. *Black Hat*, 2017:1–11, 2017.

39. Xupeng Luo, Qiao Yan, Mingde Wang, and Wenyao Huang. Using MTD and SDN-based honeypots to defend DDoS attacks in IoT. In *Computing, Communications and IoT Applications (ComComAp 2019), Shenzhen, China, October 26–28, 2019*, pages 392–395. IEEE, 2019.

40. Ramón Martínez, Nuria Vela, Abderrazak El Aatik, Eoin Murray, Patrick Roche, and Juan M Navarro. On the use of an IoT integrated system for water quality monitoring and management in wastewater treatment plants. *Water*, 12(4):1096, 2020.

41. Momina Masood, Mariam Nawaz, Khalid Mahmood Malik, Ali Javed, Aun Irtaza, and Hafiz Malik. Deepfakes generation and detection: State-of-the-art, open challenges, countermeasures, and way forward. *Applied Intelligence*, 53(4):3974–4026, 2023.

42. Volviane Saphir Mfogo, Alain B. Zemkoho, Laurent Njilla, Marcellin Nkenlifack, and Charles A. Kamhoua. AIIPot: Adaptive intelligent-interaction honeypot for IoT devices. *CoRR*, abs/2303.12367, 2023.

43. Amgad Muneer, Suliman Mohamed Fati, and Saddam Fuddah. Smart health monitoring system using IoT based smart fitness mirror. *TELKOMNIKA (Telecommunication Computing Electronics and Control)*, 18(1):317–331, 2020.

44. SR Nithin Chandra and TM Madhuri. Cloud security using honeypot systems. *International Journal of Scientific & Engineering Research*, 3(3):1, 2012.

45. Oracle and KPMG. Demystifying the cloud shared responsibility security model, 2020. Online, Accessed on May 09, 2023. www.oracle.com/a/ocom/docs/cloud/oracle-ctr-2020-shared-responsibility.pdf

46. OWASP. Open worldwide application security project. https://owasp.org/, 2022. Online, Accessed on May 09, 2023.

47. Yin Minn Pa Pa, Shogo Suzuki, Katsunari Yoshioka, Tsutomu Matsumoto, Takahiro Kasama, and Christian Rossow. IoTPOT: Analysing the rise of IoT compromises. *Emu*, 9(1), 2015.

48. Robin Rombach, Andreas Blattmann, Dominik Lorenz, Patrick Esser, and Björn Ommer. High-resolution image synthesis with latent diffusion models. In *Proceedings of the IEEE/CVF Conference on Computer Vision and Pattern Recognition*, pages 10684–10695, 2022.

49. Neil C Rowe, E John Custy, and Binh T Duong. Defending cyberspace with fake honeypots. *Journal of Computers*, 2(2):25–36, 2007.

50. Lance Spitzner. The honeynet project: Trapping the hackers. *IEEE Security & Privacy*, 1(2):15–23, 2003.

51. Lance Spitzner. *Honeypots: Tracking Hackers*, volume 1. Addison-Wesley Reading, 2003.

52. Biljana L Risteska Stojkoska and Kire V Trivodaliev. A review of internet of things for smart home: Challenges and solutions. *Journal of Cleaner Production*, 140:1454–1464, 2017.

53. Salvatore J Stolfo, Malek Ben Salem, and Angelos D Keromytis. Fog computing: Mitigating insider data theft attacks in the cloud. In *IEEE Symposium on Security and Privacy Workshops*, pages 125–128. IEEE, 2012.

54. Armin Ziaie Tabari, Xinming Ou, and Anoop Singhal. What are attackers after on IoT devices? An approach based on a multi-phased multi-faceted IoT honeypot ecosystem and data clustering. *arXiv preprint arXiv:2112.10974*, 2021.

55. Radwan Tahboub and Yousef Saleh. Data leakage/loss prevention systems (DLP). In *World Congress on Computer Applications and Information Systems (WCCAIS)*, pages 1–6. IEEE, 2014.

56. Noor Thamer and Raaid Alubady. A survey of ransomware attacks for healthcare systems: Risks, challenges, solutions and opportunity of research. In *1st Babylon International Conference on Information Technology and Science (BICITS)*, pages 210–216, 2021.

57. S Vishnu, SR Jino Ramson, and R Jegan. Internet of medical things (IoMT)-an overview. In *5th International Conference on Devices, Circuits and Systems (ICDCS)*, pages 101–104. IEEE, 2020.

58. Meng Wang, Javier Santillan, and Fernando Kuipers. Thingpot: An interactive internet-of-things honeypot. *arXiv preprint arXiv:1807.04114*, 2018.

59. Duane Wilson and Jeff Avery. Mitigating data exfiltration in Storage-as-a-Service clouds. *arXiv preprint arXiv:1606.08378*, 2016.

60. Moeka Yamamoto, Shohei Kakei, and Shoichi Saito. FirmPot: A framework for intelligent-interaction honeypots using firmware of iot devices. In *9th International Symposium on Computing and Networking Workshops (CANDARW)*, pages 405–411. IEEE, 2021.

61. Afsoon Yousefi and Seyed Mahdi Jameii. Improving the security of internet of things using encryption algorithms. In *International Conference on IoT and Application (ICIOT)*, pages 1–5, 2017.

62. Jim Yuill, Mike Zappe, Dorothy Denning, and Fred Feer. Honeyfiles: Deceptive files for intrusion detection. In *Proceedings from the 5th Annual IEEE SMC Information Assurance Workshop*, pages 116–122, West Point, NY, USA, IEEE. 2004.

63.	Mohammad Zarei, Ayoub Mohammadian, and Rohollah Ghasemi. Internet of things in industries: A survey for sustainable development. *International Journal of Innovation and Sustainable Development*, 10(4):419–442, 2016.

64.	Qian Zhu, Ruicong Wang, Qi Chen, Yan Liu, and Weijun Qin. IoT gateway: Bridgingwireless sensor networks into internet of things. In *2010 IEEE/IFIP International Conference on Embedded and Ubiquitous Computing*, pages 347–352. IEEE, 2010.

11 AI-Based Mass Screening Using Cloud-Based EHR and CXR

An Implication for India's National Strategic Plan to End Tuberculosis

Vinaytosh Mishra, Chandra Sekhar,
Jitendra Kumar, and Ashutosh Kumar Singh

11.1 INTRODUCTION

Tuberculosis (TB) remains a significant public health challenge in India, with the country bearing the highest burden of TB cases globally. India's large population, high population density, and socioeconomic factors contribute to the persistence of TB. TB has emerged as a serious threat to public health as Indians are susceptible to transmission because of poor living conditions [1].

The disease burden of TB in India has been alarming for decades. Out of the total global cases around two-thirds originate from six countries, namely India, Indonesia, China, Nigeria, Pakistan, and South Africa. India notoriously leads these countries, which is a grave concern. In 2015 the country reported 2.8 million cases, while 0.5 million deaths were due to the disease. In addition, the country has reported the maximum number of multi-drug-resistant TB patients and the number was as high as 130,000. Factors such as malnutrition, HIV co-infection, inadequate health-care infrastructure, and limited access to quality diagnostics and treatment further exacerbate the situation [2]. In recent years the Government of India has taken initiatives to fight back against TB and started some flagship programs such as the Revised National Tuberculosis Control Program and a recent National Strategic Plan for Tuberculosis Elimination [3]. Despite these continuous efforts challenges such as stigma, medication adherence, drug resistance, and late detection persist. Thus, there is a need for continued and upscaled interventions to control and eliminate this deadly disease in the country.

11.1.1 DIGITAL HEALTH MISSION

There have been significant advances in digital technologies in the last two decades. There are various use cases of digital technologies in assuring quality and efficiency

DOI: 10.1201/9781003390954-11

in healthcare. The Indian government has recognized the transformative power of these technologies in driving positive changes in various sectors, such as health, education, infrastructure, finance, agriculture, manufacturing, and governance. The National Health Policy of 2017 stresses the importance of digital technologies in addressing key gap areas in the Indian Healthcare System. Further National Digital Health Blueprint of 2019 highlighted the need for a National Digital Health Mission (NHDM) to establish a comprehensive digital health ecosystem supplementing the traditional model of care delivery. With these objectives Government of India wants to move toward the goal of providing Universal Health Coverage to its citizen. The NHDM aims to create a robust digital infrastructure, empower the workforce and patients, and streamline the delivery of health services and provide information transparently to all stakeholders [4].

The success of NDHM depends on adoption by both the central and state governments, public and private entities, and individuals, as well as analysis of international experiences to learn from their achievements and challenges.

The coronavirus disease 2019 pandemic has accelerated the need for digital transformation and prompted a reevaluation of existing healthcare practices. In response to this growing demand, India launched the Ayushman Bharat Digital Mission in September 2021. The mission has a goal to enhance the healthcare services in 4 A dimensions, namely availability, accessibility, affordability, and acceptability utilizing digital health goods. One of the objectives of establishing the digital health ecosystem is to create an integrated, effective, safe, and inclusive healthcare system in India. Collaboration among citizens, public and private healthcare providers, digital innovations, and other stakeholders is facilitated via interoperable frameworks, open protocols, and consent artifacts, resulting in equitable digitization of healthcare across the country [5].

11.1.2 AYUSHMAN BHARAT HEALTH ACCOUNT

India is building a comprehensive ecosystem in which all medical tests, procedures, and records are securely stored in the cloud and accessible from anywhere and anytime upon taking the required consent and authorization. The goal of a personal electronic health record (EHR) connected to a unique health ID, now known as the Ayushman Bharat Health Account (ABHA) number, falls in alignment with this vision.

The objective is to offer people a full and simply accessible health record that can be viewed and maintained in real time. Individuals can have a centralized repository of their medical history with an ABHA number, empowering them to take charge of their healthcare journey and allowing authorized parties to access valuable information as needed.

The Ayushman Bharat Digital Mission (ABDM) method takes the "Think Big, Start Small, Scale Fast" strategy, enabling rapid acceptance and continual learning. It emphasizes the significance of standardizing individual identity in healthcare to maintain precise medical documentation and consent-based access. The ABHA Number serves as an individual's unique proof of identity, promoting authentication

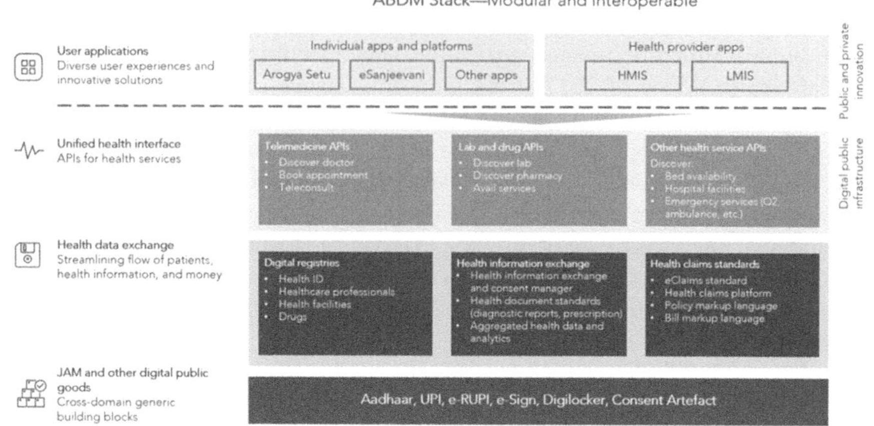

FIGURE 11.1 Ayushman Bharat Digital Health Mission's stack approach. (From Government of India, ABDM [6].)

and linkage of medical data across multiple systems and stakeholders. The Healthcare Professionals Registry is a holistic repository that links all medical professionals to India's digital health ecosystem. In a similar vein, the Health Facility Registry acts like a repository enabling medical facilities, which facilitates them to participate in the digital health system. The ABHA mobile applications work as an interface for personal health records empowering individuals to manage their health-related information. It helps in the entry, modification, and sharing of healthcare information with stakeholders effectively and securely. It enables the establishment of an ABHA address, the finding of health information, the connection of health data, and the viewing of a longitudinal record across numerous healthcare institutions. Furthermore, the app simplifies consent management for data sharing.

ABDM follows a stack approach as shown in Figure 11.1. The Stack-Based Architecture is a formal framework that focuses on object-oriented queries to databases as well as programming languages. It utilizes a combination of online portals, mobile apps, and various technologies to grant access to authorized users for health services and management of the processing of the scheme's data.

11.1.3 NEED OF COLLABORATION

Collaboration among service providers is critical for the success of a health system. ABDM provides an ecosystem for transforming healthcare services, necessitating the development and implementation of various digital tools and platforms. ABDM invites collaborators to develop solutions using the data available in the cloud. They can partner with ABDM by offering a range of services and products, such as platform development, product development, implementation, and maintenance. They provide a variety of services, including developing and implementing the digital platform, providing technical support and maintenance, providing training and support to healthcare providers, conducting data analytics and reporting, ensuring cybersecurity and data protection, and integrating various systems and services into the ABDM platform. These products and services are crucial for the delivery, security, and efficiency of the digital health ecosystem such as ABDM.

11.1.4 RATIONALE OF THE STUDY

Indian health care system has been traditionally resource deficient. There is a shortage of healthcare workforce, including doctors, nurses, and allied healthcare professionals. This shortage is even worse in rural areas where most of the TB cases are reported. The government is putting effort into developing healthcare infrastructure and training more human resources, but it is a time taking process [7]. Digital technologies such as mobile apps and artificial intelligence (AI) can help in eliminating TB. Mobile apps can be utilized for health education, identification, medical adherence, data collection, and monitoring. The generated data can help in identifying trends and devising the strategy accordingly. It can further help in devising the digital twin of the TB management programs and policies can be evaluated before launch so that they do not fail or underperform [8]. Making chest X-rays (CXR) mandatory can aid in the early detection and prompt treatment of TB, but it should be used in conjunction with other diagnostic and treatment methods to ensure that TB is detected and treated. Although TB is a perennial problem associated with the Indian healthcare system, there is a lack of studies discussing the framework to use EHRs for early detection of TB before it becomes severe and drug resistant. Recent digital transformation in Indian Health System has made it opportune to discuss how we can utilize existing cloud-based patient health records to achieve the goal of elimination of TB in India. This study attempts to fill this research gap.

The study focuses on the potential of using AI-based methods to fuse EHRs and medical imaging data to develop AI methods for clinical screening of TB. The study aims to use deep learning architectures such as convolutional neural networks, recurrent neural networks, and autoencoders to fuse the data. The study proposes an approach utilizing cloud-based EHR to provide an initiative-taking approach to managing TB in India. Thus, the objective of this research is to develop a mass screening tool for TB utilizing CXR and information available in EHRs. Apart from risk classification a specific objective of the model is to minimize the false negative.

11.2 METHODOLOGY

This section discusses the methodology to be used in the proposed research work. The study used a secondary research focus group of five experts in pulmonary disease in India to finalize the dependent and independent variables for the study [9]. The study uses multimodal data available in the cloud-based EHRs for the classification of TB in three classes described in Table 11.1. Based on the classification an alert can be sent to healthcare agencies and appropriate interventions can be planned.

TABLE 11.1
Classes for classification of tuberculosis

Class	Name	Description
C1	Pulmonary TB	This is the most common type of TB, which affects the lungs.
C2	Extrapulmonary TB	This type of TB affects parts of the body other than the lungs, such as the lymph nodes, bones, kidneys, and brain.
C3	Drug-susceptible TB	This is a type of TB that can be treated with standard first-line anti-TB drugs, such as isoniazid, rifampicin, and ethambutol.
C4	Drug-resistant TB	This is a type of TB that is resistant to one or more of the first-line anti-TB drugs.
C5	Latent TB infection	This is a type of TB in which the person has been infected with TB bacteria but does not have an active disease.

Source: Author's compilation.

The study proposes to utilize multimodal data for the development of the fusion model for the classification of subjects into the five above-mentioned cases. Again, based on focus group input and secondary research the study proposes to use five types of data sources for model building listed in Table 11.2 [10]. Thus, this study attempts to provide a framework for the prediction of TB which can be empirically tested using patient health information from cloud-based EHRs.

By combining information from these sources, the study proposes to develop a more accurate and comprehensive diagnosis of TB. This will help ensure that patients receive the appropriate treatment and care, which can improve health outcomes and reduce the spread of TB in communities.

11.2.1 Sampling Method

The number of cases required for classification in five classes using multimodal data can vary depending on several factors, including the complexity of the data, the quality of the images, the variability within the classes, and the chosen classification algorithm. A larger number of cases is better for training a robust classification model. However, there is no fixed number of cases that guarantees good performance for all cases. As a rule of thumb, the study using the approach described in this chapter will use a minimum of 200 cases per class to train a reliable model, making it 1000 cases for this study. However, this number can increase to several thousand cases per class for more complex data or to achieve higher levels of accuracy [11]. The researchers should ensure that the data set is balanced and that there is an equal number of cases in each class, to avoid bias toward the majority class. Since the proposed solution is based on a cloud based large EHR, a modeler can easily get the required number

TABLE 11.2
Probable multimodal data sources for TB classification

Data Source	Name	Description
D1	Medical records	Patient's medical history, including previous TB infections, and other underlying medical conditions.
D2	Chest X-ray	Chest X-rays are often used to diagnose TB and can provide valuable information on the extent and severity of the infection.
D3	Sputum tests	These tests can help identify the presence of TB bacteria in a patient's sputum, which can aid in diagnosis.
D4	Blood tests	Certain blood tests can help identify the presence of TB infection or determine if a patient has been exposed to TB in the past.
D5	Molecular diagnostics	These tests can help identify specific genetic markers associated with TB and can provide a more accurate and rapid diagnosis of the infection.

Source: Author's compilation.

of cases. In case the data is imbalanced, the researcher can use techniques such as oversampling or under-sampling to address this issue [12]. Simple random sampling can be used further to split the available data set into training and testing sets. The training set can be used to train the machine learning model, while the testing set should be used to evaluate the performance of the model.

11.3   RESULTS AND DISCUSSION

This study suggests using two data sources (EHR and CXR) to generate the data for the proposed AI tool. A modeler needs to retrieve the patient medical records from the EHR systems which can be accessed from different public or private agencies, whereas the CXR would be accessed from the databases or the hospitals. Following the data collection, the data samples must be annotated by the subject expert to train the AI model. Furthermore, the data size should be increased by applying the data augmentation technique if necessary. The proposed model will examine or preprocess the collected data to generate the important features from the data. Feature engineering is suggested to involve the use of various techniques such as analyzing the statistical properties of data, heatmap of feature importance, etc. More specifically, a modeler can use descriptive statistics to summarize and describe the characteristics of the data, such as mean, median, standard deviation, and percentiles for continuous data, while frequencies and proportions are used for categorical data. The relationships among variables can be identified by simple linear regression. We

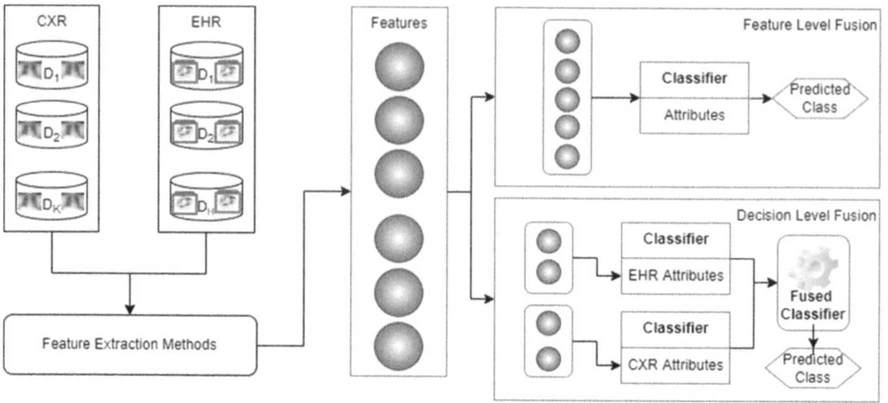

FIGURE 11.2 Data collection and feature engineering.

also suggest using clustering to group patients based on similarities in CXR and EHR data, which will be eventually helpful in identifying subgroups of similar outcomes. If required, a modeler can also perform the network analysis to establish the relationships between attributes of CXR and EHR data.

Since the features are collected from different sources, they must be combined in an effective way to get the best possible outcome. Data fusion techniques can help to improve the accuracy of TB diagnosis by combining information from different modalities. However, it is important to carefully design the fusion strategy and evaluate the performance of the model on a validation data set to ensure that it generalizes well to new data. For this purpose, we suggest a feature-level fusion and/or decision-level fusion as shown in Figure 11.2.

11.3.1 ETHICAL CONSIDERATIONS

The use of EHR data for research purposes raises important ethical considerations related to patient privacy, confidentiality, and informed consent. Ethical review boards will be required to use EHR data for clinical research. Therefore, a collaborator will need to obtain ethical approval from an institutional review board or ethics committee before using EHR data for research purposes. Similarly, for commercial applications the government needs to come up with clear guidelines to foster the collaboration for use of AI in TB prediction.

A collaborator should adhere to the ethics committee requirements, such as data de-identification or restricted access to sensitive data, to further protect patient privacy and confidentiality if required. The modeler using the proposed approach for TB detection should follow fairness, accountability, transparency, and explain ability (FATE) for product development. To ensure that AI solutions are transparent, explainable, and accountable we will strive to make the FATE model. It will reduce the risk of legal or regulatory issues and help us to stay ahead of industry trends. To achieve this objective, we have proposed a six-step strategy:

1. Model Development: When developing AI models, ensure that you are using a transparent and interpretable model architecture, rather than a black box model.
2. Data Quality: Use data from many reliable sources to ensure that the data used to train the model is of high quality, and diverse, and hence reduces the risk of biases and errors.
3. Model Interpretation: Saliency maps, activation maximization, and feature importance can be utilized to interpret the model and understand its decision-making process.
4. Model Validation: The model should be regularly validated to ensure that it makes accurate predictions and identifies any biases that creep in during training.
5. Model Documentation: The project team should keep a detailed documentation of the model, including its architecture, training data, and performance metrics, to ensure that its behavior can be easily understood and reproduced.
6. Model Deployment: Deployment includes making a provision of a layer to ensure the safety of the patients. The model deployment should be followed by getting informed consent from all stakeholders and explaining the possible risk a associated with the product.

In the last section of this chapter, we have tried to summarize the proposed approach and suggested recommendations for the implementation of the model for screening TB in India.

11.4 CONCLUSION

This chapter proposes an initiative-taking approach for the management of TB in India. The approach suggests the use of CXR and cloud-based EHR data to predict TB at an early stage of progression. The goal is to create a mass screening tool for clinical TB screening that employs AI approaches. The study advocates combining medical imaging data from CXRs with EHR data using multimodal fusion models such as convolutional neural networks, recurrent neural networks, and autoencoders. Early detection of TB through this approach can lead to timely treatment, improved patient outcomes, cost savings, and facilitation of contact tracing. The proposed model has the potential to be implemented in mobile applications accessible to primary healthcare professionals and individuals at risk, contributing to public health efforts.

Early detection of TB is important for ensuring prompt treatment and improving patient outcomes. It can also provide cost savings, facilitate contact tracing, and have a positive impact on public health. The model developed in this research can be used in developing mobile applications freely available to primary healthcare professionals and people at risk. The model developed using the proposed approach in this study should be validated in the real setting using real data and fine-tuned if required before deploying it at a large scale. In the future, this model can be mandatory and integrated with cloud-hosted EHRs to flag the risk of TB beforehand and an alarm can be sent to government agencies to take timely action.

Finally, we conclude this chapter with ten-point recommendations for the implementation strategy for the prediction system of TB in India.

1. Define the problem: The first step is to define problem related to TB which we intend to solve. Then accordingly define gaols and objectives of the fusion model.
2. Collect and preprocess data: The next step is to collect and preprocess the multimodal data that will be used to train the learning model.
3. Split the data: Once the data has been collected and preprocessed, it is typically split into training and testing sets.
4. Select the deep learning algorithm: Depending on the problem and the data, a suitable machine learning algorithm should be selected.
5. Train the model: The selected algorithm is then used to train the model on the training set. This involves fitting the model to the data and adjusting the model parameters to optimize performance.
6. Evaluate the model: Once the model has been trained, it is evaluated on the testing set to determine its performance on various measures.
7. Improve the model: Based on the evaluation results, the model may be improved by adjusting the algorithm, modifying the features, or changing other model parameters.
8. Deploy the model: Once the model has been developed and evaluated, it can be deployed for use in real-world applications.
9. Monitor and update the model: Finally, the model should be monitored for performance over time and updated as needed.
10. Think about next: Once the technology is standardized, think about solving another problem related to TB and repeat steps 1 to 9.

Thus, this research work has contributed to research and policy in two prongs. First, the research identifies independent and dependent variables for the risk prediction system for TB using multimodal data stored in the cloud. Second, it provides recommendations for implementing the same. In totality, the research provides a framework for early detection of TB using CXR and EHRs. One of the limitations of the study is that it does not venture into model development and suggestion of the best model for achieving the objectives of the framework. This can be addressed in the future direction of the study. Second, the variables identified in the study are specific to India and thus the findings can be generalized to India or countries having socioeconomic and climate comparable with that of India. The findings of the study could also be generalized to other South Asian Association for Regional Cooperation countries with slight adjustments. A future study can utilize the existing cloud-based electronic records available under ABHA ID. The approach suggested in this chapter is useful for policymakers, and collaborators related to digital health missions in India. The approach can also be customized for other chronic diseases such as diabetes, chronic obstructive pulmonary disease asthma, etc.

11.5 ACKNOWLEDGMENTS

The authors are grateful to the expert participating in the focus group discussion for decision-making related to the identification of dependent and independent variables for the approach. They are also grateful to the Gulf Medical University, Ajman, UAE, and the National Institute of Technology, Kurukshetra, for providing infrastructural support.

REFERENCES

1. Sathiyamoorthy, R., Kalaivani, M., Aggarwal, P., & Gupta, S. K. (2020). Prevalence of pulmonary tuberculosis in India: A systematic review and meta-analysis. *Lung India: Official Organ of Indian Chest Society*, 37(1), 45–52.

2. Laxmeshwar, C., Stewart, A. G., Dalal, A., et al. (2019). Beyond 'cure' and 'treatment success': Quality of life of patients with multidrug-resistant tuberculosis. *The International Journal of Tuberculosis and Lung Disease*, 23(1), 73–81.

3. Yadav, J., John, D., & Menon, G. (2019). Out-of-pocket expenditure on tuberculosis in India: Do households face hardship financing? *Indian Journal of Tuberculosis*, 66(4), 448–460.

4. Manju, E. (2022). National digital health mission. *TNNMC Journal of Community Health Nursing*, 10(2), 27–30.

5. Sharma, R. S., Rohatgi, A., Jain, S., & Singh, D. (2023). The Ayushman Bharat Digital Mission (ABDM): The making of India's Digital Health Story. *CSI Transactions on ICT*, 11(1), 3–9.

6. National Health Authority (2023), Ayushman Bharat Digital Health Mission, https://abdm.gov.in/abdm (Retrieved May 17, 2023).

7. Karan, A., Negandhi, H., Hussain, S., et al. (2021). Size, composition, and distribution of health workforce in India: Why, and where to invest? Human Resources for Health, 19(1), 1–14.

8. Santosh, K. C., Antani, S., Guru, D. S., & Dey, N. (Eds.). (2019). *Medical Imaging: Artificial Intelligence, Image Recognition, and Machine Learning Techniques*. CRC Press.

9. Sathitratanacheewin, S., Sunanta, P., & Pongpirul, K. (2020). Deep learning for automated classification of tuberculosis-related chest X-Ray: Dataset distribution shift limits diagnostic performance generalizability. *Heliyon*, 6, e04614. www.cell.com/heliyon/pdf/S2405-8440(20)31458-4.pdf

10. MacLean, E., Broger, T., Yerlikaya, S., Fernandez-Carballo, B. L., Pai, M., & Denkinger, C. M. (2019). A systematic review of biomarkers to detect active tuberculosis. *Nature Microbiology*, 4(5), 748–758.

11. Kaur, H., Pannu, H. S., & Malhi, A. K. (2019). A systematic review on imbalanced data challenges in machine learning: Applications and solutions. ACM Computing Surveys (CSUR), 52(4), 1–36.

12. Fiorentini, N., & Losa, M. (2020). Handling imbalanced data in road crash severity prediction by machine learning algorithms. *Infrastructures*, 5(7), 1–24.

12 Defending Medicare

Unleashing the Power of Anti-Vortex Technology against Hackers in Cloud Environment

Sataditya Jana, Anitya Kumar Gupta,
Nitesh Kumar Vashisht, Lokraj, Aman Jolly,
Vikas Pandey, and Indrasen Singh

12.1 INTRODUCTION

In the modern era of the Internet, every biotech and healthcare company collects, preserves, and stores various kinds of data in their gigantic databases. Among this data, there are also some valuable and sensitive data. Valuable data includes research and development documents, discoveries of drugs and technologies, business plans, marketing strategies, and other things, whereas sensitive data contains patient records, genetic data, clinical trial results, and others. In this high-tech age, valuable data needs security, whereas sensitive data needs privacy [1]. But sometimes these companies fail to maintain their security and privacy due to malicious cyber activities. Cyber attacks are most common attacks that cause security breaches in the Medicare industries, whereas rouge employees or insiders breach the privacy of the company intentionally.

In the modern era of the Internet, biotech and healthcare companies heavily rely on the collection, preservation, and storage of vast amounts of data in their databases [2, 3, 8]. This data encompasses a wide range of information, including valuable and sensitive data. Valuable data refers to critical assets that hold strategic importance for the company. It includes research and development documents, groundbreaking drug and technology discoveries, business plans, marketing strategies, and intellectual properties. Protecting this valuable data is crucial to maintain a competitive edge, safeguarding proprietary information, and securing future advancements. On the other hand, sensitive data within the healthcare industry encompasses highly personal and confidential information. This can include patient records, genetic data, clinical trial results, medical history, and other personally identifiable information. To preserve patient data confidentiality, adhere to legal obligations (such as the Health Insurance Portability and Accountability Act), and foster trust with patients and stakeholders, it is crucial to maintain the privacy and security of sensitive data [2, 4, 5]. Unfortunately, despite the advancements in cybersecurity measures, biotech

DOI: 10.1201/9781003390954-12

and healthcare companies are still vulnerable to malicious cyber activities and insider threats. Cyberattacks are a prevalent threat, targeting these industries due to the value and sensitivity of the data they hold. Hackers and cybercriminals employ various tactics, such as data breaches, ransomware attacks, and phishing scams, to gain unauthorized access to databases and exploit vulnerabilities in security systems [6, 7]. Moreover, insider threats pose an additional risk to the privacy of company data. Rogue employees or anybody with access may intentionally or accidentally release sensitive data, resulting in privacy violations. These breaches can occur due to various reasons, including personal gain, revenge, or negligence. Insider breaches can have severe consequences, as individuals within the organization have intimate knowledge of security measures and access to critical data [9]. The consequences of security and privacy breaches in the Medicare industry can be detrimental. Financial losses are incurred through the destruction of cybersecurity systems, the theft of valuable data, and the subsequent costs of investigating and recovering from the attack. Competitively, the leakage of confidential data can provide a significant advantage to competitors, undermining a company's research and development efforts or revealing proprietary information. Furthermore, regulatory damages are incurred due to the breach of trust among customers, the company, and shareholders. Breaches of patient privacy can result in legal and regulatory penalties, loss of reputation, and erosion of customer trust. These damages can have long-lasting effects on the company's standing within the industry and its relationships with stakeholders or accidentally leaking it. This costs the company financial losses for the destruction of cybersecurity systems and theft of valuable data, competitive damages due to leakage of confidential data, and regulatory damages through the mistrust among customers, other companies, and shareholders [8, 9]. Even though Medicare companies have advanced technologies like encryption, firewalls, access control, and others, they are not strong enough to handle systematic and physical breaches single-handedly. While patching cybersecurity gaps, insider attacks take place, whereas for stopping insider attacks, companies have to lower the cybersecurity devices, which can leave a gap for cyber attackers. Our hypothetical study, analysis, and research uses emerging technologies like block chain, analytical AI, federated learning, and others to provide a probable solution to this problem. If it is successfully implemented in the industries, it can help to protect the databases with minimal data loss.

12.2 PREVIOUS TECHNOLOGIES FOR DATA SECURITY AND PRIVACY

12.2.1 DATA MASKING

Conceptually, data masking is the technique for hiding sensitive data such as age, sex, race, genetic records, or confidential data like research documents, models, new technologies or drugs, business plans, marketing strategies, and other data for maintaining privacy [10]. The data masking method follows a few processes listed below.

1. Masking through Suppression: Organizations use this technique to suppress data features related to statistical reliability, confidentiality, or data quality,

following mainly two types: Record suppression and Attribute suppression [10].

2. Masking through Perturbation: Perturbation is a purposeful manipulation of the attribute values in data records while reducing the risk of unauthorized disclosure or re-identification [10].

3. Masking Micro-Data through Sampling: The sampling technique is a masking technique that entails releasing a probability sample of the micro-data as the data product to reduce the level of detail or granularity of the data [10].

4. Masking Micro-Data through Aggregation: Aggregation combines the attribute values for specific data records or other attributes, sometimes both [10].

5. Synthetic Micro-Data: Synthetic micro-data is a data masking technique which involves mimicking the statistical properties or patterns of the original micro-data while protecting the privacy of the data set [10]. Though this method hides data from outsiders, it is completely vulnerable to cyberattacks or insider threats. As there is no security input in this method, it is easy to break the system with cyberattacks. As for insider threats, insiders have access to these hidden data and can easily breach these data [10].

12.2.2 ACCESS CONTROL

Access control is a cybersecurity method that restricts the permission of accessing or using a specific data set previously set by the owner. It follows mainly three types of process: Discretionary Access Control, where the user has access to use and freely copy the data; Mandatory Access Control, where the transfer of data follows unidirectionally in a lattice of security labels; and Role-Based Access Control, where the user can use a certain amount of data based on their respective roles [11]. Although this method can protect from unauthorized access, it is vulnerable to high-level cyberattacks, which generate the access code in the attacker-device; Denial of Service (DoS) attacks, where attacks disrupt the system preventing users from accessing the system or its resources; and insiders who can access the data from the system and easily breach the data for their personal benefits [12].

12.2.3 INTRUSION DETECTION AND PREVENTION SYSTEM

Intrusion detection and prevention system, combines two techniques – IDS Intrusion detection system, which analyzes and monitors the network of traffic and the behavior of devices, and intrusion prevention system, which helps prevent various types of network intrusion, attacks, and malicious activities. Unlike the first three cited systems, it can protect the system from distributed Denial of Service attacks, privacy leaks, Man in the Middle attacks, rouge gateway attacks, privilege escalation attacks, service manipulation attacks, and injecting information. Though it is a strong cybersecurity system, it cannot deal with insider threats and act against the data breaches caused by them. It is also vulnerable to developing DoS attacks and mutated cybersecurity attacks [13].

TABLE 12.1
Virus detection methods and their limitations [14]

No.	Type of Method	Usages in Virus Detection	Limitations
1	Simple Signature Scanning	Detects simple viruses that are spread and their copy in the system byte-by-byte	Cannot detect unknown viruses
2	Generic Signature Scanning	Swiftly identifies all the viruses in the same family using the pattern observed in the family of viruses	Cannot detect new viruses of the same family
3	Integrity Checking	Compares a file's hash value against that of an equivalent, clean copy of the file to find infections	Has no use if the device is already infected
4	Heuristic Scanning	Analyzes a target program's source code to see whether it contains any signs of a virus	Only recognizes a few virus-like operations
5	Behavior Monitoring	Tries to detect any kind of virus activity	Has limited coverage and produces FN and FP

12.2.4 ANTIVIRUS

Antivirus is a type of program that detects viruses in the system and tries to protect the system by removing the virus-written codes and restoring the original system. These are the base-level security programs given to protect the system from cyber threats. The working processes of antivirus and their flaws are given in Table 12.1. Although antivirus is good for protecting the system against malware or viruses, it has no use in securing the system from high cyberattacks. Also, it does not have any operations to protect the system against rogue employees or insiders, making it completely vulnerable to high-level threats [14].

12.2.5 DATA LOSS PREVENTION

Data Loss Prevention (DLP) refers to the tactics, tools, and procedures created to protect sensitive or confidential data at rest and in motion from loss, leak, or unauthorized access through deep content analysis. It has various uses in file system protection, network protection, and application protection from cyber threats and rouge employees. Although DLP is strong enough for the cybersecurity framework, it cannot withstand evolving cybersecurity attacks and fails to detect or prevent data loss caused by authorized insiders [15].

12.2.6 FIREWALLS

Firewalls refer to a set of impenetrable components placed between two network systems in such a position that all traffic must go through it and only authorized

TABLE 12.2
Working Process of Firewalls and Their Limitations [17]

No.	Type of Firewall	Working Process	Flaws
1	Packet Filter	Receives packets and evaluates them according to listed rules	1. Cannot detect attacks from out of the list 2. Cannot handle DoS attacks 3. Cannot deal with insider threats
2	Application-Gateway	Stops every connection at gateway and connects the devices if only the connection is permitted	1. Cannot deal with high-quality cyberattacks 2. Gives FP and FN results 3. Vulnerable to insider threats
3	Circuit-Gateway	Drops some connection and accepts the permitted ones but always creates a new connection whenever a connection is accepted	1. Cannot deal with application-layer cyberattacks 2. Cannot protect from DoS attacks 3. Cannot secure the device from insider attacks
4	Hybrid Firewalls	Combines features and functions of different types of firewalls into a single integrated system	1. Gives FN and FP 2. Cannot protect the system from advanced cyberattacks 3. Cannot deal with insider threats
5	Virtual Private Network	Establishes a safe connection from any individual machine to any other using end-to-end encryption	1. Has performance and latency issues 2. Vulnerable to cyberattacks 3. Can cause connection leaks 4. Vulnerable to insiders

traffic is allowed to pass. These are the basic security system that tries to protect the system against cyberattacks. A few types of firewalls and their flaws are shown in Table 12.2. Although firewalls provide great security for lower-level threats, they cannot handle high-level cybersecurity threats for being basic-level security. It also lacks the ability to protect against rouge employees and insider threats [16, 17], as referred in Table 12.2.

12.2.7 ENCRYPTION

Data encryption refers to the process of transforming plain, readable data into an unintelligible format called ciphertext to protect sensitive and confidential data from theft or leakage. This technique can prevent cyberattacks and rogue employees effectively from stealing or breaching the data [18]. It follows a few algorithms as given.

1. RSA: Named after its creators Rivest, Shamir, and Adelman, RSA is one of the most known public key crypto systems used for exchanging keys, digital signatures, or encryption of data blocks. It is an asymmetric algorithm which uses a public key for encryption and a private key for decryption; the key size varies from 1024 to 4096 bits [18].

2. Data Encryption Standard (DES): Recommended by the National Institute of Standards and Technology (NIST) in 1977, DES is a symmetric key block cipher. Developed by IBM in 1974, this algorithm uses a key length of 56 bits and a block size of 64 bits, and it can switch to different modes, namely cipher block chaining (CBC), electronic codebook (ECB), cipher feedback (CFB), and output feedback (OFB), making it flexible to use [18].
3. AES: AES is a variable bit block cipher developed by Joan Daemen and Vincent Rijmen and recommended by the NIST, which replaced DES in 1998. AES is a safe, fast, and flexible algorithm with variable key lengths of 128, 192, and 256 bits. It has various implementations, especially in small devices [18].
4. BLOWFISH: Designed by Bruce Schneier in 1993, BLOWFISH is a symmetric key block cipher. It contains a variable key length from 32 to 448 bits and a block size of 64 bits and has a Fiestel network. This algorithm consists of two parts – a key expansion part, which converts a key with a maximum of 448 bits into many sub-key arrays with a total size of 4168 bytes, and a data encryption part, which is only appropriate for applications where the key does not frequently change (such as communications links and automated file encryption) and is carried out through a 16-round Feistel network. DES has the advantage of having its compactness optimized in hardware applications, and when used with 32-bit microprocessors with big data caches, it becomes noticeably quicker [18].

Encryption is one of the strongest cybersecurity systems nowadays, but it still has some flaws. If the encryption algorithm is weak, then cyberattacks can break the system. But mainly, with the access key, authorized insiders can leak, steal, or breach valuable and sensitive data. So, even if the encryption is strong, it cannot deal with insider attacks [18].

12.3   PROPOSED SOLUTION EVALUATING EMERGING TECHNOLOGIES

12.3.1   BLOCKCHAIN TECHNOLOGY

Block-chain technology has gained significant attention and recognition for its secure and decentralized nature. At its core, blockchain is a digital ledger that enables the transparent and tamper-evident recording of assets and transactions. By eliminating the need for a reliable third party, it provides a platform for various applications such as smart contracts, supply chain management, and payments, offering a multitude of benefits. A blockchain consists of a series of blocks, each containing transaction data and a unique identifier called a hash. These blocks are linked together in a chronological chain. The decentralized nature of blockchain technology is one of its main benefits. Blockchain runs on a network of nodes, as opposed to conventional centralized systems, and each node has a copy of the complete blockchain. This decentralization enhances security by eliminating a single point of failure. It makes the blockchain more resistant to corruption and hacking attempts, as altering the data would require controlling a majority of the network's computing power, which is highly improbable in a large and well-established blockchain network. Transparency is another crucial aspect of block chain technology. Every transaction recorded on the blockchain is visible to all participants in the network. This transparency ensures accountability and prevents fraudulent activities. Each participant can verify the transactions and ensure that no unauthorized changes have occurred. This feature makes blockchain particularly suitable for industries such as supply chain management, where tracking and verifying the origin and movement of goods are crucial. Immutability is a fundamental characteristic of blockchain. Once a block is added to the chain, it becomes virtually impossible to alter or delete the data it contains. This immutability is achieved through cryptographic hashing and consensus mechanisms. Each block in the chain contains a reference to the previous block's hash, creating a link that extends throughout the entire chain. Any attempt to modify a block would require recalculating the hash of that block and all subsequent blocks, making it computationally infeasible and immediately evident to the network. Furthermore, blockchain technology offers enhanced security in transactions. With traditional systems, there is a reliance on centralized authorities or intermediaries to validate and authorize transactions. In the blockchain, transactions are verified by consensus algorithms executed by network participants, known as miners or validators. This distributed validation process ensures the integrity of transactions without the need for intermediaries, reducing costs and eliminating single points of failure. Additionally, blockchain technology has the potential for faster transaction processing compared to traditional systems. While traditional financial transactions may involve multiple intermediaries and can take days to settle, blockchain-based transactions can be executed in near real time. The decentralized nature of the blockchain eliminates the need for intermediaries, streamlines processes, and enables direct peer-to-peer interactions, resulting in faster transaction speeds. However, blockchain technology has a great potential for security services. In future, it can be implemented in vast fields of cybersecurity [19, 20]. The working process of blockchain technology is discussed in Figure 12.1.

12.3.2 Artificial Intelligence

Artificial intelligence (AI) has emerged as a powerful tool in the field of cybersecurity, offering significant potential for detecting and mitigating threats in an increasingly complex digital landscape. AI has the potential to improve cybersecurity systems in a variety of ways due to its capacity to analyze enormous volumes of data and spot trends. One notable application of AI in cybersecurity is fraud detection. By leveraging complex algorithms and machine learning techniques, AI can identify anomalous behavioral, locational, or working patterns that may indicate fraudulent activities. For instance, AI systems can analyze user behavior and transaction patterns to identify suspicious activities in real time, enabling organizations to take proactive measures to prevent financial losses and protect sensitive information. Furthermore, AI is capable of detecting large-scale botnet attacks, which involve a network of compromised computers controlled by malicious actors. By leveraging advanced algorithms, AI systems can analyze network traffic, identify

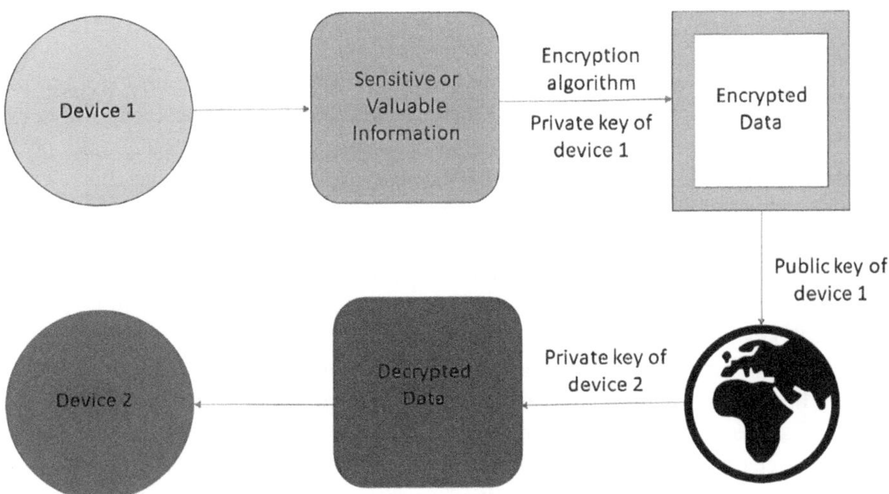

FIGURE 12.1 Working process of blockchain technology.

patterns associated with botnet activity, and take immediate action to mitigate the threat. This proactive approach helps organizations prevent the spread of malware, protect their networks, and safeguard critical data. AI's ability to analyze systems and detect patterns has tremendous potential in the future of cybersecurity. Traditional rule-based techniques frequently find it difficult to keep up with the complexity and number of cyber threats. AI, on the other hand, can adapt and learn from new data, enabling it to detect emerging threats and zero-day vulnerabilities that may evade traditional security measures. By continuously analyzing and correlating vast amounts of data, AI systems can identify potential vulnerabilities, predict attacks, and provide actionable insights to cybersecurity professionals. Moreover, AI can be utilized in the field of predictive analytics, where it can analyze historical data to identify trends and predict future attacks. This can help organizations prioritize their security efforts and allocate resources effectively, strengthening their overall cybersecurity posture. AI's ability to automate certain cybersecurity tasks also improves operational efficiency. For example, AI-powered systems can autonomously monitor networks, detect and respond to incidents, and even perform automated vulnerability assessments. By reducing the manual workload on cybersecurity professionals, AI allows them to focus on more complex and strategic tasks [21]. However, it is important to note that AI is not a silver bullet for cybersecurity. It has its limitations and can be vulnerable to adversarial attacks where malicious actors attempt to manipulate AI systems. Therefore, a combination of AI and human expertise is crucial to effectively defend against evolving cyber threats. Implementing a federated learning system for AIs can be a good approach to train the AIs in human system and with human resources [22].

12.3.3　EDGE COMPUTING

Edge computing is a paradigm that involves deploying computing and storage resources, referred to as edge devices, to the edge of the network. These edge devices, which can include micro data centers, fog nodes, and cloudlets, are strategically positioned between the Internet of Things (IoT) system and the devices or systems located at the network's edge. By placing edge devices in proximity to the devices or systems they serve, edge computing enables efficient and effective monitoring and data processing. Each edge device acts as a connection point between the IoT system and the devices or systems on the network's edge, forming distributed network architecture. One of the key advantages of edge computing is its ability to facilitate high-quality monitoring of every device or system within the network. By deploying multiple edge devices throughout the network, each device or system can be seamlessly connected to the IoT system, allowing for comprehensive monitoring and control. This ensures that no device or system is left unmonitored, providing enhanced visibility and management capabilities. Furthermore, edge computing enables faster data processing and real-time responses. With edge devices near the devices or systems they serve, data can be processed locally without the need to transmit it to a centralized cloud server. This significantly reduces the latency associated with sending data over long distances, resulting in near-instantaneous response times. The local processing capability of edge devices also helps alleviate the processing load on IoT devices. Instead of relying on resource-constrained IoT devices to perform complex computations, edge devices can handle the bulk of the processing tasks. This offloading of processing tasks to edge devices improves the overall efficiency and performance of the IoT system. In addition, the proximity of edge devices to the devices or systems at the network's edge ensures that there is no communication gap. Real-time responses from edge devices enable seamless and continuous interaction between the IoT devices and the main devices or systems at the edge of the network. This ensures smooth and uninterrupted operation of the network, even in situations where connectivity may be limited or intermittent [23]. The working methodology of edge computing is shown in Figure 12.2.

12.4　ALTERNATIVE HYPOTHESIS MODEL

This model was hypothesized by taking ideas from blockchain technology, analytical AI, and edge computing, and the hypothesis combined the ideas and modified it to the extent at which the model will become much safer in case of privacy from insider breaches and much more durable in case of security from cyberattacks. These emerging technologies are selected because of their notable pros and much lesser cons, such as – for blockchain technology:

1. Blockchains are immutable so that none can alter or delete the system.
2. Blockchain technology can also be implemented with encryption keys or access codes.
3. It is easier to share data forming a network using a blockchain system.

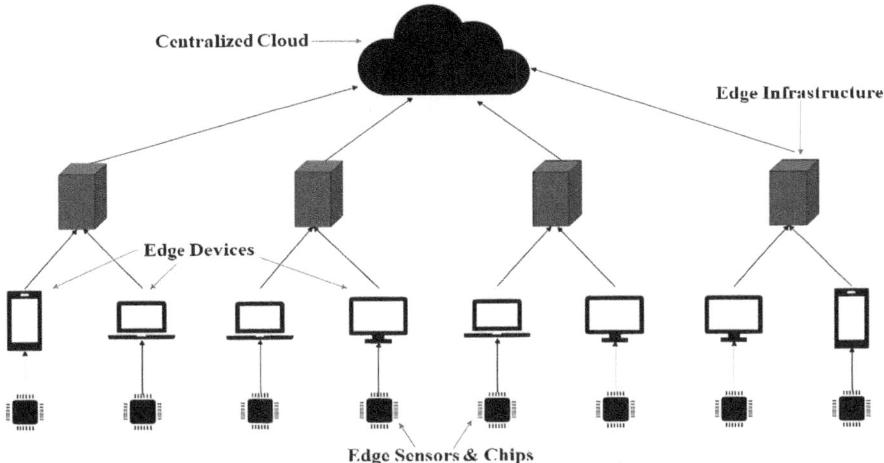

FIGURE 12.2   Working process of edge computing.

For analytical AI:

1. It can be used to analyze the system and identify potential risks.
2. It can identify potential flaws and the points of system failures in blockchain technology during cyberattack or insider threats.
3. With the help of federated learning, AI can help our system to improve more, so that it can face evolving cyberattacks.

For edge computing:

1. It can analyze the local system with the help of analytical AI.
2. It enables faster processing power in the system.
3. It gives real-time response about the system.
4. It can store local data for a short time reducing the load of traffic in the platform.
5. It leaves no communication gap in the system.

Hypothesized Model: The model has four main parts:

1. Main Unit Cell: In this hypothesis, the confidential datasheets containing intellectual and sensitive data are organized in a specific structure referred to as a "cubic data frame." This structure is designed to be the core part of the system, making it the primary target for cyberattackers and rogue insiders. To create this cubic data frame, the datasheets are interconnected layer by layer, forming a three-dimensional structure. Each datasheet is linked to others through open-access pathways, creating a network of interconnected information. The specific method used for this linking process involves complex coding techniques. The cubic data frame allows for efficient organization

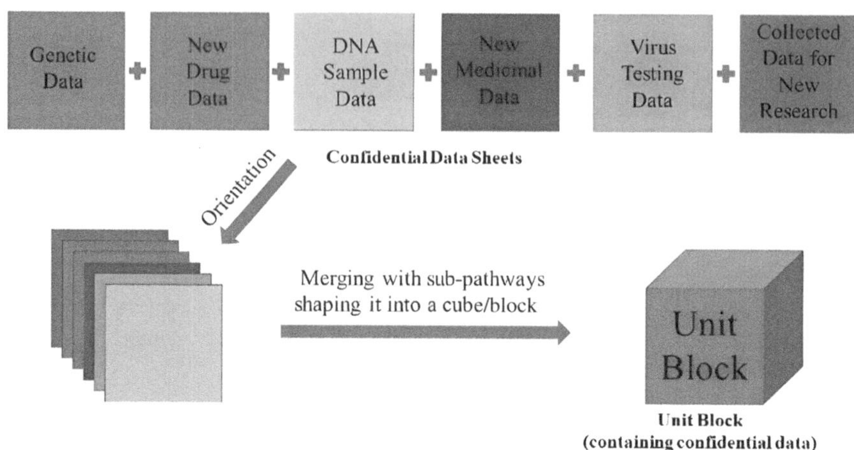

FIGURE 12.3   Main Unit Cell.

and retrieval of confidential data. It is important to note that the size of each cell within the cubic data frame can vary, but it is recommended to keep the cell size as minimal as possible. This minimization helps reduce the potential data loss during cyberattacks, ensuring that only a small portion of data is compromised if an attack occurs. The structure of the Main Data Cell is shown in Figure 12.3.

2. Core Security Structure: The base layer of the main security system is designed to safeguard the Main Unit Cell, which is a crucial component. This security system comprises one Main Unit Cell and six Access Point Cells of same size as the Main Unit Cell, making it a compact three-dimensional structure. The purpose of the Access Point Cells is to connect the Main Unit Cell with six two-way pathways, where each Access Point Cell is equipped with a two-way pathway (users can both enter and exit through them) to control user access from the open access pathway. In simpler terms, the Access Point Cells serve as gateways or checkpoints that regulate entry into the Main Unit Cell. The complex access control system implemented in these Access Point Cells ensures that only authorized individuals can enter the Main Unit Cell. It employs sophisticated security measures to authenticate and validate users, preventing unauthorized access. The core security system is shown in Figure 12.4.

3. Unit System: The system is the smallest security unit within a larger platform, which includes 20 False Data Cells of the same size and appearance as the Access Point Cells, with the primary purpose of concealing the core security structure by employing data masking techniques. These False Data Cells have been strategically placed in a way that the entire security structure takes on the shape of a Rubik's cube-like structure. Essentially, they create a diversion for potential attackers, making it more difficult for them to identify and target the actual security measures. Importantly, there is no access system

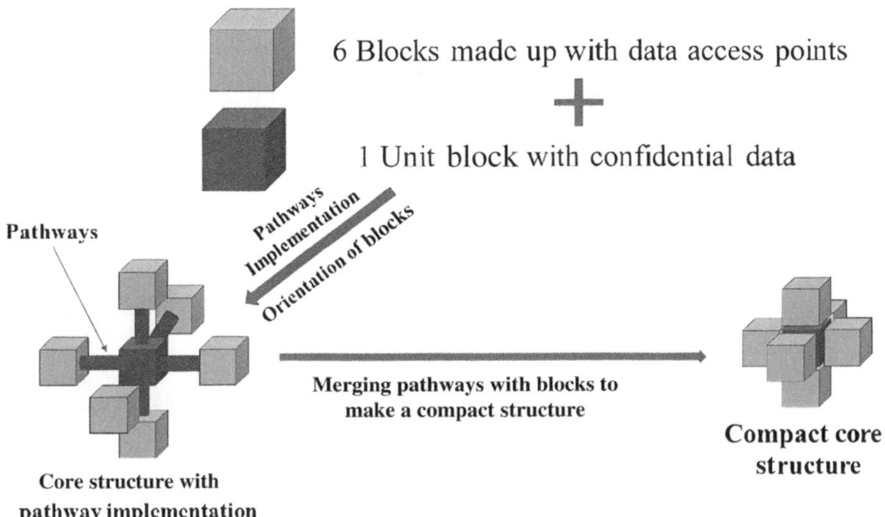

FIGURE 12.4   Core security structure.

provided to interact with these False Data Cells, further enhancing their role as a deceptive layer. Each system in the security unit features an L-shaped pathway, which is implemented with a complex access control system, ensuring that only authorized individuals can navigate through the pathway and gain access to confidential data. The primary function of this security system is to hide the Main Data Cell, which contains sensitive and confidential information. By incorporating the False Data Cells while implanting the L-shaped pathway, the system achieves the goal of making the Main Data Cell accessible exclusively through the designated Access Point Cells. The unit system is shown in Figure 12.5.

4. L-Shaped Pathway: This a complex type of pathway hypothetically that helps the system to protect against cyberattacks and even insiders. This structure has the plan to sacrifice a minimal database to protect the whole data system in mean time detecting the failure or attacked point in the system. The structure of this pathway system is shown in Figure 12.6.

This pathway system contains four pathways joined with three access points. The first pathway (1, 2) joins two systems with each other via access point 1 in between them. This access point operates on two codes for one time access in only one of two pathways. When the access code for pathway 1 is given, the access to pathway 1 is granted while pathway 2 locks itself simultaneously and vice versa. Then the user goes to pathway (1, 3) where the given code grants access to pathway 3 while locking pathway 1 simultaneously. Hypothetically, if the user is an insider, he will surely try to access pathway 4 for stealing or leaking data. But when he puts the code for pathway 4 in pathway (3, 4), pathway 3 and pathway 4 get terminated and pathway

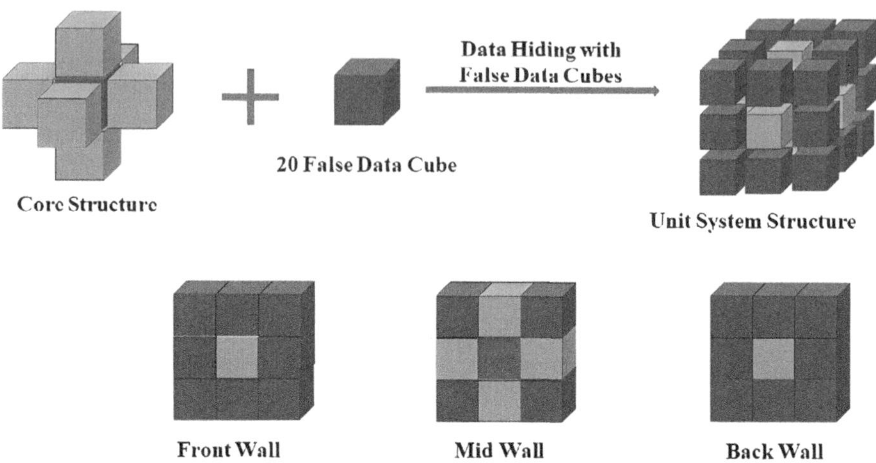

FIGURE 12.5 Unit system.

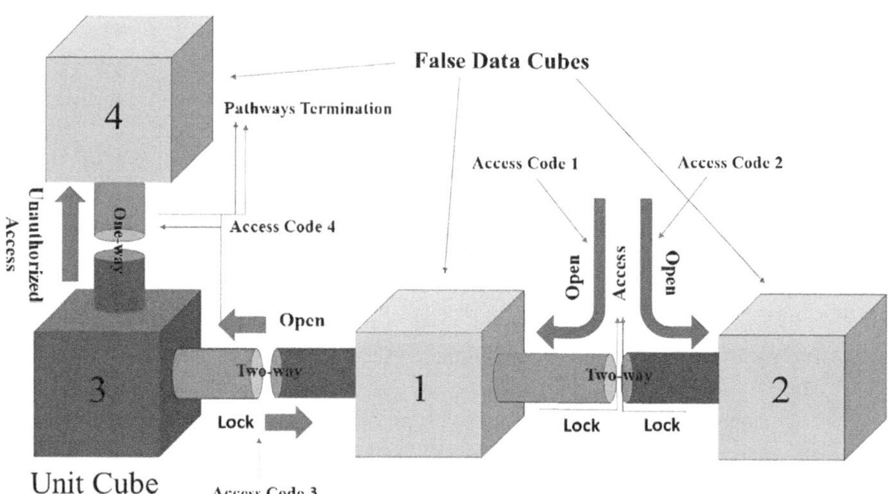

FIGURE 12.6 L-shaped pathway.

1 remains lock, causing that particular cube system to collapse, trapping the insider or attacker in that particular data sphere.

Working Process: The security system begins by collecting confidential data from various platforms and sorting it into intellectual and sensitive datasheets using data analytics. These datasheets are then connected with an open access pathway, forming a three-dimensional cubic data frame known as the Main Unit Cell. The Main Unit Cell contains only confidential data and serves as the central component of the security system. Surrounding the Main Unit Cell, there are six Access Point Cells implanted with access points. These cells are interconnected to the Main Unit Cell through pathways, creating a compact three-dimensional structure called the Core Security Structure. This structure enhances the security of confidential data by organizing it cohesively. To further protect the Core Security Structure from public exposure, 20 False Data Cells are utilized. These cells employ data masking techniques to cover the entire structure, creating a Rubik's cube-like shape. This masking hides the true core security structure, providing an additional layer of security and preventing unauthorized access. Each Unit System within the security system is equipped with four L-shaped pathways. These pathways regulate user access by allowing horizontal access to datasheets while blocking vertical access through the Main Unit Cells. The pathways operate by locking the opposite access gate while opening one. If anyone attempts to gain unauthorized vertical access through a Main Unit Cell, the connected pathways are terminated, rendering the unit system defective or unusable. The entire security system is protected by three-dimensional firewalls, safeguarding all unit systems from cyberattacks targeting the access points in the Access Point Cells. These firewalls act as barriers, monitoring and filtering network traffic to prevent unauthorized access and potential security breaches.

Certain independent unit systems are connected through specified main pathways, which serve as open-access routes, enabling linear user access between two columns of unit systems. Every set, consisting of a number of pathways, is connected to an edge device equipped with an edge sensor. The access gate of each edge device is protected by firewalls. The edge sensors utilize analytical AI programming to analyze user behavior by recording their working time, habits, and procedures; detect defects or errors or unauthorized access or multiple accesses from a system at an instant caused by malfunctions, cyberattacks or insider breaches within the unit systems; and provide real time responses at a local level. This analytical AI can also detect the location of the breach in the server and the device responsible for the breach, and it analyzes the results to decide if the server can work for now or needs to be closed immediately while detecting the source of the breach or the device or ID of the hacker or the insider in real time by trapping their devices in a non-escapable cyber-trap by creating an anti-vortex structure of the attack within the area of breach in the server. Additionally, these sensors store all user data within the pathway continuously with the help of analytical AI directly from the server, facilitating comprehensive monitoring and analysis. This type of analytical AI implementation is very effective to prevent DoS attacks, advanced persistent threats, and both unauthorized and authorized insider attacks by locally managing the unprotected server by cyber-traps and pathway manipulation which does not interrupt the function of the whole server as all the unit systems are independent. The edge devices can establish wireless connections or cloud-based connectivity with IoT devices, ensuring seamless communication within the platform and enabling faster data processing. This connectivity extends to satellite devices, creating a comprehensive network for the company and simplifying data sharing. The full system is discussed in Figure 12.7.

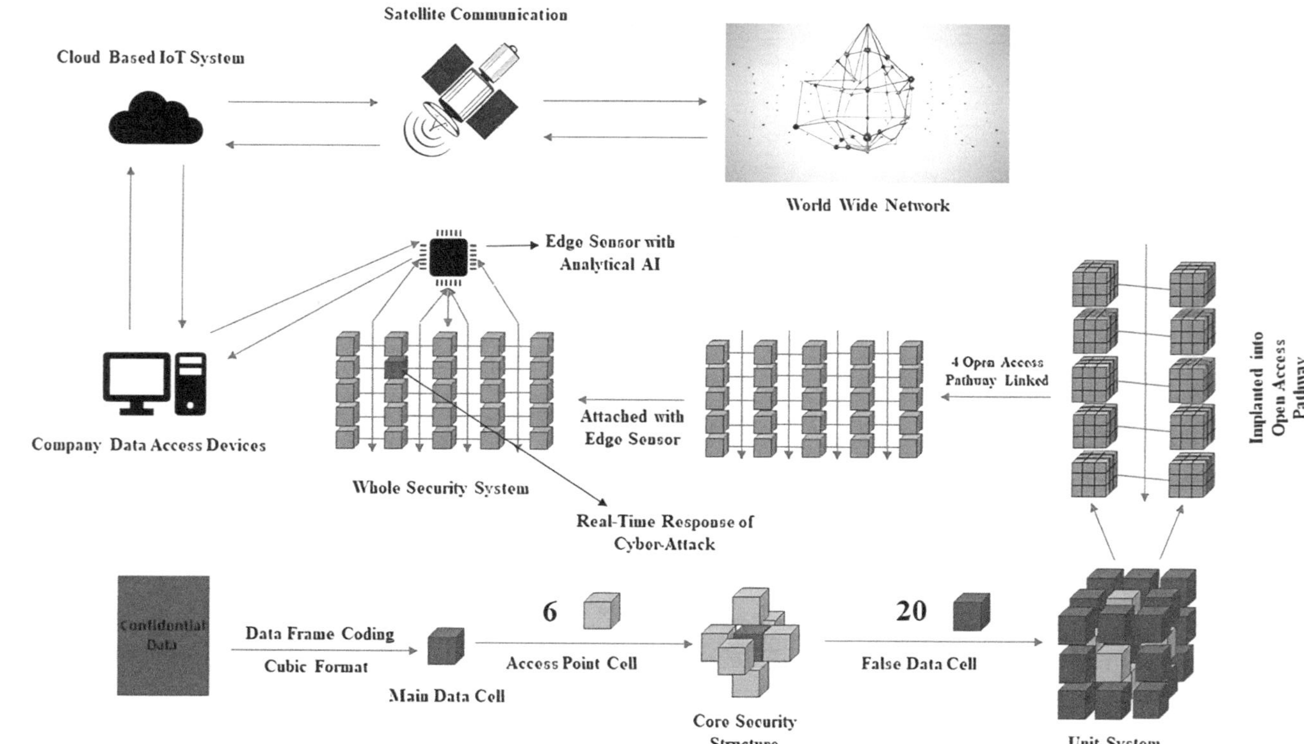

FIGURE 12.7 Complete working process of hypothesized model (anti-vortex system).

12.5   RESULTS AFTER BASIC EVALUATION OF MODEL

Following the hypothesis, some experimental coding was conducted with the sole purpose of making a complex pathway system. After experimenting with numerous different codes, a complex code was found that can be implemented into the pathway if modified properly. This code is still in experimental state by our team and can be vulnerable to cyberattacks if not implemented properly. But as of now, it showed the possible outcomes of that pathway in the result section. The L-shaped pathway model system was tested with the possible access codes and the results are shown in Tables 12.3, 12.4, and 12.5. The results of normal access, invalid access, and insider attack are shown in Table 12.6

12.6   DISCUSSION AND LIMITATIONS

The increasing prevalence of cyberattacks and insider threats has highlighted the critical need for robust network security systems. This research explores a hypothetical cybersecurity model designed to protect the server or its confidential data from

TABLE 12.3
Results for pathways 1 and 2

Access Code	Pathway 1	Pathway 2
Any code except access codes	Lock (Access invalid)	Lock (Access invalid)
Access code 2	Lock	Open
Access code 1	Open	Lock

TABLE 12.4
Results for pathways 1 and 3

Access Code	Pathway 1	Pathway 3
Any code except access codes	Lock (Access invalid)	Lock (Access invalid)
Access code 1	Open	Lock
Access code 3	Lock	Open

TABLE 12.5
Results for pathways 3, 4, and 1

Access Code	Pathway 3	Pathway 4	Pathway 1
Any code except access codes	Lock (Access invalid)	Lock (Access invalid)	Lock (Access invalid)
Access code 4	Delete	Delete	Lock
Access code 3	Lock	Lock	Open

TABLE 12.6

Access code check against normal user vs. rouge employee vs. insider threats

Normal Access	Invalid Access	Insider Attack
3650	9317	Enter your access code: 3650
Opening pathway pathway_1	Locking pathway pathway_2	Access code is valid!
9819	1545	Enter your access code: 1545
Opening pathway pathway_3	Locking pathway pathway_1	Access code is valid!
2091	3751	Enter your access code: 2091
Opening pathway pathway_1	Locking pathway pathway_3	Locking pathway pathway_4
Access code is valid!	6187	0723
Enter your access code: 6187	Opening pathway pathway_1	Locking pathway pathway_2
Access code is valid!	0245	3662
Enter your access code: 3662	Opening pathway pathway_3	Locking pathway pathway_1
Access code is valid!	8613	0931
Enter your access code: 3457	Access code is invalid!	7500
1496	Enter your access code: 7500	Opening pathway pathway_1
Locking pathway pathway_2	Access code is valid!	3702
7623	Enter your access code: 7623	Opening pathway pathway_3
Locking pathway pathway_1	Access code is valid!	8561
1613	Deleting pathway pathway_4	Deleting pathway pathway_3
Locking pathway pathway_1	Access code is valid!	

unauthorized or authorized access and cyberattacks like DoS attacks or advanced persistent threats. The model incorporates various techniques and security measures to ensure the safety of sensitive information. To begin with, the model focuses on safeguarding confidential data by employing encryption, access controls, and secure storage methods. Encryption techniques help ensure that data remains hidden and in accessible to unauthorized individuals. Access controls restrict system access based on user authentication and authorization, preventing unauthorized entry. Secure storage methods, such as data segregation and restricted access, add an extra layer of protection to sensitive information. Moreover, the model operates on a unique algorithm that incorporates a system of separation between different units. By compartmentalizing units, the model minimizes the risk of cyberattacks that could potentially compromise the entire system. When one unit is accessed or opened, it triggers the locking mechanism for other units, effectively isolating them from the active unit. This separation mechanism enhances the overall security posture, as any breach or compromise in one unit does not automatically affect the others. In addition to protect against external threats, the model addresses the issue of insider threats. It possesses the capability to confuse rogue employees or insiders who attempt to access or misuse confidential data. By implanting false or misleading information within the system, the model creates a sense of uncertainty or mistrust for unauthorized individuals. This serves as a deterrent and aids in the detection and prevention of potential insider threats. Furthermore, the model incorporates the concept of a blockchain framework and AI analysis to detect and identify cyberattacks or data breaches in real time. Cubic data-frames provide a decentralized and tamper-resistant framework for storing and verifying data. By leveraging AI analysis, the model continuously monitors and analyzes various system activities and behaviors, searching for anomalies or patterns indicative of cyberattacks or unauthorized access. This real-time detection enables swift response and mitigation measures to minimize potential damage and data loss. Despite its promising features, it is essential to acknowledge the limitations and research gaps that exist in the presented model. As with any complex system, further research, development, and testing are required to refine and enhance its functionality, security, and effectiveness. Research gaps may include optimizing the system performance, addressing potential vulnerabilities, and ensuring compatibility with existing industry standards and regulations. This ongoing research and development process is crucial to ensure the system's reliability and resilience against evolving cyber threats. Moreover, implementing such a sophisticated cybersecurity system requires careful coding practices. The complexity of the model necessitates the incorporation of various security measures and algorithms. Developers must consider factors such as encryption, access controls, error handling, and the seamless integration of edge computing and AI technologies. Proper coding practices, rigorous testing methodologies, and collaboration between domain experts and cybersecurity professionals are crucial to ensuring the system's robustness and reliability. Lastly, it is important to recognize that cyberattacks continually evolve and become more sophisticated over time. Future advancements in hacking techniques, vulnerabilities, or new attack vectors may pose challenges to the system's security. Therefore, a proactive approach to cybersecurity is essential. This includes continuously updating and enhancing the system's defenses, staying informed about emerging threats and industry best practices, and conducting regular security audits and penetration testing. Ongoing research and development efforts are vital to adapt the system to address evolving cyber threats effectively.

12.7　FUTURE SCOPES

Based on the hypothesis and future predictions, the described security model showcases potential advancements and applications in various industries. A breakdown of the possibilities by incorporating emerging technologies is presented here:

Enhanced Defense against Cyberattacks: With significant system implementation and continuous improvement, this security model has the potential to provide robust protection against cyberattacks, minimizing data loss and ensuring the integrity of stored information. As future technologies evolve, such as advancements in AI, machine learning, and anomaly detection algorithms, the model can further strengthen its defense mechanisms and adapt to emerging threats.

Application in Healthcare and Research: The model's ability to maintain the full privacy of data makes it a suitable candidate for implementation in healthcare and research industries. Medical institutions and research organizations often deal with sensitive and confidential patient data, clinical trials, and research findings. By utilizing this security system, they can safeguard valuable data from unauthorized access, protecting patient privacy and research integrity.

Military and Defense Applications: The model's robust security features make it applicable in the military and defense sectors. These industries handle highly classified and sensitive information, including military strategies, weapon systems, and intelligence data. By implementing this security system, military organizations can ensure the confidentiality of critical data, protecting national security interests and preventing unauthorized access from external threats or rogue insiders.

Mitigating Insider Threats: The complex access control system integrated into the model can help prevent data breaches caused by rogue insiders within organizations. By carefully managing and controlling access permissions, the system can minimize the risk of unauthorized data access or leakage. This provides an additional layer of protection against internal threats, ensuring data privacy and maintaining trust within the company.

Integration with Emerging Technologies: By integrating quantum computing, data visualization, the metaverse, human AI, and virtual reality tools, a technology similar to "Holographic Technology" could emerge. This transformative advancement would revolutionize the infrastructure of cybersecurity and cyber networking systems. Quantum computing would enhance encryption techniques, while data visualization would enable efficient analysis of complex cybersecurity data. The metaverse would provide a collaborative platform for real-time threat monitoring and response. Human AI would assist in decision-making and automate security processes. Virtual reality tools would create immersive environments for visualizing and interacting with cyber networks. Collectively, these technologies would reshape cybersecurity by improving speed, accuracy, collaboration, and situational awareness.

It's important to note that while these future predictions and potential applications are plausible, their realization will depend on various factors, including technological advancements, implementation challenges, and industry-specific requirements.

12.8 CONCLUSION

Evaluating the previous technologies and improving the concept of blockchain technology, edge computing, and analytical AI, this chapter hypothesizes a model, made up of cubic data-frames and L-shaped security pathways, which can effectively handle high-level cyberattacks such as DoS attacks, advanced persistent threats, and both unauthorized and authorized insider threats, even sometimes physical data thefts, while detecting the location of breach or the system responsible for breach or even the insider or the hacker in the meantime. Having the potential to be merged with any type of technology and capable of containing huge amounts of data, it can also provide the ease of use to normal users rather than complex security systems, mainly in the fields of Medicare, research, and defense. Implementing this technology in cybersecurity fields can boost the levels of security in a great way and save companies or sectors from huge data loss by sacrificing minimal data.

REFERENCES

1. Malin, B. A., Emam, K. E., & O'Keefe, C. M. (2013). Biomedical data privacy: Problems, perspectives, and recent advances. *Journal of the American Medical Informatics Association*, 20(1), 2–6.

2. Gao, Y. L., Zhang, L., & Wei, W. (2021). The effect of perceived error stability, brand perception, and relationship norms on consumer reaction to data breaches. *International Journal of Hospitality Management*, 94, 102802.
3. Kim, S. H., Wang, Q. H., & Ullrich, J. B. (2012). A comparative study of cyberattacks. *Communications of the ACM*, 55(3), 66–73.
4. Seh, A. H., Zarour, M., Alenezi, M., Sarkar, A. K., Agrawal, A., Kumar, R., & Ahmad Khan, R. (2020, May). Healthcare data breaches: Insights and implications. In *Healthcare*, 8(2), 133.
5. Barona, R., & Anita, E. M. (2017, April). A survey on data breach challenges in cloud computing security: Issues and threats. In *2017 International Conference on Circuit, Power and Computing Technologies (ICCPCT)* (pp. 1–8). IEEE.
6. Liu, S., & Cheng, B. (2009). Cyberattacks: Why, what, who, and how. *IT Professional*, 11(3), 14–21.
7. Agrafiotis, I., Nurse, J. R., Goldsmith, M., Creese, S., & Upton, D. (2018). A taxonomy of cyber-harms: Defining the impacts of cyberattacks and understanding how they propagate. *Journal of Cybersecurity*, 4(1), tyy006.
8. Whitler, K. A., & Farris, P. W. (2017). The impact of cyberattacks on brand image: Why proactive marketing expertise is needed for managing data breaches. *Journal of Advertising Research*, 57(1), 3–9.
9. Probst, C. W., Hunker, J., Bishop, M., & Gollmann, D. (Eds.). (2010). *Insider Threats in Cyber Security* (Vol. 49). Springer Science & Business Media.
10. Duncan, G., & Stokes, L. (2009). Data masking for disclosure limitation. *Wiley Interdisciplinary Reviews: Computational Statistics*, 1(1), 83–92.
11. Sandhu, R., & Samarati, P. (1996). Authentication, access control, and audit. *ACM Computing Surveys (CSUR)*, 28(1), 241–243.
12. Gauthier, F., Lavoie, T., & Merlo, E. (2013, December). Uncovering access control weaknesses and flaws with security-discordant software clones. In *Proceedings of the 29th Annual Computer Security Applications Conference* (pp. 209–218). Annual Computer Security Applications Conference by ACM.
13. Raponi, S., Caprolu, M., & Di Pietro, R. (2019). Intrusion detection at the network edge: Solutions, limitations, and future directions. In *Edge Computing–EDGE 2019: Third International Conference, Held as Part of the Services Conference Federation, SCF 2019, San Diego, CA, USA, June 25–30, 2019, Proceedings 3* (pp. 59–75). Springer International Publishing.
14. Mishra, U. (2010). Methods of virus detection and their limitations. Available at SSRN 1916708.
15. Torsteinbø, T. (2012). Data loss prevention systems and their weaknesses. Master's Thesis. University of Agder.
16. Bellovin, S. M., & Cheswick, W. R. (1994). Network firewalls. *IEEE Communications Magazine*, 32(9), 50–57.
17. Schultz, E. E. (2014). Types of Firewalls. Internet: www.ittoday.info/AIMS/DSM/83-10-41.pdf [Nov. 5, 2014].
18. Panda, M. (2016, October). Performance analysis of encryption algorithms for security. In *2016 International Conference on Signal Processing, Communication, Power and Embedded System (SCOPES)* (pp. 278–284). IEEE.
19. Golosova, J., & Romanovs, A. (2018, November). The advantages and disadvantages of the blockchain technology. In *2018 IEEE 6th Workshop on Advances in Information, Electronic and Electrical Engineering (AIEEE)* (pp. 1–6). IEEE.
20. Shrier, D., Wu, W., & Pentland, A. (2016). Blockchain & infrastructure (identity, data security). *Massachusetts Institute of Technology-Connection Science*, 1(3), 1–19.

21. Dalave, C. V., & Dalave, T. (2022). A review on artificial intelligence in cybersecurity. In *Proceedings of the 6th International Conference on Computer Science and Engineering (UBMK)* (pp. 304–309). IEEE.
22. Li, T., Sahu, A. K., Talwalkar, A., & Smith, V. (2020). Federated learning: Challenges, methods, and future directions. *IEEE Signal Processing Magazine*, 37(3), 50–60.
23. Cao, K., Liu, Y., Meng, G., & Sun, Q. (2020). An overview on edge computing research. *IEEE Access*, 8, 85714–85728.

13 An Automated Approach for Migration of Microservices-Based Applications to Serverless Architecture

Vinay Raj and Uma Sankararao Varri

13.1 INTRODUCTION

Distributed systems have evolved rapidly from a monolithic style of client-server applications to today's trending serverless architectures. The demand for quick design and deployment of services with low downtime of the applications has made the evolution of different architectural styles, such as service-oriented architecture (SOA), microservices, and serverless architectures [1]. SOA has been the most popular style for designing large enterprise applications with the implementation of web services. It has been one of the successful architectural styles for designing Internet-based applications post the era of Internet. However, due to the tight coupling with Enterprise Service Bus and the services in SOA tending toward monolithic in size lead to the evolution of the new architectural style, microservices [2]. The definition of microservices clearly states that every service should implement only one business goal, i.e., every service should follow the single responsibility principle [3]. The major benefit of microservices over SOA is that the services are deployed in cloud containers instead of traditional servers at the developer end. The cloud containers are lightweight and they have the feature of auto-scalability based on the demand of user requests. Currently microservices is one of the trending styles of software design because of its diverse benefits and advantages of cloud platforms.

On the other hand, serverless computing is an emerging and potential cloud computing paradigm which has gained popularity in recent times for its constantly available servers driving the web application behavior owing to its large influence in reducing costs of computing and storage space, decreasing latency, improving scalability, and eliminating server-side management [4]. The main intention behind the concept of serverless computing is to completely abstract away servers from the developers. The term "serverless" has a misconception that the servers are not available, which is incorrect. The actual servers exist and all the development, testing, and deployment are performed by the developer but at the cloud provider. The complete infrastructure, hardware, and software configurations are taken by the provider and

DOI: 10.1201/9781003390954-13

the developer just needs to have good Internet connection. In this computing, developers merely need to write functions in high-level languages such as Java, Python, etc., define a few simple attributes, and then upload these functions to a serverless platform. The resultant Application Programming Interface (API) or Hyper Text Transfer Protocol (HTTP) requests might then be used to conduct their well-defined computing tasks [5, 6]. In contrast to serverful computing models, developers employing serverless computing do not need to worry about managing infrastructure resources because platforms manage such things on their behalf. The heart of serverless computing is the cloud functions which are written by developers and invoked as units of execution via the Internet. Serverless computing may also be seen from the perspective of developers as Function-as-a-Service (FaaS), which enables developers to create and execute their applications (or functions) without having to deal with the complexity of creating and managing the underlying infrastructure. On the other hand, serverless service providers always provide their clients Backend-as-a-Service (BaaS), or application-dependent services, including Database and Object Storage Service. Accordingly, from a broad perspective, serverless computing is the merging of BaaS and FaaS to provide clients with a single service model [7].

Generally, microservices are deployed in cloud containers using the Docker software [8]. However, with the advent of serverless platforms, it has become the alternative for deployment of microservices [9, 10]. In a recent study, it has been found that the performance of microservices is better when deployed in serverless platforms compared to containerized microservices [11, 12]. Additionally, microservice architectural style is a popular alternative for creating applications for cloud since the updation, scaling, and upgrading can be done individually for each microservice. Considering these facts, application development and cloud infrastructure management were merged into what is today known as DevOps. However, there is growing interest in using FaaS and serverless (CaaS) technologies for refactoring microservices-based applications because of the attention the serverless computing technology has gained and its many benefits, including no infrastructure management, a pay-per-use billing policy, and on-demand fine-grained autoscaling [13].

Because of the significant advantages of serverless computing, it has earned positive attention in the industry. In a recent survey, it is expected that the global industries will adopt serverless platforms by 2025. There has been a transition from microservices deployment in a containerized architecture to a serverless architecture over time. Despite these advantages, the time and money necessary to modify existing code restrict the accessibility of these applications and impede attempts to develop serverless computing platforms [14]. Supporting the migration of current applications to serverless platforms will help developers while also broadening the scope of serverless computing. Furthermore, there is no effective algorithm for migrating microservices-based applications to serverless automatically [15]. However, migrating systems to serverless involves a number of challenges, including (1) not understanding the impact of migration and (2) not having enough material on automated migration strategies [16, 17]. These difficulties prompted us to research and develop approaches for migrating microservices-based applications to a serverless architecture. Based on many surveys in the literature, it is found that Amazon

Web Services (AWS) Lambda is the most popularly used serverless platform and hence, in our work, AWS Lambda is considered as the serverless platform. Any term referring to serverless indicates AWS Lambda in this study.

The remaining part of the chapter is structured as follows. The motivation behind proposing an automated approach considering the existing works is presented in Section 13.2. An automated approach for migration is proposed in Section 13.3 and experimental analysis of the proposed work is presented in Section 13.4. The comparison and discussion on the proposed approach is presented in Section 13.5 and Section 6 concludes the work.

13.2 LITERATURE REVIEW

The existing literature on both microservices and serverless computing has focused more on the software engineering activities of the applications designed using those styles. The research on both microservices and serverless is trending as both the styles have emerged recently. In this section, the existing and related works towards the evolution of serverless computing are discussed.

The authors in [18] have presented the challenges in implementation of serverless applications and suggested few possible solutions for such problems. Also, automated migration process is highlighted as an open problem. An automated process for migration of FaaS from one serverless to another platform is highlighted as a challenge. In [19], the authors present an empirical analysis of how migrating to serverless reduced the hosting costs between 66% and 95%. It speeds us time to market for delivery of new features. However, the authors have selected legacy monolithic applications for migration to serverless. The authors present in [20] how a FinTech application is migrated to serverless and also the performance analysis of pre- and post-migration of the application is presented. A tool, ToLambda, for automatic conversion of Java monolithic application code into AWS Lambda Node.js microservices is proposed in [21]. A semiautomated approach for migrating monolithic to microservices is proposed in [22], where different approaches for transforming code are presented. All the above approaches for migrating to serverless have focused on migrating legacy monolithic applications. However, our focus is on migrating microservices-based applications to serverless. Hence, in our work, an automated approach for migrating microservices to serverless is proposed.

A similar work of migrating microservices to serverless is proposed in [23] where a complex IoT platform application based on microservices is migrated to Open-Whisk (OW) and Google Cloud Run (GCR). However, the proposed approach is specific to only OW and GCR. The authors in [24] have presented a comparative study between different serverless platforms in terms of cost and performance by deploying a microservices application. It is also highlighted that there is a need for automatic approaches for migrating microservices to serverless and also migrating the applications from one serverless to other platforms. They have not proposed any mechanism for migration which is considered as a major contribution in our work. A partial migration of monolithic application to microservices and serverless platform is proposed in [25]. However, the serverless platform is only used to deploy the application. Also, finding an optimal automatic migration solution for existing legacy systems is an interesting research direction [26]. Additionally, only a few applications have been migrated to serverless either because of no proper mechanism for migration or because of lack of awareness of the benefits of serverless. In our work, an automated and generalized approach is proposed to migrate to any serverless platforms.

13.3 AUTOMATED MIGRATION OF MICROSERVICES TO SERVERLESS PLATFORMS

Serverless computing has changed the way applications are designed and deployed. It has been a best alternative for deploying microservices applications [27]. The applications which are migrated to microservices from SOA using the service graph-based approach [28] are deployed in containers using Docker. In this section, an automated approach for extracting the required serverless yml file from containerized microservices and deployment into serverless platforms using the yml file is presented. The complete process of migrating the containerized microservices till the deployment in serverless platforms is presented in Figure 13.1. The component which plays a major role in the automated migration approach is the generation of yml file. The serverless.yml file is a Yet Another Markup Language (YAML) configuration file that defines the functions, resources, plug-ins, and other configuration information for our serverless application. As an initial study, AWS Lambda is considered as the serverless platform in the entire chapter. The term "serverless platform" refers to AWS Labmda in this chapter. The structure and the configurations included in the yml file of the AWS Lambda are discussed here.

13.3.1 STRUCTURE OF YML FILE

This file is used to configure a service and contains information about your functions, the events that cause them, and the AWS resources you should use [29]. Each service

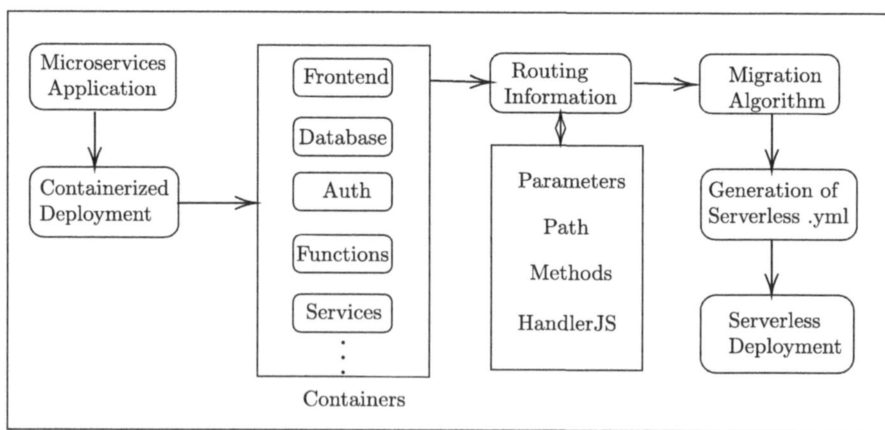

FIGURE 13.1 Migration of microservices to serverless platforms.

configuration is managed in the file. Following are the main features that should be considered in yml file:

- Define a serverless service.
- Provide the cloud platform details to which the service will be deployed, like Google Cloud, AWS, Azure, etc.
- Write one or more functions.
- Specify the events that will trigger each function (e.g. HTTP requests).
- Define any plug-in to use that extends the behavior of serverless frameworks.
- Define a set of AWS resources to create.
- Allow events listed in the events section to automatically create the resources required for the event upon deployment.
- Allow flexible configuration using variables.

13.3.2 Automated Algorithm

This is the most critical phase in the whole process of migration, also called the automation phase. A file named serverless.yml needs to be created which contains all the functions and infrastructure resources. It acts as a service configuration. This creation is automated by a Python script that extracts parameters for serverless.yml file. Every microservice project has one such file containing all the routes and requests along with other parameters of interest like path, handler function, etc. We are assuming this file is in the root folder of the project. Algorithm 13.1 contains the procedure that parses "routes.js" file first of all to get the methods of requests that exist, their paths that are used by API in the web browser, and their respective handlers defined where the main logic of microservice rests. All these extracted values are populated in serverless.yml file under the functions section. This automated algorithm which inputs the routing information of the containerized microservices generated a yml file which can be directly deployed in serverless platform.

13.3.3 Serverless Deployment

Once the serverless.yml file is generated and handler functions are ready, the next step is to deploy the modified microservices into serverless platforms, such as AWS. In the process of deployment, different lambda functions are created from individual microservices, starting with the least critical part first. For each microservice, separate HTTP gateways are created and lambda functions are invoked using URLs. The general process of serverless deployment is shown in Figure 13.2.

13.4 EXPERIMENTAL ANALYSIS

In this section, a containerized microservice is considered as a case study application and the proposed approach is illustrated using this application. Also, a comparative analysis of performance parameters including response time and throughput is conducted for both containerized microservices and the ones migrated to serverless, the details of which are presented in next sections.

Algorithm 13.1 serverless.yml generation

Input: Routing information file "routes.js"
Output: Serverless.yml file
  1: **Begin**
  2: Initialize routes file to zero
  3: Initialize each route to zero
  4: Input the AppType
  5: **if** AppType == NodeJS **then**
  6:     Search for "package.json" in directory
  7:     **if** Directory found **then**
  8:         Search for string "scripts" and "start"
  9:         Set path to the value of filename written next to start
10:     **end if**
11:     Search for routes file in the path found by above statement
12:     **if** lines contain "get" or "post" or "put" or "delete" **then**
13:         Insert values of method, path and handler js function in each route.
14:     **else**
15:         Discard lines
16:     **end if**
17: **else**
18:     print "Not supported"
19: **end if**
20: Export each route to "serverless.yml" file **return** .yml file
21: **End**

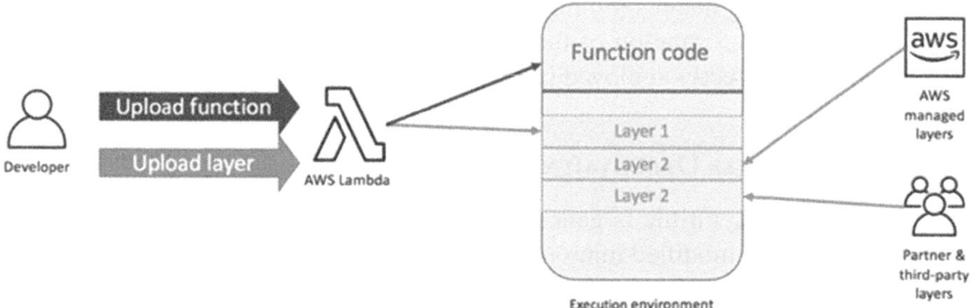

FIGURE 13.2   Serverless deployment framework.

13.4.1 CASE STUDY APPLICATION

A simple NodeJS microservices application is considered to be migrated to server-less architecture. The application is a To-Do Manager which is divided into a set of services specialized in doing specified tasks using a certain set of protocols. The services communicate with each other over a network. A diagram showing all the microservices involved is shown in Figure 13.3.

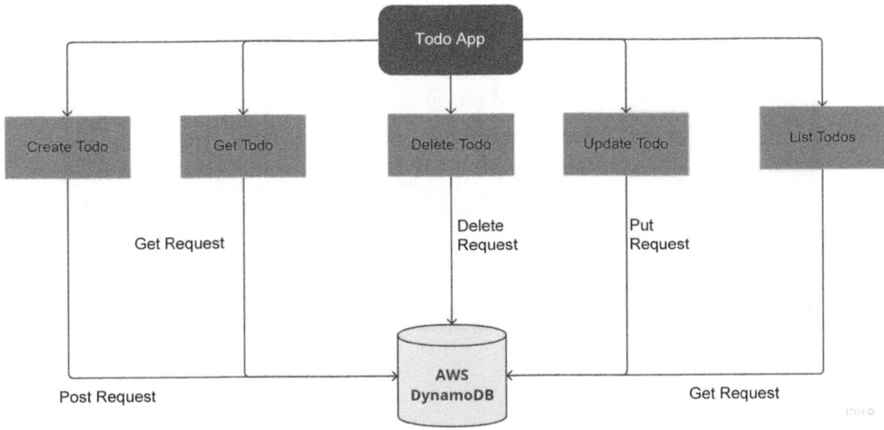

FIGURE 13.3   Microservices of a To-Do application.

The modules in Figure 13.3 illustrate main functionalities of the application. However, there are multiple services involved in each of these modules which are running in separate containers like Frontend, Database, Authentication services, etc. For easy understanding and implementation of the application, a demo skeleton of a basic microservice is created, which includes different functions as listed below.

1.   **Hello**: This represents just a home page of the application.
2.   **Create Todo**: A function to create a new to-do in the database by user.
3.   **Get Todo**: A function that helps to fetch any existing to-do from the database.
4.   **Delete Todo**: This function allows users to delete any existing to-do, permanently erasing from the database.
5.   **Update Todo**: This function allows to update any kind of details for created to-do's and store the updated details back to the database.
6.   **List Todos**: This function lists down the complete set of created to-do's fetching them from the database.

13.4.2   MIGRATION RESULTS

By applying the algorithm on the case study application discussed in Section 13.4.1, the following output is generated with the method and the handler function. The handler functions are the methods in the code that activates the events upon request and once a particular handler returns a response, it activates the other events to satisfy the request. As mentioned in Sections 13.1 and 13.3, AWS Lambda is considered as the serverless platform in this chapter. The below handler functions are generated suitable for AWS platform.

```
+----------+--------------+----------------------+
| Method   |     Path     |   Handler Function   |
+----------+--------------+----------------------+
|   get    |    /hello    |    handler.hello     |
|   post   |  /createTodo |  createTodo.handler  |
|  delete  |  /deleteTodo |  deleteTodo.handler  |
|   get    |   /listTodo  |   listTodo.handler   |
|   put    |  /updateTodo |  updateTodo.handler  |
|   get    |   /getTodo   |   getTodo.handler    |
+----------+--------------+----------------------+
```

13.4.3 EMPIRICAL ANALYSIS

Generally, to migrate applications from one architecture to another or from one programming language to another, it is very important to study the behavior of the applications before and after the migration. In this regard, since the application is migrated from containers to serverless platform, an empirical analysis is carried out to assess the performance of the application post-migration. Performance metrics such as response time and throughput are considered for analysis and since the applications are based on services, SoapUI is the suitable tool for analysis. SoapUI is a tool for testing Web Services; these can be the SOAP Web Services as well RESTful Web Services or HTTP-based services. SoapUI is an Open Source and completely free tool with a commercial companion – ReadyAPI – that has extra functionality for companies with mission-critical Web Services. It can be used to do functional testing, performance testing, interoperability testing, regression testing, etc. Additionally load testing is performed for the chosen application to check latency, utilization, and various other parameters.

In order to analyze the parameters, different test scenarios are defined. Initially, the microservices deployed in containers undergo load testing and then the serverless applications are tested for cold start latency and load testing. Finally, a comparison between both the styles of deployment is presented.

13.4.3.1 Load Testing for Containerized Microservices

The To-do applications deployed in containers are tested by sending HTTP requests to the microservices (API Calls) using SoapUI. These requests are sent in random order by a user and this test is aimed at estimating the performance of the application with throughput and response time as metrics. The average response time captured for the containerized microservices is presented in Table 13.1.

13.4.3.2 Cold Start Latency

The most important factor of discussion is cold start latency in serverless platforms, especially in AWS [30, 31]. It is the time taken by the serverless functions to load all the required configurations and warming the instances. The functions of the chosen case study application are tested with different loads and the cold start latency is

TABLE 13.1
Average response time for microservices deployed in containers

API Call	Average Response Time
Hello_api	3 ms
create_todo	91 ms
update_todo	122 ms
delete_todo	129 ms
get_todo	129 ms
list_todos	105 ms

TABLE 13.2
Cold start latency for each API call

API Call	Maximum Value	Average Value
Hello_api	9 ms	6 ms
create_todo	354 ms	250 ms
update_todo	349 ms	200 ms
delete_todo	310 ms	233 ms
get_todo	355 ms	261 ms
list_todos	296 ms	145 ms

TABLE 13.3
Load testing results with 500 and 1,000 users

Performance metric	500 users	1,000 users
Average response time (ms)	4620.6	8126
Throughput (transactions/sec)	4.06	0.68

observed for each of the microservices. To analyze the performance of serverless functions with containerized microservices, cold start latency is observed and the results are presented in Table 13.2.

13.4.3.3 Load Testing on Serverless Functions

Using SoapUI tool, API calls are submitted for the case study application deployed in serverless platform. In order to analyze the results in detail, different loads with 500 and 1,000 users are submitted through SoapUI and the performance metrics such as response time and throughput are captured. The results are presented in Table 13.3.

TABLE 13.4
Comparison of metrics for serverless and containerized deployment

Performance Metric	Serverless	Containerized
Average response time (ms)	200	389
Throughput (transactions/sec)	156.85	17.06

13.5 COMPARISON AND DISCUSSION

The results of performance testing conducted on both containerized microservices and corresponding serverless functions are compared in terms of average response time and throughput. The comparison is shown in Table 13.4. It is clear from the results that the serverless deployment model has better response time and more number of transactions are executed successfully. This states that the serverless platforms are best suited for deploying microservices-based applications instead of containers.

With the current buzz about the serverless computing in both industry and academia, and with the comparative results, serverless computing will attain high attention and most of the Internet applications will be designed and deployed using serverless. Though container-based approach for microservices is running fine for applications, looking into the factors affecting the customer satisfaction, serverless will dominate the other existing cloud solutions.

13.5.1 THREAT TO VALIDITY

Though it is claimed that the proposed approach works for migrating containerized microservices to any serverless platform, Amazon AWS is considered for implementation and experimental analysis. As serverless.yml file is common for all serverless platforms, it is assumed that the proposed approach will be suitable for all serverless platforms. Also, the experimental analysis is carried out by considering only one toy example and it cannot be claimed that the performance of containerized microservices is poor compared to serverless model. A detailed empirical analysis is required with more case study examples to prove that serverless deployment of microservices performs better than containerized microservices.

13.6 CONCLUSION

With the evolution of new technologies and software design methodologies, it has become a challenging task for software developers, software architects, and project managers to select the appropriate design model for their new applications. After the Internet age, SOA has been considered the most suitable architecture with its web services implementation and since Martin Fowler coined the term "microservices," it has become the buzzword. However, on the other hand, serverless platforms have become popular in providing robust cloud services and as the apt model for deployment of microservices. Considering these evolutions, an automated approach for migrating containerized microservices to serverless platforms is proposed. A comparative study of containerized microservices and services deployed in serverless platforms in terms of performance parameters is also presented. The study suggests that the theoretical aspects of serverless computing are true as it exhibits better performance compared to containerized microservices. However, AWS Lambda is considered for the migration of microservices and for comparison. Many other serverless providers exist in the market, such as Microsoft Azure, Google Cloud, Open Whisk, etc. Proposing a generalized approach suitable for all such platforms and also shifting applications from one serverless platform to an other can be considered as a future study.

REFERENCES

1. Raj, V. and Sadam, R. 2021 Sep. Evaluation of SOA-based web services and microservices architecture using complexity metrics. *SN Computer Science*, 2:1–0.
2. Raj, V., 2021. Framework for migration of SOA based applications to microservices architecture. *Journal of Computer Science & Technology*, 21, pp. 196–198.
3. Raj, V. and Ravichandra, S. 2018. Microservices: A perfect SOA based solution for enterprise applications compared to Web Services. *In 2018 3rd IEEE International Conference on Recent Trends in Electronics, Information & Communication Technology (RTEICT)* 2018 May 18 (pp. 1531–1536). IEEE.
4. Baldini, I., Castro, P., Chang, K., Cheng, P., Fink, S., Ishakian, V., Mitchell, N., Muthusamy, V., Rabbah, R., Slominski, A. and Suter, P., 2017. Serverless computing: Current trends and open problems. *Research Advances in Cloud Computing*, pp.1–20. Springer Singapore.
5. Cassel, G.A.S., Rodrigues, V.F., da Rosa Righi, R., Bez, M.R., Nepomuceno, A.C. and da Costa, C.A., 2022. Serverless computing for Internet of Things: A systematic literature review. *Future Generation Computer Systems*, 128, pp. 299–316.
6. Kaushik, N., Kumar, H. and Raj, V., 2023. Empirical evaluation of microservices architecture. communication and intelligent systems: *Proceedings of ICCIS 2022*, Volume 2, 89, p.241.
7. Li, Z., Guo, L., Cheng, J., Chen, Q., He, B. and Guo, M., 2022. The serverless computing survey: A technical primer for design architecture. *ACM Computing Surveys (CSUR)*, 54(10s), pp. 1–34.
8. Capuano, R. and Muccini, H., 2022, March. A systematic literature review on migration to microservices: A quality attributes perspective. In *2022 IEEE 19th International Conference on Software Architecture Companion (ICSA-C)* (pp. 120–123). IEEE.
9. Li, Y., Lin, Y., Wang, Y., Ye, K. and Xu, C., 2022. Serverless computing: State-of-the-art, challenges and opportunities. *IEEE Transactions on Services Computing*, 16(2), pp.1522–1539.
10. Lloyd, W., Ramesh, S., Chinthalapati, S., Ly, L. and Pallickara, S., 2018, April. Serverless computing: An investigation of factors influencing microservice performance. In *2018 IEEE International Conference on Cloud Engineering (IC2E)* (pp. 159–169). IEEE.
11. Jin, Z., Zhu, Y., Zhu, J., Yu, D., Li, C., Chen, R., Akkus, I.E. and Xu, Y., 2021, August. Lessons learned from migrating complex stateful applications onto serverless platforms. In *Proceedings of the 12th ACM SIGOPS Asia-Pacific Workshop on Systems* (pp. 89–96).

12. Adouth, V. and Rajagopal, E., 2023. Blockchain-based certificateless public auditing with privacy-preserving for cloud-based cyber-physical systems. *Concurrency and Computation: Practice and Experience*, 35(12), p.e7690.
13. Jambunathan, B. and Yoganathan, K., 2018, March. Architecture decision on using microservices or serverless functions with containers. In *2018 International Conference on Current Trends Towards Converging Technologies (ICCTCT)* (pp. 1–7). IEEE.
14. Wen, J., Chen, Z., Liu, Y., Lou, Y., Ma, Y., Huang, G., Jin, X. and Liu, X., 2021, August. An empirical study on challenges of application development in serverless computing. In *Proceedings of the 29th ACM Joint Meeting on European Software Engineering Conference and Symposium on the Foundations of Software Engineering* (pp. 416–428).
15. Wen, J., Chen, Z., Jin, X. and Liu, X., 2023. Rise of the planet of serverless computing: A systematic review. *ACM Transactions on Software Engineering and Methodology*, 32(5): 1–61.
16. Lloyd, W., Vu, M., Zhang, B., David, O. and Leavesley, G., 2018, December. Improving application migration to serverless computing platforms: Latency mitigation with keep-alive workloads. In *2018 IEEE/ACM International Conference on Utility and Cloud Computing Companion (UCC Companion)* (pp. 195–200). IEEE.
17. Nupponen, J. and Taibi, D., 2020, March. Serverless: What it is, what to do and what not to do. In *2020 IEEE International Conference on Software Architecture Companion (ICSA-C)* (pp. 49–50). IEEE.
18. Yussupov, V., Breitenbücher, U., Leymann, F. and Müller, C., 2019, December. Facing the unplanned migration of serverless applications: A study on portability problems, solutions, and dead ends. In *Proceedings of the 12th IEEE/ACM International Conference on Utility and Cloud Computing* (pp. 273–283).
19. Adzic, G. and Chatley, R., 2017, August. Serverless computing: Economic and architectural impact. In *Proceedings of the 2017 11th Joint Meeting on Foundations of Software Engineering* (pp. 884–889).
20. Goli, A., Hajihassani, O., Khazaei, H., Ardakanian, O., Rashidi, M. and Dauphinee, T., 2020, April. Migrating from monolithic to serverless: A fintech case study. In *Companion of the ACM/SPEC International Conference on Performance Engineering* (pp. 20–25).
21. Kaplunovich, A., 2019, May. ToLambda–Automatic path to serverless architectures. In *2019 IEEE/ACM 3rd International Workshop on Refactoring (IWoR)* (pp. 1–8). IEEE.
22. Osman, M.H., Saadbouh, C., Sharif, K.Y. and Admodisastro, N., 2022. From monolith to microservices: A semi-automated approach for legacy to modern architecture transition using static analysis. *International Journal of Advanced Computer Science and Applications*, 13(10): 907–916.
23. Chadha, M., Pacyna, V., Jindal, A., Gu, J. and Gerndt, M., 2022, November. Migrating from microservices to serverless: An IoT platform case study. In *Proceedings of the Eighth International Workshop on Serverless Computing* (pp. 19–24).
24. Menéndez, J.M., Gayo, J.E.L., Canal, E.R. and Fernández, A.E., 2023. A comparison between traditional and Serverless technologies in a microservices setting. arXiv preprint arXiv:2305.13933.
25. Bajaj, D., Bharti, U., Goel, A. and Gupta, S.C., 2020, May. Partial migration for re-architecting a cloud native monolithic application into microservices and FaaS. In *International Conference on Information, Communication and Computing Technology* (pp. 111–124). Springer, Singapore.
26. Hassan, H.B., Barakat, S.A. and Sarhan, Q.I., 2021. Survey on serverless computing. *Journal of Cloud Computing*, 10(1), pp. 1–29.
27. Heikkinen, J., 2023. Serverless and microservice architecture in modern software development. *Masters thesis*. JAMK University of Applied Sciences.

28. Raj, V. and Ravichandra S., 2022 Jul. A service graph based extraction of microservices from monolith services of service-oriented architecture. *Software: Practice and Experience*, 52(7):1661–1678.

29. Casale, G., Artač, M., Van Den Heuvel, W.J., van Hoorn, A., Jakovits, P., Leymann, F., Long, M., Papanikolaou, V., Presenza, D., Russo, A. and Srirama, S.N., 2020. Radon: Rational decomposition and orchestration for serverless computing. *SICS Software-Intensive Cyber-Physical Systems*, 35, pp.77–87.

30. Lin, P.M. and Glikson, A., 2019. Mitigating cold starts in serverless platforms: A pool-based approach. arXiv preprint arXiv:1903.12221.

31. Vahidinia, P., Farahani, B. and Aliee, F.S., 2020, August. Cold start in serverless computing: Current trends and mitigation strategies. In *2020 International Conference on Omni-layer Intelligent Systems (COINS)* (pp. 1-7). IEEE.

14 Cloud Defender

Developing a Machine Learning Framework for Real-Time Detection and Mitigation of DDoS Attacks in Cloud Computing Environments

Om Prakash Suman and Mohit Kumar

14.1 INTRODUCTION

In this chapter we presents a detailed overview of the research topic, which focuses on using machine learning (ML) to improve intrusion detection systems (IDS) and address distributed denial of service (DDoS) attacks in cloud environments. It covers important areas such as different types of classifiers used in IDS and the significance of feature selection techniques. Additionally, it highlights the security requirements in cloud environments and discusses the motivation behind the research, emphasizing the need for a strong defense mechanisms against DDoS attacks and the importance of real-time detection and response capabilities offered by ML. The chapter outlines the research goals and emphasizes the significant contributions of the research, including better ML algorithms, innovative approaches to detect and mitigate zero-day attacks, efficient feature selection methods, scalable detection systems, and lightweight algorithms for resource-constrained environments. Furthermore, the introduction part establishes the foundation for the research study, highlighting the vital role of ML in enhancing IDS to counter DDoS attacks in cloud environments.

14.1.1 CLOUD COMPUTING

Cloud computing is a technology that has changed the way data and software applications are accessed and stored. It allows users to access their information from any location with an Internet connection, instead of having it on their personal computer or local server. The technology works by using a network of remote servers that process and store data, enabling users to access their information and applications whenever they need to.

DOI: 10.1201/9781003390954-14

The advantages of cloud computing include increased flexibility, scalability, and cost-efficiency. It also provides better data security, as data is stored in secure data centers that can be easily backed up and recovered in case of a disaster.

There are three types of cloud computing: public, private, and hybrid. Public cloud services are accessible to anyone with an Internet connection and are provided by third-party providers. Personal cloud services are used by a single organization and are not accessible to the public. Hybrid cloud services combine public and private cloud services, providing the best of both worlds [1].

However, like any technology, cloud computing has security risks. The main concern is network-based attack (DDoS) that impacts data privacy, as unauthorized parties can access sensitive information if proper security measures are not in place. It is important for businesses and individuals to have robust security protocols and ensure that their cloud service provider complies with industry standards and regulations.

14.1.1.1   Security Issue in Cloud Computing and Their Types

Network-based cloud computing has become increasingly prominent in recent years due to its cost efficiency and flexibility. Security issues in cloud computing have emerged as a paramount concern, demanding immediate attention. With the rise of data breaches, compromised transparency, relentless network attacks like DDoS, and intricate challenges surrounding identity and access management, the vulnerability of cloud environments cannot be ignored. Fortunately, recent research has brought forth promising solutions to these predicaments. Cutting-edge encryption techniques, robust access controls, and transparent mechanisms have been developed, bolstering network security and fortifying identity management protocols. Furthermore, ongoing efforts focus on enhancing security standards, certifications, and auditing procedures, alongside real-time detection of DDoS attacks and secure virtualization [2]. Yet, the quest for staying ahead of emerging threats persists, as safeguarding cloud computing environments remains a paramount goal to ensure data integrity, confidentiality, and trustworthiness.

14.1.2   Distributed Denial-of-Service Attack

A DDoS attack is a type of cyberattack that aims to overwhelm or disable a computer system or network by flooding it with traffic from multiple sources. This type of attack is more sophisticated than a DoS attack because it involves a network of compromised devices called a botnet. To carry out a DDoS attack, the attacker uses malware or other techniques to take control of the devices and then uses them to launch a coordinated attack against the target system or network. The impact of a DDoS attack can be severe, leading to extended service disruption, data loss, and harm to the organization's reputation. DDoS attacks can take several forms, including amplification attacks, reflection attacks, and application-layer attacks. The consequences of a DDoS attack can range from temporary inconvenience to significant financial or operational loss. To reduce the risk of a DDoS attack, organizations should have a robust defense strategy that includes measures such as network segmentation, intrusion detection and prevention systems, content filtering, and load balancing. An incident response plan is also critical to detect and contain the attack quickly. Regular security testing and assessments can help identify vulnerabilities and enhance defenses against DDoS attacks [3].

14.1.2.1 Classification of DDoS Attack

DDoS attacks can be classified based on various criteria, including the type of traffic used, the technique used to launch the attack, and the target of the attack. Here are some common classifications [4]:

a. **Based on Traffic Type:** DDoS attacks can be categorized based on the type of traffic they employ. Volumetric attacks involve flooding the target with an overwhelming amount of traffic, which exceeds its capacity and renders it unresponsive. Protocol attacks exploit vulnerabilities in network protocols, depleting server resources and resulting in system crashes. Application-layer attacks specifically target web applications, consuming resources and causing them to fail. SYN flood attacks take advantage of Transmission Control Protocol/Internet Protocol (TCP/IP) vulnerabilities, blocking legitimate traffic from accessing the target. User Datagram Protocol (UDP) flood attacks overload the target with a large volume of UDP packets, disrupting services. Ping of Death attacks exploit weaknesses in networking systems, leading to system crashes.

b. **Based on Target:** DDoS attacks can be classified based on the target they focus on. Network layer DDoS attacks aim to disrupt the target's network connectivity by overwhelming network infrastructure components like routers, switches, and firewalls with excessive traffic. Transport layer DDoS attacks specifically target transport layer protocols (TCP and UDP), causing disruptions in the communication between network devices or consuming server resources. Application layer DDoS attacks directly target web applications that rely on protocols such as HTTP and HTTPS. Attackers flood the server with a high volume of requests, leading to resource exhaustion. These attacks pose a challenge for detection and mitigation as they often imitate legitimate traffic patterns.

14.1.3 Intrusion Detection System

In the realm of cloud security, an IDS refers to a security solution specifically tailored to monitor and detect suspicious activities or potential intrusions within a cloud computing environment. These IDS systems play a vital role in ensuring the protection of cloud infrastructure, applications, and data from various security threats. Continuously monitoring network traffic, application behavior, and user activities within the cloud environment, their primary objective is to identify and respond to any unauthorized access attempts, malicious activities, or abnormal patterns that may compromise the overall security of the cloud infrastructure or the integrity of the stored data. Employing advanced detection techniques and leveraging threat intelligence, IDS systems enable real-time alerts and facilitate prompt incident response, thereby reducing risks and bolstering the overall security stance of cloud environments [5].

IDS can be categorized into various classes based on the specific activities they monitor and the detection methods they employ. This classification allows for a systematic understanding of IDS functionalities [6].

14.1.3.1 Classification of IDS by Analyzed Activities

IDS can be classified into three categories according to the type of activities that are analyzed [7]: network-based intrusion detection system (NIDS), host-based intrusion detection system (HIDS), and application-based intrusion detection system (AIDS).

a. **Host-Based IDS:** HIDS is a software that is designed to monitor and analyze specific hosts, such as files, processes, and system logs, to identify suspicious or unauthorized activities. It compares system snapshots, monitors indicators like failed logins and high CPU usage, and can detect potential attacks by examining system calls and changes to binaries. HIDS has a lower traffic volume compared to NIDS, reduces overhead, and provides detailed information about the host system's activities.

b. **Network-Based IDS:** NIDS uses sensors placed throughout the network to analyze traffic, either locally or remotely through a central controller. It is more scalable and cross-platform than HIDS, making it a preferred option for securing a company's IT infrastructure. NIDS and HIDS can be used together for a better level of protection. NIDS can analyze a wide range of application protocols and provide a strong response against outsider attacks. However, NIDS is not able to analyze encrypted traffic, which is a significant disadvantage.

c. **Application-Based IDS:** AIDS is a security solution that operates at the application layer of a network to detect and prevent unauthorized access and suspicious activities. It analyzes data and transactions within the application itself, focusing on application-layer attacks like SQL injection and XSS attacks. This IDS can be customized for specific applications, providing flexible and effective security.

14.1.3.2 Classification of IDS by Detection Method

IDS can also be classified into three categories according to detection method: signature-based IDS (SIDS), anomaly-based IDS, and hybrid-based IDS (HIDS).

a. **Signature-Based IDS:** It is a security mechanism that identifies malicious activities by examining network traffic for established patterns or signatures. It compares incoming traffic against a predetermined set of signatures stored in a database, generating alerts if a match is found. While proficient in detecting recognized threats, it may not effectively identify novel or unfamiliar threats.

b. **Anomaly-Based IDS:** It is a security mechanism that recognizes potential threats by analyzing deviations from normal behavior. It establishes a standard baseline of expected activity and contrasts current behavior against it to detect any unusual or anomalous patterns that could indicate a security threat. It has the capability to detect previously unknown threats, such as zero-day exploits, although it can generate false positives due to infrequent but harmless behavior.

   c. Hybrid IDS: It is an IDS that combines signature-based and anomaly-based approaches. It uses known signatures for recognized threats and ML algorithms to detect abnormal behavior. This hybrid approach improves threat detection for known and unknown threats, reducing false positives and negatives. Ongoing research focuses on updating the signature database and adapting to emerging threats.

14.1.4 FIREWALL VS. INTRUSION DETECTION SYSTEMS

Firewalls and IDS are two essential components of network security that differ in their approaches to detecting and mitigating DDoS attacks. A firewall acts as a protective barrier between an internal network deemed trustworthy and an external network. It diligently monitors and regulates incoming and outgoing traffic by employing pre-established rules. While firewalls can offer a certain degree of protection against DDoS attacks by impeding suspicious traffic patterns or IP addresses, their core functionality lies beyond the realm of specific DDoS detection. Conversely, IDS systems are meticulously crafted to identify and meticulously scrutinize network anomalies, encompassing DDoS attacks within their purview. IDS stands sentinel, vigilantly examining real-time network traffic, seeking telltale signs of aberrant behavior indicative of a potential DDoS onslaught. It astutely recognizes the distinctive patterns associated with DDoS attacks, promptly triggering alarms and dutifully notifying administrators to expedite further action. Thus, IDS systems epitomize a more targeted, proactive methodology toward DDoS detection, surpassing the capabilities of firewalls, which predominantly focus on traffic filtration and access control [8].

14.1.5 MACHINE LEARNING IN THE FIELD OF INTRUSION DETECTION SYSTEM

ML has become a crucial tool in the field of IDS, which is used to identify possible threats or attacks on a network. These systems, commonly used in security setups, monitor network traffic and system activities to detect potential intrusions or security threats. In the past, most IDS relied on signature-based methods. These methods compared new network traffic with preexisting signatures of known attacks in order to identify threats. However, this approach became ineffective as new attacks and similar threats emerged that couldn't be recognized by the existing signatures. To overcome this limitation, ML techniques have been implemented for real-time detection and prevention of potential security threats. ML algorithms enable IDS to learn the patterns, behaviors, and flow of normal network activity. With this understanding, the algorithms can identify anomalies in network activity that may indicate an intrusion or attack. This detection method is more effective than signature-based approaches because it can identify new potential threats and reduce false positive alerts [9].

There are different types of ML algorithms used in IDS. Some commonly used ones include supervised learning, unsupervised learning, and deep learning [10].

a. **Supervised Learning Algorithms:** These algorithms use labeled training data that consists of input and output pairs. They discover patterns that represent normal network behavior and then evaluate new data to determine if it is normal or suspicious. These algorithms can accurately detect variations in communication patterns and trigger an alert when any deviation is detected.

b. **Unsupervised Learning Algorithms:** These algorithms identify normal network activity by monitoring communication protocols, traffic volumes, and other network indicators. They enable IDS to identify abnormal behavior by flagging potential threats as "outliers."

c. **Deep Learning Algorithms:** These algorithms are designed to handle complex data and can detect sophisticated cyber threats. They diagnose large amounts of data, find patterns, and learn to detect previously unknown threats.

The use of ML in IDS has strengthened the security of many organizations and reduced potential attacks while minimizing false alarms. As attackers continue to develop new and more sophisticated threats, the adoption of ML is crucial for the ongoing health and resilience of networks and security systems (see Table 14.1).

14.1.5.1　Types of Classifiers Based on ML

There are different types of classifiers used in the classification approach using ML techniques. These classifiers can be classified as single, ensemble, or hybrid, based on how they work together to solve a problem.

a. **Single Classifiers:** In this section, IDS models are designed using a single technique such as clustering, classification, or association. Recent research has utilized classifiers like Support Vector Machine (SVM), Neural Networks (NN), Decision Trees (CART, C4.5, and J48), Naive Bayes (NB), and K-Nearest Neighbors (KNN) to create single classifier-based IDS models.

b. **Hybrid Classifiers:** In this category, IDS models combine two or more functional components to improve performance compared to a single classifier approach. The implementation of hybrid classifiers involves two stages. The first stage focuses on optimizing the learning performance through parameter tuning. In the second stage, intermediate results are used to predict the final output. One example of a hybrid classifier design is the DENFIS algorithm [1].

c. **Ensemble Classifiers:** Ensemble models enhance performance by combining the opinions of multiple learners. These models are particularly efficient for training and testing due to their parallel nature and access to multiple processors. There are two ways to implement ensemble models.

TABLE 14.1

Comparative analysis of proposed ML models for DDoS detection

References	Objective	Techniques	Exp. Setup	Limitation
[18]	Detect DDoS attacks promptly using ensemble machine learning techniques with high accuracy rates	Ensemble ML techniques, Feature Selection Algorithm	Weka	Reliance on the quality and diversity of training data for optimal performance
[13]	To propose a system utilizing ML techniques for accurate detection and classification of DDoS attacks	ML- and DL-based algorithm, Optimizer, and Boosting Algorithm	Google Colab	Limited scope of attack data, specific algorithm evaluation, lack of real-time implementation
[14]	Develop a method for detecting DDoS attacks using unsupervised data mining techniques and entropy	Unsupervised data mining techniques, CURE algorithm	Mininet, and POX	Limited effectiveness in dealing with sophisticated DDoS attacks that can evade unsupervised techniques
[15]	Develop a hybrid method for proactive detection and prevention of DDoS attacks	Hybrid method combining k-means clustering and Standardization	MATLAB	Reliance on the quality and relevance of the CIADA dataset for accurate comparison
[17]	Develop a statistical approach for DDoS detection using the FFSc metric	Statistical approach using the FFSc metric, correlation measures	TFN2K, and TopDump	Difficulty in differentiating attack traffic closely mimicking normal network traffic
[11]	Investigate network traffic patterns, identify potential attacks using statistical attributes	TRA method, statistical attribute selection, machine learning-based prediction	nmeta2	susceptibility to false positives/negatives due to statistical attributes and potential bias
[18]	Establish correlation between feature selection methods and ML classification algorithms	Filter-based feature selection methods and ML algorithm	Jupyter Notebook	Limited to the effectiveness of the selected FS and ML classification algorithms
[16]	Effectively detect and prevent DDoS attacks using statistical information from VMs and cloud servers approaches	Statistical analysis of VM activities ML	Jupyter Notebook	Reliance on accurate and comprehensive statistical information from VMs and cloud servers

When it comes to learning techniques in ML, they are generally classified as supervised or unsupervised, depending on the presence or absence of labeled data and the goal of prediction.

14.2 MOTIVATION

Cloud Defender, a cloud security mechanism, is primarily motivated by the need to safeguard cloud infrastructure from DDoS attacks. As the business world increasingly relies on cloud services, ensuring the security and integrity of the underlying infrastructure is becoming more important than ever before. Cloud Defender aims to create a secure and reliable cloud ecosystem using advanced defense mechanisms that can detect and neutralize DDoS attacks promptly. These attacks can cause significant financial losses for businesses, including direct costs, loss of customer trust, and reputation damage. By preventing these financial burdens with real-time detection and response, Cloud Defender can also create an enhanced user experience by guaranteeing high service availability, uninterrupted access to cloud resources, and data protection confidence for customers. Cloud Defender's integration with various ML classifiers strengthens the entire security posture and supports it by adding a layer of transparency and interpretation ability to the detection and mitigation process. This mechanism provides deeper insights into attack patterns and allows security professionals to fine-tune their defense strategies while also contributing to wider research advancements.

14.3 PROBLEM SCOPE AND RESEARCH OBJECTIVES

The problem statement in the research paper highlights that detecting DDoS attacks using ML is hindered by inaccurate ML algorithms, imbalanced data sets, and the difficulty of identifying new attack patterns. Additionally, selecting relevant features and managing high-dimensional data present additional obstacles. Scalability and real-time detection become challenging in scenarios with high network traffic volumes, while resource-constrained environments necessitate lightweight ML algorithms. This chapter emphasizes that addressing these challenges can improve the accuracy and efficiency of DDoS detection systems across different network environments.

14.4 SIGNIFICANT CONTRIBUTIONS

There are several ways in which significant progress can be made in addressing the challenges mentioned in the problem statement:

- Develop better ML algorithms for detecting DDoS attacks, improving prediction accuracy, and handling imbalanced datasets.
- Create a new DDoS-FD-23 data set as a solution to address the scarcity of data sets pertaining to DDoS attacks.
- Innovate techniques for detecting and mitigating zero-day DDoS attacks, including anomaly detection, behavioral analysis, and leveraging threat intelligence.

- Design methods for efficient feature selection, reducing data set complexity while analyzing network traffic data accurately and quickly. Consider using correlation coefficient and recursive feature elimination (RFE).
- Build scalable and high-performance detection systems that can process and analyze large volumes of network traffic in real time. Utilize distributed computing, parallel processing, and optimized algorithms, like the StormShield DDoS detection system.
- Develop lightweight ML algorithms for resource-constrained environments, like edge computing or IoT networks. Ensure accurate detection while minimizing computational and memory requirements.

By implementing these contributions, along with advancements from other researchers and organizations, we can significantly improve DDoS detection. These enhancements will improve prediction accuracy, address data set limitations, detect new attack types, optimize feature selection, ensure real-time detection, and cater to resource-constrained environments. Ultimately, this will enhance the security and reliability of cloud computing environments.

14.5 LITERATURE SURVEY

This section provides an overview of intrusion detection research: ML algorithms, feature selection, ensemble methods, performance comparison, and valuable resources for further research.

14.5.1 SEARCH STRATEGY

The search strategy proposed for the literature survey in this research study involves carefully selecting databases and search engines like IEEE Xplore, Digital Library, and Google Scholar. Specific keywords such as "DDoS detection," "intrusion detection," "machine learning," "neural networks," and "anomaly detection" are used to guide the search effectively. The focus is narrowed down by specifying the desired study type or method, prioritizing papers from reputable conferences or journals, and considering recently published papers. The chapter should also include a comprehensive evaluation of the proposed methods, including a comparison with state-of-the-art techniques, highlighting limitations and challenges for a balanced perspective.

14.5.2 RESEARCH QUESTIONS

- How can ML effectively detect and mitigate DDoS attacks in cloud networks?
- What are the important features and metrics used in ML models for accurate DDoS detection in cloud environments?
- What are the limitations and challenges of using machine ML for DDoS detection in cloud networks, and how can they be overcome?
- How does the performance of ML-based DDoS detection compare to traditional rule-based approaches in cloud computing?

- What are the recommended practices for training ML models for DDoS detection in cloud networks, including data preprocessing, feature selection, and model optimization?

14.5.3 RELATED WORK

Most of the suggested methods since the development of ML for DDoS detection may be divided into four main categories: clustering, classification, statistics, and hybrid.

The authors in [12] proposed ensemble ML techniques, specifically RF, histogram-based gradient boosting, and adaptive boosting classifiers. Their main objective was to detect DDoS attacks promptly using the CIC-DDoS2019 dataset, aiming to achieve high accuracy rates collating to prior research endeavors. In their study, several steps were followed, including data preprocessing, feature extraction, feature labeling, training, and testing the ensemble ML models on the dataset. Compared with existing literature, the study's maximum accuracy score of 91% outperformed previous approaches by 0.5487 points. The results and findings section presents an evaluation of the ensemble models, highlighting their effectiveness in accurately detecting and classifying DDoS attacks.

The authors in [13] explored the issue of DDoS attacks and proposed a system that utilizes ML techniques for their detection and classification. They employed various algorithms, including extreme gradient boosting (GB), KNN, NB, and CNN, to accurately identify and classify these attacks. Through their study, they found that XGBoost achieves the highest accuracy, followed closely by CNN and KNN. The paper suggests some future enhancements, such as including a wider variety of attack data and developing a comprehensive system for attack prevention. The research highlights the effectiveness of ML techniques in combating DDoS attacks and emphasizes the importance of ongoing efforts to improve network security.

14.5.4 CLUSTERING-BASED TECHNIQUES

The authors in [14] proposed a method for detecting DDoS attacks using unsupervised data mining techniques and introduced the concept of entropy and applied it to windowed incoming packets using Clustering Using Representative (CURE) as a cluster analysis method. The proposed approach was evaluated using real-world datasets and compared to existing approaches, demonstrating its superiority in terms of accuracy, detection rate, false alarm rate, F-measure, and Phi coefficient. The paper also discusses related works in the field of intrusion detection and presents the EM-CURE cluster analysis algorithm. Data reduction using entropy windows is employed to preprocess the datasets.

The authors in [15] proposed a hybrid method called centroid-based rules for proactive detection and prevention of real-world DDoS attacks. Their approach utilized k-means clustering techniques along with proactive rules to analyze the Center for Applied Internet Data Analysis (CAIDA) attack dataset. By randomly selecting one million packets from each dataset and normalizing them, they created a final dataset that resembled real-life scenarios. The proposed "Proactive DDoS Attack Detection System" consisted of several steps, including feature selection, data transformation, standardization using Shannon's entropy method, and splitting the data into training and testing sets. The k-means clustering algorithm was then applied to generate centroids and establish max-min rules for each cluster. Outliers were handled using a shrink factor, and the accuracy model was contrasted to a baseline centroid-based method.

14.5.5 STATISTICS-BASED TECHNIQUES

When it comes to identifying DDoS attacks, statistical techniques have emerged as valuable tools. By analyzing key statistical properties of incoming network packets, such as the source and destination IP addresses, as well as the packet rate, experts have discovered reliable measures for identifying and distinguishing DDoS attacks from legitimate network traffic. These statistical approaches have gained widespread popularity in the field due to their ease of implementation and computational efficiency.

The authors in [16] proposed an innovative statistical approach to DDoS detection. The researchers introduced a statistical metric called Feature Feature Score (FFSc) that enables the analysis of multiple variables to differentiate between attack traffic and legitimate traffic effectively. They observed that when an attack is initiated by a botnet, the attack traffic tends to exhibit secure correlations between its samples. This correlation arises from the fact that the attacker employs the same attack statistics during the attack-initiating process. Leveraging this insight, the researchers proposed a correlation measure to distinguish attack packets from regular ones.

However, it should be noted that some attackers generate attack traffic that closely mimics normal network traffic, making it challenging to differentiate between the two solely based on correlation measures. To address this, the authors in [17]. explored multiple network traffic and devised an analytical approach that accounts for changes in individual feature values that may reflect broader alterations in the overall network traffic sample. To demonstrate the feasibility of their suggested strategy, the researchers conducted experiments using widely recognized datasets for DDoS research, namely the CAIDA and Defense Advanced Research Projects Agency (DARPA) datasets. They employed a feature selection step that involved extracting and calculating various attributes, including entropy of source IP addresses, variation of source IP addresses, and packet rate. For instance, the entropy of source IPs was computed using a specific equation outlined in the research paper.

14.5.6 HYBRID TECHNIQUES

The authors in [11] introduced a hybrid approach that combines statistical concepts with ML algorithms to address the challenge of detecting DDoS attacks. This unique method, known as Traffic Rate Analysis (TRA), presents a novel approach to investigating network traffic patterns and identifying potential attacks. By computing the ratio of TCP flags to the total number of TCP packets, they aimed to uncover irregularities that might indicate the occurrence of an attack. To accomplish this, an ML algorithm was employed to compile state-action rules that establish connections between TCP flag rates and the likelihood of a DDoS attack.

The TRA method entails the collection of TCP, UDP, and Internet Control Message Protocol (ICMP) packets from the network traffic. However, only the TCP packets are retained for subsequent analysis. The payload is discarded, and the focus is placed on the TCP headers. These headers contain six possible flags, such as SYN, FIN, RST, ACK, PSH, and URG. Whenever any of these flags are set within a packet, they are quantified and aggregated. The TCP flag rates considered the primary metric are then calculated using a specific equation as described in the research paper. By examining TCP flag rates, their method enables the identification of patterns and anomalies that could indicate the presence of a DDoS attack. The state-action rules derived from ML algorithms facilitate the establishment of a connection between specific flag rate patterns and the likelihood of an attack. By harnessing the strengths of statistical analysis and ML, the proposed hybrid technique offers a comprehensive and precise approach to detect and mitigate DDoS attacks.

14.5.7 Classification-Based Techniques

In the realm of cybersecurity, researchers have been exploring advanced techniques to combat the growing threat of DDoS attacks. One prominent approach gaining recognition is the utilization of ML, specifically classification and clustering algorithms, which have demonstrated remarkable potential in detecting and mitigating these attacks. These methods have shown superiority over traditional detection mechanisms, boasting not only greater speed but also significantly enhanced accuracy.

A paper by Osanaiye et al. [18] delves into the analysis of various feature selection methods and ML classification algorithms. The primary objective of their study was to establish a correlation between these methods and identify the most effective combination for achieving a higher rate of DDoS detection. To evaluate their proposed approach, they employed the well-known Knowledge Discovery in Databases (KDD) Cup dataset, which encompasses 41 feature sets specifically designed for DDoS attack detection experimentation and testing. The initial phase of their research involved processing the data using filter-based feature selection methods. By applying techniques such as information gain, gain ratio, and chi-squared, the authors aimed to extract the most pertinent features from the complete feature set. These selected features play a crucial role in accurately identifying and distinguishing DDoS attacks from legitimate network traffic.

Another noteworthy approach, introduced by in [16] takes advantage of statistical information obtained from the virtual machines (VMs) and cloud servers. By leveraging this data, their objective was to effectively detect DDoS attacks and prevent malicious network packets from propagating to the external network. To illustrate their proposed system, the authors presented an architectural framework where multiple VMs are rented by an attacker and subsequently transformed into botnets. The

Virtual Machine Manager (VMM) collects relevant information from the VMs and forwards it to an ML engine, designed to detect any signs of malicious behavior. Upon detecting suspicious activity across multiple VMs, the system proceeds to disconnect the network connection of all implicated VMs, effectively thwarting the DDoS attack.

14.6 METHODOLOGY

In our research, we propose a method to create a DDoS-FD-23 dataset by utilizing published datasets from CIC Canada. The raw data is transformed into a structured dataset with clear features and labels. We enhance the dataset's quality through various data preprocessing techniques, including addressing missing values, outliers, normalization, duplicate removal, and feature engineering. Data pruning is also performed to refine the dataset by eliminating unnecessary columns, handling singleton values, and reducing dimensionality. The preprocessed and pruned dataset is then divided into training, validation, and testing sets.

Furthermore, our research incorporates a range of ML models to achieve accurate DDoS attack detection. We employ classification algorithms such as Random Forest, Linear Regression, and Decision Trees, as well as ensemble methods like Ada Boost and Extreme Gradient Boosting. To capture intricate patterns within the data, these models are trained using the training set and evaluated using the validation set. We optimize the model performance by fine-tuning the hyperparameters through techniques such as cross-validation or grid search. Based on evaluation metrics, including accuracy, we select the best-performing model based on the Voting algorithm, which is then tested on an independent testing set to assess its generalization ability. Our ultimate objective is to achieve the highest possible accuracy in detecting DDoS attacks by evaluating multiple ML and DL models on the dataset (see Figure 14.1).

14.6.1 DATASET

We propose a data acquisition process to generate a new dataset named as DDoS Fusion Dataset of 2023 (DDoS-FD-23) for model analysis. Our focus is on obtaining datasets from open sources, particularly the CSE-CICIDS2017, CSE-CICIDS2018AWS, and CSECICIDS2019 [19] datasets published by the Canadian Institute of Technology. These datasets are recognized as valuable resources for cybersecurity research and IDS development. In order to create a balanced dataset, we randomly sampled 70% of benign flows and 30% of DDoS flows from the previously extracted 6.4 million data points. This sampling process resulted in a final imbalanced dataset consisting of 6.5 million data points with 82 features.

Additionally, we address the limited availability of recent DDoS attack datasets by creating a new dataset called DDoS-FD-23 (see Table 14.2). This dataset is a combination of four open datasets and undergoes data preprocessing, pruning, and normalization. It serves as a comprehensive resource for studying DDoS attacks and improving intrusion detection systems, containing a vast volume of network traffic data.

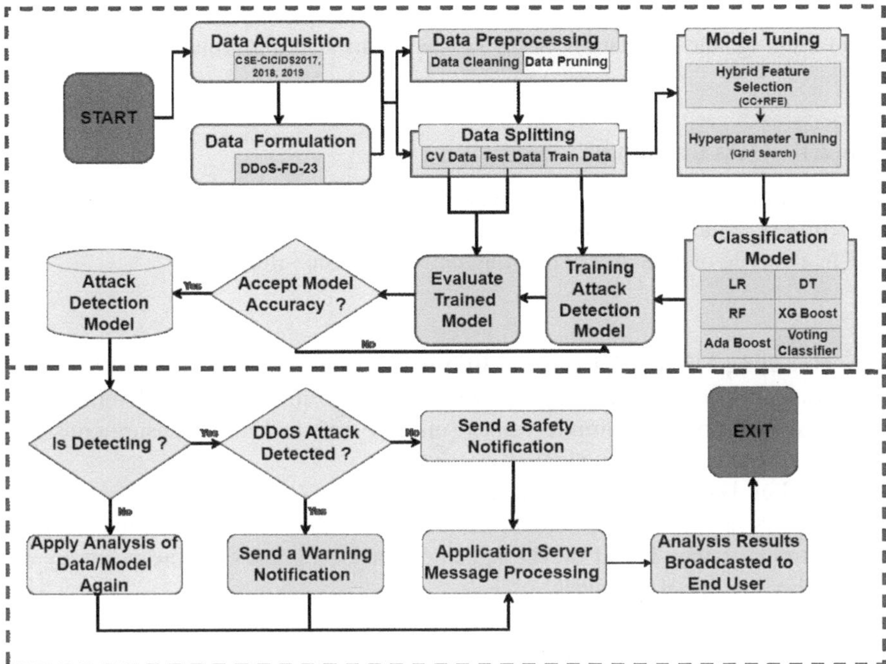

FIGURE 14.1 Proposed methodology.

TABLE 14.2
Formulate DDoS_FD_23 dataset description

Dataset Property	Description
Dataset Name	DDoS Fusion Dataset of 2023 (DDoS-FD-23)
Size	3.7 GB
Source	Canadian Institute for Cybersecurity (CIC)
Year(s) Captured	2016, 2017, 2018, 2019
Original Datasets	CSE-CIC-IDS2017, CSE-CIC-IDS2018-AWS, CSE-CIC-IDS2019 [19]
Data Points	6.4 million
Features	82
Imbalance Ratio	70% Benign flows, 30% DDoS flows

14.6.2 Data Prepossessing

Data preprocessing is vital for preparing DDoS datasets for analysis and modeling. These datasets are collected to study and comprehend cyberattacks that seek to overwhelm online services and hinder availability. By applying data preprocessing techniques, researchers and analysts can improve the quality, usability, and reliability of the DDoS dataset. This leads to more accurate analysis, robust modeling, and valuable insights into the nature of DDoS attacks, facilitating the development of effective countermeasures. Popular tools like TensorFlow are used for data preprocessing, model training, and evaluation. Some key techniques are presented here.

14.6.2.1 Data Cleaning

Data cleaning is an essential step in data preprocessing that involves identifying and correcting or removing errors, inconsistencies, and inaccuracies in a dataset. Let's take an example of a dataset containing network traffic information.

a. **Handling Missing Values and Infinite Values:** In our network traffic dataset, we identified two columns, "Flow Byts/s" and "Flow pkts/s," with missing values. We converted infinite values (inf and -inf) to NaN to ensure consistent handling during subsequent processing.

b. **Outlier Detection and Treatment:** Using the Z-score method, we detected outliers in the "Packet Length" column. We either removed or replaced these extreme values based on the specific analysis, improving the accuracy and reliability of data analysis.

c. **Data Normalization/Scaling:** To prevent the dominance of certain features, we normalized the "Packet Size" column using min-max scaling. This transformed the values to a range between 0 and 1, allowing for unbiased analysis and improved model performance.

d. **Removing Duplicates:** Based on the values in the "Source IP," "Destination IP," and "Timestamp" columns, we identified and removed duplicate records from the dataset. This preserved data integrity and improved the reliability of the analysis.

e. **Handling Data Discrepancies:** We resolved inconsistencies in the "Protocol" column by converting all values to a consistent format. This ensured accurate analysis and interpretation.

f. **Handling Skewed Data:** The "Packet Size" column exhibited a highly skewed distribution. To address this, we applied a logarithmic transformation, reducing the skewness and improving the suitability of the data for analysis.

g. **Feature Engineering:** To enhance the predictive power of the dataset, we created a new feature called "Total Bytes Transferred" by summing the "Incoming Bytes" and "Outgoing Bytes" columns. This new feature provides additional information and improves model performance.

h. **Data Validation and Verification:** By performing data validation and verification, we ensured the accuracy, completeness, and adherence to expected patterns or constraints in the dataset. This helps identify and rectify any anomalies or errors.

i. **Checking Column Data Types:** We examined the column data types and identified four columns with object data type ("Flow ID," "Src IP," "Dst IP," "Timestamp," "Label"), 35 columns with int64 data type, and 42 columns with float64 data type. Understanding data types informs appropriate transformation and analysis techniques.

By following these data cleaning steps, we addressed issues such as missing values, improper data types, and redundant columns, preparing the dataset for further analysis or modeling.

14.6.3 Data Pruning

To address the challenge of analyzing the massive DDoS attack dataset called DDoS-FD23, which consists of approximately 10 million data points and over 85 features, we employed data pruning techniques. Data pruning in the context of DDoS attacks refers to the process of reducing the size or complexity of the dataset used for analyzing and detecting DDoS attacks. The aim is to create a lightweight model that could efficiently detect DDoS attacks with high accuracy, reducing computational requirements while retaining relevant information. Given the scale and complexity of the dataset, we employed the following techniques tailored specifically for the DDoS-FD-2323 dataset:

a. **Handling Singleton Values and Categorical Features:** In our dataset, we identified columns with singleton values, such as the "Fwd URG Flags" column, which contains only a single unique value. We can safely drop this column. Additionally, we identified categorical features like the "Label" column, which may require special encoding or transformation for numerical models.

b. **Trimming Unnecessary Columns:** To streamline the dataset and enhance computational efficiency, we removed irrelevant or redundant columns. For example, the "Flow ID" column can be represented by combining "Src IP," "Dst IP," "Src Port," "Dst Port," and "Protocol," so we dropped the "Flow ID" column. Similarly, we trimmed other irrelevant columns like "Fwd URG Flags" and "Bwd URG Flags."

c. **Feature Extraction:** To leverage underlying patterns and relationships, we converted the "Src IP" and "Dst IP" columns from strings to numerical features. For example, we removed the dots in IP addresses and combined the resulting digits into a new numerical feature. (Example: Dst IP and Src IP - 172.31.69.25 to 172316925)

d. **Correlation-Based Dimension Reduction:** We applied Pearsons correlation-based feature selection to identify the most relevant features for distinguishing between normal and attack traffic in a DDoS attack dataset. Highly correlated pairs of columns were dropped to remove redundant information and improve model performance. We calculated the correlation matrix and set a threshold value. For pairs of columns with correlation coefficients exceeding the threshold, one of the features was removed. The remaining features were ranked based on their correlation with the target variable, and the top-ranked features were selected.

e. **RFE-Based Feature Selection:** After applying correlation-based feature selection and selecting the top 46 features, RFE can be used to further refine the feature subset. RFE is an iterative technique that eliminates the least important features by training models repeatedly until a desired number of features is reached. This helps identify the best 29 subsets of features for the analysis.

By applying these techniques, we address singleton values, categorical features, and unnecessary columns, and perform dimension reduction to create a refined feature subset for subsequent analysis or modeling.

14.6.4 Data Splitting

Here we separate the target variable "y" as a label from the input features to facilitate model training. The dataset is split into a training set, a validation set, and a testing set. The training set of 60% of the dataset is used to train the model, the validation set of 20% of the dataset aids in model evaluation and selection, and the testing set of 20% of the dataset is used to assess the final performance. This separation ensures unbiased evaluation, prevents overfitting, and measures the model's ability to generalize to new data.

14.6.5 Model Construction

Model construction for DDoS detection involves developing ML models to accurately classify network traffic as attacks or normal. Algorithms like Random Forest, Logistic Regression (LR), Decision Tree, XG Booster, Ada Booster, and Voting Classifier are used, with hyperparameter tuning and ensemble methods to enhance performance. Models are trained on a training set, and evaluated on a validation set using metrics like accuracy, precision, and recall. The best model is selected for testing on an independent set to assess generalization. Overall, model construction includes algorithm selection, tuning, and evaluation to build accurate DDoS detection models.

14.6.5.1 Logistic Regression

LR is a statistical model used for classifying data into two categories. In the context of DDoS attacks, it helps differentiate between normal and malicious network traffic by analyzing patterns and features in the data. In this work, the LR model is instantiated with the parameter C, which takes values of [0.01, 0.1, 10, 100]. Smaller C values indicate stronger regularization. To optimize the model's performance, a Randomized Search Cross Validation is employed, performing a randomized search over the specified hyperparameter space. The scoring metric used is the F1 score, combining precision and recall. The model achieves an accuracy of 97% and F1 scores of 96.939% on the training set. It achieves an F1 score of 96.958% on the test set, with training and testing times of 6.58 seconds and 7.84 seconds, respectively. The cross-validated results show a mean test F1 score of 96.926% and a mean test macro F1 score of 98.147%. Overall, the LR model with a C value of 100 achieves high accuracy and F1 scores on the given dataset, and the training and evaluation processes are computationally efficient.

14.6.5.2 Decision Tree

The Decision Tree Classifier is a machine learning model used for classifying data, specifically in detecting DDoS attacks. It employs a decision tree structure with parameters such as "max depth" and "max features" to control the depth and feature selection in the tree. The model's performance is optimized using Randomized Search CV, with ten different combinations of parameter values using two fold cross-validation and evaluating the F1 score. The result achieves high accuracy and F1 scores, with a perfect F1 score of 99.987% on the training data and an F1 score of 99.973% on the test data. The mean test F1 score of 99.972% indicates high accuracy in predicting both positive and negative classes, while the mean test macro F1 score of 99.983% indicates strong overall performance.

14.6.5.3 Random Forest

The Random Forest Classifier is an ML algorithm commonly used for detecting DDoS attacks. It combines multiple decision trees through ensemble learning, making accurate predictions. Each tree is built using random subsets of data and features, ensuring diversity and reducing correlations. By aggregating the outputs of all the trees, the classifier reaches a final prediction. Random Forest Classifiers handle noisy data, generalize well, and effectively handle high-dimensional feature spaces. In the proposed model, hyperparameters such as the number of estimators (trees) and maximum depth are optimized using Randomized Search CV. The F1-score is used as the evaluation metric, measuring accuracy based on precision and recall. The model achieves high performance, with a perfect F1 score on the training data and approximately 99.997% on the test data. Cross-validation demonstrates consistent performance, with a mean F1 score of approximately 99.988% and a mean macro F1 score of approximately 99.993%. The training and evaluation processes are computationally efficient, with a total wall time of approximately 12 minutes and 16 seconds.

14.6.5.4 XG Boost

Extreme Gradient Boosting (XG Boost) is a highly performant ML algorithm known for its accuracy in classification and regression tasks. It utilizes ensemble learning by combining weak prediction models, typically decision trees, to create a strong predictive model. In the proposed methodology, the XG Boost classifier undergoes a randomized search for hyperparameter optimization, enhancing its performance. The training phase involves learning from the provided dataset to identify patterns relevant to DDoS attack detection. The testing phase evaluates the trained model, demonstrating a high F1 score of approximately 99.999%, indicating the accurate classification of network flows. Cross-validation confirms the model's consistency, with a mean test F1 score of approximately 99.998% and a mean test macro F1 score of approximately 99.998%. These results emphasize XG Boost's effectiveness in detecting DDoS attacks, providing robust and reliable predictions to safeguard network systems.

14.6.5.5 Ada Boost

The Adaptive Boosting (Ada Boost) classifier is an ML algorithm designed for DDoS attack detection by combining weak classifiers into strong ones. It iteratively trains these classifiers, assigning weights to samples and focusing on accurately classifying challenging instances. In our proposed methodology, we utilize Randomized Search CV to find optimal hyperparameter values, such as "n estimators" set to 800, to train the Ada Boost classifier model. The model achieves exceptional performance during training, with an impressive F1 score of 99.999%. When evaluated on a separate test dataset, the model demonstrates perfect accuracy with an F1 score of 99.999%. Cross-validation confirms consistent high performance, with mean test F1 score and macro F1 score values of 99.999% and 99.999%, respectively. The integration of Ada Boost classifier with Randomized Search CV in our proposed model offers a robust solution for DDoS attack detection, enhancing system and network security.

14.6.5.6 Voting Classifier

A voting classifier combines predictions from multiple ML models to detect DDoS attacks in network traffic. It improves accuracy and robustness by leveraging diverse models and using majority voting or weighted averaging for final classification. Our experiments using Ada Boost in the voting classifier achieved nearly 100% accuracy.

Proposed Correlation Algorithm:

1.	Start
2.	Calculate correlation matrix (n=6.5M, m=73) (c=nxm).
3.	Set correlation threshold (T=0.7.)
4.	Remove highly correlated features (i, j) where i != j, except Label.
5.	Rank remaining features by correlation with binary target variable y.
6.	Select the top 46 features with the highest correlation coefficients.
7.	Drop redundant and multicollinear columns based on correlation analysis.
8.	End.

14.7 RESULTS AND DISCUSSION

Based on our proposed research, we aimed to address the crucial issue of DDoS attacks in cloud environments. Our research focused on designing and implementing the ML-based system, Cloud Defender, to provide real-time detection and mitigation of these attacks (see Table 14.3).

We evaluated multiple ML models including LR, Decision Tree, Random Forest, XG Boosting, and Ada Boost. Performance assessment was done using various metrics such as accuracy, precision, recall, F1 score , true positives (TP), false positives (FP), false negatives (FN), and true negatives (TN).

TABLE 14.3
Selected features of DDoS-FD-23 datasets based on CC and RFE

Final Selected Features

Fwd Seg Size Min	Fwd Pkt Len Std	CWE Flag Count	Src Port
Fwd Pkt Len Mean	Dst Port	Init Bwd Win Byts	Tot Fed Pkts
Totlen Fwd Pkts	Fwd Pkt Len Std	TotLen Fwd Pkts	Tot Bwd Pkts
Fwd Seg Size Avg	Pkt Len Max	Init Fwd Win Byts	Subflow Fwd Pkts
Fwd Seg Size Min	Dst IP	ACK Flag Cnt	Subflow Bed Pkts
Fwd Act Data Pkts	Src IP	Fwd Header Len	
Bwd Pkt Len Min	Bwd Header Len	Pkt Len Var	
Fwd IAT Tot	Down/Up Ratio	Subflow Fwd Bytes	

To ensure the dataset's quality and suitability, we conducted comprehensive data preprocessing. This involved data cleaning, handling missing values, outlier detection, data normalization/scaling, removing duplicates, data pruning, feature engineering, and feature extraction. Additionally, we conducted correlation-based dimension reduction and RFE-based feature selection.

Furthermore, the dataset was divided into training, cross-validation, and testing sets. Approximately 60% of the data was allocated for training, allowing the models to learn patterns and relationships effectively. An additional 20% was set aside for cross-validation to evaluate the models' generalization capabilities and fine-tune hyperparameters. The remaining 20% was dedicated to the testing set, enabling the assessment of the model's performance on unseen data.

To evaluate the effectiveness of Cloud Defender, we conducted comprehensive assessments using multiple datasets, including CSE-CICIDS2017, CSECI-CIDS2018AWS, and CSE-CICIDS2019, and our own created DDoS-FD-23. The results obtained from our model (Tables 14.4 and 14.5) were compared to previously published research papers, highlighting the superior performance of Cloud Defender, particularly when compared to the CIC datasets and DDoS-FD-23 dataset that we developed. From the comparison results it is evident that our created DDoS-FD-23 dataset yielded superior results for both model training and testing.

We utilized a voting classifier, which combined predictions from multiple ML models. Notably, our experiments with the voting classifier, specifically using Ada Boost, demonstrated outstanding accuracy of 100% with 0 FP and 0 FN.

The research presented in this chapter reflects our efforts in addressing the challenges posed by DDoS attacks in cloud environments, showcasing the effectiveness and potential of Cloud Defender, which was developed as a result of our proposed research.

14.8 CONCLUSION AND FUTURE SCOPE

In this chapter, we tackle the significant issue of DDoS attacks in cloud environments. We propose an IDS that utilizes ML techniques to analyze performance and enhance the security of cloud computing. Through extensive testing on the different

TABLE 14.4
Comparative analysis of proposed ML models for CIC datasets

Model Name	Datasets	Accuracy (%)	Precision (%)	Recall (%)	F1 Score (%)	TP (%)	TN (%)	FP (%)	FN (%)
Logistic	CSE-CICIDS2017	86.23	86.28	83.32	84.74	83.33	98.03	1.97	16.67
Regress	CSE-CICIDS2018	82.53	82.75	87.27	84.95	87.27	82.72	17.27	12.72
ion	CSE-CIC-IDS2019	86.42	86.49	83.33	84.99	83.33	79.51	20.49	16.67
Decision	CSE-CICIDS2017	94.49	94.87	88.7	91.63	88.71	99.26	0.74	11.29
Tree	CSE-CICIDS2018	90.99	91.37	85.48	88.33	85.48	80.51	19.48	14.51
	CSE-CIC-IDS2019	85.25	85.93	79.67	82.74	79.67	85.93	14.07	20.33
Random	CSE-CICIDS2017	95.99	96.63	90.39	93.47	90.32	99.50	0.50	9.68
Forest	CSE-CICIDS2018	92.50	92.72	86.46	89.49	86.46	81.53	18.46	13.53
	CSE-CIC-IDS2019	92.99	93.73	83.61	88.43	83.61	92.73	7.27	16.39
XG	CSE-CICIDS2017	98.18	98.29	91.93	94.91	91.88	99.75	0.25	8.12
Boost	CSE-CICIDS2018	94.47	94.82	88.23	91.42	88.23	83.76	16.23	11.76
	CSE-CIC-IDS2019	94.19	94.55	83.87	90.48	83.87	94.55	5.45	16.13
Ada	CSE-CICIDS2017	98.97	99.45	96.6	97.8	96.81	99.88	0.12	3.19
Boost	CSE-CICIDS2018	95.93	96.55	93.33	94.92	93.33	88.66	11.33	6.66
	CSE-CIC-IDS2019	96.96	97.30	86.94	92.79	86.94	97.30	2.70	13.06

TABLE 14.5

Proposed ML models evaluation on DDoS_FD_23 dataset

Model Name	Accuracy (%)	Precision (%)	Recall (%)	F1 Score (%)	TP (%)	TN (%)	FP (%)	FN (%)
LR	99.27	99.31	99.45	99.38	99.45	96.56	3.44	0.55
DT	99.94	99.99	99.99	99.99	99.99	99.99	0.01	0.0011
RF	99.99	100.00	99.99	100	99.99	99.999	0.0004	0.00079
XG Boost	99.99	100	100	100.00	100	99.998	0.0015	0
Ada Boost	100	100	100	100	100	100	0	0

datasets and employing feature selection methods, we achieve impressive accuracy rates in detecting various types of DDoS attacks. Importantly, our model outperforms existing approaches, as demonstrated by comparisons with other research papers. We emphasize the crucial role of ML in countering DDoS attacks and discuss potential future developments in the field. Our ML-based system, Cloud Defender, is designed to establish a secure cloud ecosystem by employing advanced defense mechanisms, real-time attack detection, and swift mitigation measures. By integrating diverse ML classifiers, we strengthen the overall security posture and contribute to broader research advancements. Our research makes a valuable contribution to improving DDoS detection systems and enhancing the security and resilience of cloud computing environments.

The study is also limited to DDoS detection in cloud environments, potentially limiting its applicability to other types of cyber threats or network settings. While efforts were made to improve the dataset's quality, it may not encompass the full range of real-world attack scenarios. The generalizability of the findings is therefore constrained. Furthermore, the evaluation of the ML-based system, Cloud Defender, was done in a controlled experimental environment, requiring further verification of its real-world implementation and scalability in different cloud environments.

Future research should aim to address limitations by adapting ML algorithms for emerging threats, exploring advanced feature selection techniques, enhancing scalability and real-time detection capabilities, optimizing ML algorithms for resource-constrained environments, and extending research beyond DDoS attacks to encompass other cyber threats in cloud computing environments.

REFERENCES

1. Voorsluys, W., Broberg, J., and Buyya, R. (2011). Introduction to cloud computing. *Cloud Computing: Principles and Paradigms*, pages 1–41. Wiley.
2. Singh, A. and Chatterjee, K. (2017). Cloud security issues and challenges: A survey. *Journal of Network and Computer Applications*, 79:88–115.
3. Deshmukh, R. V. and Devadkar, K. K. (2015). Understanding DDoS attack & its effect in cloud environment. *Procedia Computer Science*, 49:202–210.
4. Mirkovic, J. and Reiher, P. (2004). A taxonomy of DDoS attack and DDoS defense mechanisms. *ACM SIGCOMM Computer Communication Review*, 34(2):39–53.
5. Roschke, S., Cheng, F., and Meinel, C. (2009). Intrusion detection in the cloud. In *2009 eighth IEEE International Conference on Dependable, Autonomic and Secure Computing*, pages 729–734. IEEE.
6. Pradhan, M., Nayak, C. K., and Pradhan, S. K. (2020). Intrusion detection system (IDS) and their types. In *Securing the Internet of Things: Concepts, Methodologies, Tools, and Applications*, pages 481–497. IGI Global.
7. Hung-Jen Liao, Chun-Hung Richard Lin, Ying-Chih Lin, Kuang-Yuan Tun. (2013). Intrusion detection system: A comprehensive review. *Journal of Network and Computer Applications*, 36(1):16–24.
8. Cavusoglu, H., Raghunathan, S., and Cavusoglu, H. (2009). Configuration of and interaction between information security technologies: The case of firewalls and intrusion detection systems. *Information Systems Research*, 20(2):198–217.

9. Om Prakash, Kumar, M. (2023). Machine learning based theoretical and experimental analysis of DDoS attacks in cloud computing. In *2023 International Conference on Device Intelligence, Computing and Communication Technologies, (DICCT)*, pages 526–531.

10. Salman, T., Bhamare, D., Erbad, A., Jain, R., and Samaka, M. (2017). Machine learning for anomaly detection and categorization in multi-cloud environments. In *2017 IEEE 4th international conference on cyber security and cloud computing (CSCloud)*, pages 97–103. IEEE.

11. Noh, S., Lee, C., Choi, K., and Jung, G. (2003). Detecting distributed denial of service (DDoS) attacks through inductive learning. In Liu, J. et al., editors, *IDEAL 2003*, volume 2690, pages 286–295. Springer-Verlag Berlin Heidelberg.

12. Amaad, Hafiz Mughal, a. H. (2023). Experimenting ensemble machine learning for ddos classification: Timely detection of DDoS using large scale dataset. In *2023 4th International Conference on Advancements in Computational Sciences (ICACS)*, pages 1–7. IEEE.

13. Usha, G., Narang, M., and Kumar, A. (2021). Detection and classification of distributed DoS attacks using machine learning. In *Computer Networks and Inventive Communication Technologies: Proceedings of Third ICCNCT 2020*, pages 985–1000.

14. Bhaya, Wesam EbadyManaa, M. (2017). DDoS attack detection approach using an efficient cluster analysis in large data scale. In *2017 Annual Conference on New Trends in Information & Communications Technology Applications (NTICT)*, pages 168–173. IEEE.

15. Mahadi, Eabaadi Manie, A. (2016). A proactive DDoS attack detection approach using data mining cluster analysis.

16. He, Z., Zhang, T., and Lee, R. B. (2017). Machine learning based DDoS attack detection from source side in cloud. In *2017 IEEE 4th International Conference on Cyber Security and Cloud Computing (CSCloud)*, pages 114–120. IEEE.

17. Hoque, N., Bhattacharyya, D. K., and Kalita, J. K. (2016). FFSc: A novel measure for low-rate and high-rate DDoS attack detection using multivariate data analysis. *Security and Communication Networks*, 9(13):2032–2041.

18. Osanaiye, O., Choo, K.-K. R., and Dlodlo, M. (2016). Analysing feature selection and classification techniques for DDoS detection in cloud. *Proceedings of Southern Africa Telecommunication*, pages 198–203. SATNAC.

19. CIC-Datasets. University of New Brunswick (Canadian Institute for Cybersecurity). www.unb.ca/cic/datasets/index.html, accessed on December 01, 2023.

15 GBDT-Based Approach for DDoS Attack Prevention in SDN-Based Cloud Virtual Machine

Sudakshina Mandal, Danish Ali Khan, and Rakesh D. Raut

15.1 INTRODUCTION

The foundation of the cloud computing has created contemporary and compelling opportunities in the world of information and communication technology (ICT) environment. Individuals, small and medium-sized businesses, and large corporations can now lease processing, storage, and network resources on an as-needed basis. Whether it can be the deployment of any service or it can be any software-based services as well as any server-oriented needs or data storage, cloud computing has enormously replaced the conventional system with distributed nature. One of the significant attributes is to serve as a data center with high computational capacity. Reliability and availability are two significant parameters when it comes to the topic of cloud data centers. However, the access of the data center is distributed in a cloud computing nature irrespective of many countries and many networks where the individual network is exclusive and unforeseeable by its characteristics. This is made happen due to advancements in virtualization technologies; precisely the VMs can be considered as the rudimentary element for providing services to the consumers from the data centers. Deployment of cloud VM is very significant for maintaining the quality of the services of cloud data centers. Individual consumers are associated with the individual or several distributed VMs in cloud architecture [1]. VM abbreviates the ongoing current system and creates an abstract for the users by hiding the complexities of underlying physical hardware details. These VMs can be allocated in a dynamical and statistical way across dispersed cloud data centers infrastructure to improve application performance for paying customers while efficiently utilizing the service provider's physical resources and alleviating bottlenecks. By focusing to improve the server-side resources, a live VM migration strategy has been used where the researchers must look into the better utilization of the resources and power consumption [2]. A consolidation technique can be proposed by minimizing the number of network switches, which can be turned on in any stipulated time [3]. However, the VM allocation affects the traffic dynamics in this regard.

The traditional computing network of cloud systems faces numerous issues regarding the arrangement of networking resources as it requires human intervention

DOI: 10.1201/978100339054-15

and overdue efforts. The emerging paradigm of Software-Defined Network (SDN) framework is considered as a significant option for operating in highly dynamic environments which makes the network more configurable and programmable. SDN can be easily accommodated in cloud data centers due to the concept of wide network abstraction where the logic of the control plane has been hidden for fast service, virtualization, and deployment. SDN framework comprises the strategy of network control logic into centralized software components by enabling policies, configuring, and helping in network resource management which can be programmed in short timescales. Openflow switches of SDN interconnect the controller and the different VMs in the network which defines the rules of dynamic architecture of networking needs to be updated into the physical cloud network hardware layer in order to change the configuration periodically [4]. SDN-based VM strategy significantly reduces the network traffic between the VMs and hosts due to the dynamic placement of VMs over the network topology. Conscientious placement of VMs significantly reduces network traffic [5]. On the other hand, the SDN framework is considered as a network-centric architecture where the VMs or hypervisors cannot reach out to the SDN directly for information of the temporal network state [6]. This framework can be used to secure a large amount of resources without any disruption of network architecture or any other threats. Existing cloud data centers prove their dynamicity in this regard in distributed scenarios by enhancing the scalability through introducing SDN architecture [7].

As the cloud data centers provide hosting services of server-centric equipment to the consumers, they should maintain a strong security paradigm in order to engage tenants without exposing potential information through some data leakage. While cloud data centers are faced with different security issues, however, in the current cyber scenario, the entire world is affected by numerous cybersecurity attacks based on different network infrastructures. This violates the availability of the network and by creating unnecessary network traffic by flooding with huge volumes of data packets in VMs, the attacker creates false traffic and bottlenecks the network bandwidth. This network data packet consumes the VMs potential resources of networking information such as storage and computing information. The CIA triad of network information such as confidentiality, integrity, and availability of the network packets also got hampered by these attacks. Distributed denial of service (DDoS) attack is considered as the most significant attack which is used to obstruct the network availability by violating several source machines together. DDoS attacks are eminent malicious cyber attacks which are straightforward to launch, cause catastrophic service outages, and are difficult to trace back to the source of the attackers. In this chapter, network-based DDoS attacks in cloud-based VMs have been focused on. With the incorporation of the SDN framework in the designing of the network topology of VMs in cloud data centers, traditional DDoS prevention approaches have been transformed with "Cyber Agility and Defensive Maneuver" methods.

Numerous studies are being conducted to detect DDoS attacks on SDN networks [8]. Different machine learning algorithms have also been employed for the detection of DDoS attacks on the server side of cloud data centers. Support Vector Machine (SVM) is a comparatively used method that has quite a good result for the detection of DDoS attacks, proving less false-positive rate and faster detection. Apart from SVM, random forest is also considered one of the significant supervised learning methods to be used for detection.

However, SVM has certain constraints while implementing the DDoS detection mechanism. Some network traffic attributes are multivalued attributes when the SVM algorithm is used to detect DDoS attacks on an SDN network. The SVM, on the other hand, was initially designed for binary classification. As a result, multiclass classification is a significant issue when using SVM. The second issue is the SVM algorithm's lengthy training and testing time. The SVM classifier performs well in terms of classification accuracy and false-positive rate. However, the SVM algorithm requires more time to train and test attack detection. Random Forest takes less time to prepare and test the traffic to detect the attack in this scenario. Apart from this, both these technologies have become obsolete where the advanced machine learning technologies have taken part.

By addressing the above-stated issues, this chapter designed a DDoS attack prevention approach based on SDN cloud VM by using Gradient Boosting Decision Tree (GBDT) along with principal component analysis (PCA). For dimensionality reduction and extraction of essential features from the dataset, PCA is used. It is proven in this chapter that GBDT needs less time than other popular machine learning techniques, and its accuracy is better than others. As the proposed method is deployed in the server-side SDN-based cloud data centers, it helps to sustain the availability and reliability of the distributed cloud data centers efficiently by analyzing the network traffic and eliminating DDoS attack-producing malicious traffic which in turn can affect the different VMs allocated in a distributed cloud network.

The contributions are illustrated below.

1. Proposing of machine learning-based DDoS attack prevention approach to shield the SDN-based cloud virtual machines.
2. GBDT has been used as a machine learning algorithm.
3. PCA has been used along with GBDT for dimensionality reduction and extraction of the essential features from the dataset.
4. This approach proves to have high accuracy (98.884%) and outperforms all the existing approaches in the literature.

The rest of the chapter is organized as follows: Section 15.2 describes the preliminaries required for constructing the chapter, while Section 15.3 illustrates the literature review. Section 15.4 describes the proposed framework, while section 15.5 presents and analyzes experimental results and discussion. Finally, Section 15.6 concludes the chapter.

15.2 PRELIMINARIES

In this section, some of the prerequisite algorithms and their workflows have been described, which will be further used by the proposed method.

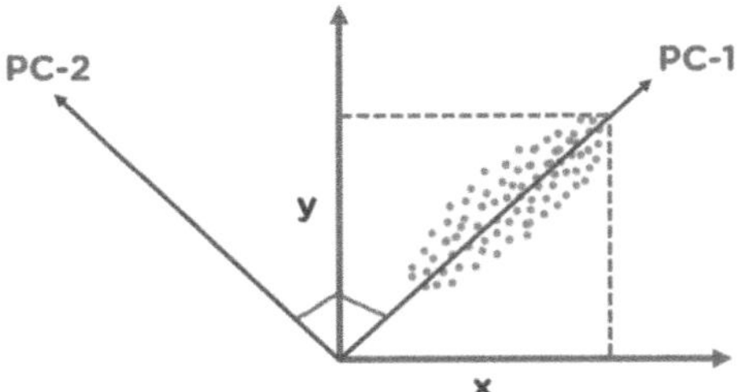

FIGURE 15.1 Maximum variance of PCA [9].

15.2.1 SIGNIFICANCE OF USING PRINCIPAL COMPONENT ANALYSIS IN DIMENSIONALITY REDUCTION

To improve the efficiency of SVM classification, PCA, an unsupervised learning technique, has been used for dimensionality reduction, which minimizes the dataset's complexity by losing significant properties. PCA is considered one of the important methods for extracting features from network traffic. PCA helps to increase interpretability and minimizes information loss. PCA helps find the significant features in a dataset and makes the dataset easy for plotting in a two-dimensional (2D) or three-dimensional (3D) plane. It reveals the relationship between the huge amount of sample datasets and multivariate datasets. The generic definition of PCA is as follows.

Let random variables be $X1, X2, , Xp$ and sample deviations be $S1, S2, , Sp$. The standard definition is as follows:

$$C_j = a_{j1}x_1 + a_{j2}x_2 + + a_{jp}x_p \quad where\, j = 1, 2, , p.$$

1. Now, if the value of $C1$ satisfies the above equation and the var$(C1)$ is largest, then $C1$ will be considered as the first component of PCA.
2. If the value of $C2$ satisfies the above equation and the var$(C1)$ is largest, with $(a21, a22, ..., a2p)$ being perpendicular to $(a11, a12, ..., a1p)$, then $C2$ will be considered as the second component of PCA.

From Figure 15.1, it is observed that many points can be plotted in the $2D$ plane. Two principal components can be found in the diagram. The primary component is $PC1$ which produces the maximum variance between data, whereas $PC2$ is considered as another significant component which is orthogonal to $PC1$.

The procedure of finding the optimum values using PCA is explained below [10].

1. **Standardization of Dataset**

   PCA is utilized for the identification of components with the maximum variance and the contribution of each variable to a component is based on the magnitude of variance. Normalization is required to scale up the unscaled dataset with different measurement units which distorted the relative comparison of variance across features. It ensures that the respective feature has a mean = 0 and variance =1. Equation 15.1 describes this step here.

$$Z = \frac{Value - mean}{standard\ deviation} \tag{15.1}$$

2. **Calculation of Covariance Matrix**

    In this step, the relationship between the input datasets has been obtained for calculating covariance among the input datasets and plotting them in a covariance matrix shown in equation 15.2 [11]. Interpreting the cumulative variance percentage calculation captured by the respective principal components is a significant part of reducing the features of the datasets.

$$\begin{bmatrix} cov(p,p) & cov(p,q) & cov(p,r) \\ cov(q,p) & cov(q,q) & cov(q,r) \\ cov(r,p) & cov(r,q) & cov(r,r) \end{bmatrix} \tag{15.2}$$

3. **Finding of Eigenvectors and Eigenvalues**

    Eigenvectors and eigenvalues are the components of linear algebra that we need to compute from the covariance matrix to obtain the principal components of the datasets. Eigenvalues are scalars by which we multiply the eigenvector of the covariance matrix.

4. **Feature Vector**

    This step computes the eigenvectors and order them by their eigenvalues in descending order for finding the principal components.

    Here either all components have been chosen or discarded those whose values are lesser significant and create feature vector Y by using the remaining components. The feature vector is considered as a matrix whose values are taken from the columns of the eigenvector. This is the first step of the dimensionality reduction procedure. If P eigenvector has been chosen out of N for the final dataset, it contains only P dimensions.

Here, D represents the features of the original dataset, whereas K represents the dataset containing only the reduced features after deducing the dimensionality using PCA. PCA helps to find the correlation between the variance and the input data of the dataset. The ratio of k and δ has been represented by T here in equation 15.3. Fewer values of K result in the maximum reduction of features in the final dataset.

$$T = K/\delta \tag{15.3}$$

The variance of the dataset gets decreased when the value of K decreases. By using the above equation in this case the values of K is represented as the maximum variance in the dataset.

15.2.1.1 Finding a Suitable Value for K

Here $P(i)$ is considered as the original features in the dataset and $p_{approx}^{(i)}$ is the projected features. So, the squared projection error can be represented by equation 15.4.

$$1/m \sum_{i=1}^{m} P(i) - p_{approx}^{(i)} ||2 \qquad (15.4)$$

The total variation of the dataset will be represented by equation 15.5.

$$1/m \sum_{i=1}^{m} x^{i} ||2 \qquad (15.5)$$

Choosing K as 99% by keeping the data percentage of variance high is a wise decision in this regard.

15.2.2 GRADIENT BOOSTED DECISION TREE

The name of the technology signifies two sub-terminologies: gradient and boosting. The GBDT is considered one of the boosting techniques. As the boosting technique suggests, it is nothing but a concept where the weak learners can be modified to perform better [12]. It modifies the concept of boosting by a numerical optimism problem. Decreasing the loss function of the model is the primary objective of boosting. To find the minimum differentiable function, a gradient descent algorithm has been used which is considered as a first-order iterative algorithm.

It is considered an additive stage-wise model which is used to generate learners during the training phase. The objective of the weak learner to the ensemble is based on the gradient descent optimization process [13]. Minimizing the overall error of the strong learner is the primary contribution here. GBDT does not modify the sample distribution as weak learners train on the remaining residual errors of a strong learner. Giving the training to the model's residuals gives more importance to misclassified observations [14]. Here, new weak learners are added for concentrating on the areas where the existing learners are performing not systematically. The working of the GBDT has been illustrated in Figure 15.2.

To summarize, it consists of three modules:

1. Optimization of the loss function.

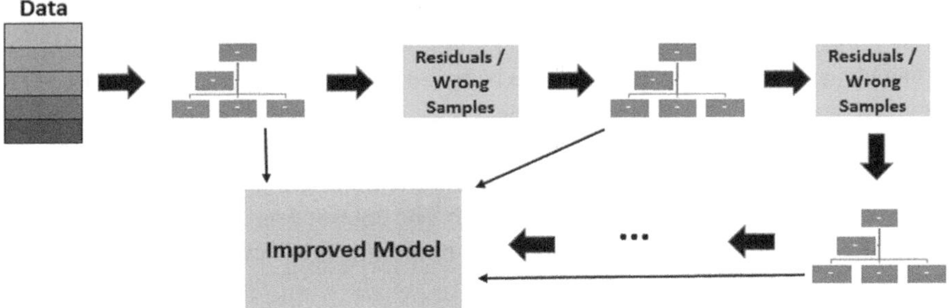

FIGURE 15.2 Workflow of GBDT [15].

2. Make predictions based on the weak learners' set.
3. Minimization of loss function by adding the model to the set of the weak learner.

15.3 LITERARURE REVIEW

In this section, we rigorously investigate existing works in the literature and some of the significant works have been briefed here for correlation with our research. It is found from the literature that DDoS attack prevention methodology in SDN-based cloud VM framework is merely investigated. Most of the existing methods are based on the client-side architecture in a cloud environment. Moreover, the existing piece of work follows SVM kernel, logistic regression, random forest, and decision tree as the conventional machine learning-based algorithms. The concept of SDN-based VM is comparatively new, and we did not find much work in this particular domain in the literature. However, the use of a GBDT brings innovation and keeps the accuracy high in the field of detecting DDoS attacks in source-side SDN-based VM.

Zecheng et al. [16] proposed a DDoS attack detection method based on Random Forest. Different supervised and unsupervised learning algorithms have been compared here to prove the best accuracy of their model. A real cloud platform simulation provides a practical idea of a DDoS attack. This method is based on the source-side cloud platform. Different types of DDoS attacks such as Transmission Control Protocol (TCP) SYNC attacks, Internet Control Message Protocol (ICMP) flooding attacks, Secure Shell (SSH)-brute-force attacks and Domain Name System (DNS) reflection attacks have been experimented by this model and proven the accuracy of 99.7%.

Based on the SDN-based framework, Jehad et al. [7] proposed a DDoS attack prevention approach. They used SVM as a machine learning technique. PCA was used to classify the incoming IP traffic as per the requirement. They proved the robustness of the algorithm based on the DARPA dataset [17]. They demonstrated the efficiency of their model by comparing it with the normal SVM technique and proved that the accuracy of their model is around 95. The training timing in their study was 120 sec.

Another significant DDoS attack prevention method is proposed by Myo Myint et al. [8] by introducing an advanced SVM in the SDN framework. Only the flooding-based DDoS attack was tested by the SDNTrafficDS dataset [8]. The proposed method was implemented in the mininet emulator as an emulator. The accuracy was 97% for this algorithm. The flooding DDoS attack tested User Datagram Protocol (UDP) flooding attacks, TCP flooding attacks, and SYN flooding attacks.

Hakak et al. [18] proposed an anomaly-based DDoS attack detection method. The KDD CUP dataset [19] was used in this paper. This dataset used a twofold classification method where 20% of the original dataset was used for training and 10% was used for testing the dataset. Artificial neural network (ANN) was used as the machine learning technology for the proposed model. The dataset was chosen uniformly from the original dataset set so that regular and malicious instances were approximately the same in number in each dataset.

Another significant work is developed by Wahab et al. [20] based on cloud VM by focusing on the mitigation of DDoS attacks. An innovative approach was used by the researchers here. A trust model was built up and Bayesian inference was used for execution of the model and to build up trust between hypervisors and VMs. By using the trust-based maximin game model the research maximized the detection rate of DDoS attacks and experimentally proved that it takes 4.4 milisecond time to detect DDoS attacks in a 50 co-hosted VM-based cloud system.

By maintaining a similarity, another significant work is proposed by Nguyen et al. [21] based on SDN-driven Moving Target Defense (MTD)-based network obfuscation scheme in a cloud VM platform. Firstly, spawning has been down to create VM replica across the IP subnets and, secondly, it has been exhibited that optimal VM selection improves the unpredictability of DDoS attack. This approach was implemented in the GENI cloud testbed by using open vSwitches over an SDN-based controller and proved the accuracy by comparing with static resource-based defense models; it performs two times better than others.

The SDN-based VM placement strategy named SDNVMP was adhered by Anderson et al. [22] in cloud data centers. Based on the model, it reduced the security risk associated with distributed VMs in cloud data centers and reduced the network traffic also. This research was experimented with contemporary works such as first fit (FF) and energy-efficient and QoS-aware algorithm, EQVMP. Experimentally the authors have proven that SDNVMP outperforms all other approaches.

The SDN-based VM in a cloud system is often flooded with Transmission Control Protocol (TCP) SYN flooding attacks. Shah et al. [23] proposed a statistical anomaly detection methodology EDOS-TCP SYN which is used to detect SYN flooding attacks by examining individual network traffic. This method was compared with existing literature by the researcher and it was found that the method generates a large number of false negatives and proves efficiency in source SDN-based cloud.

Ahalawat et al. [24] proposed a low-rate DDoS attack prevention method established on Renyi Entropy with Packet Drop technique (REPD) where probabilistic distributions were taken into consideration while measuring fluctuations of network traffic. Experimental results and comparative results proved the accuracy and robustness of this research here.

By taking the limitations from the above-mentioned pieces of literature, we designed a GBDT-based DDoS attack prevention method based on SDN cloud VM, which in turn protects the cloud data centers.

15.4 PROPOSED FRAMEWORK

In this chapter, a novel GBDT-based DDoS prevention methodology has been proposed for protecting cloud data centers. As the cloud infrastructural details of cloud data centers have been shifted to an SDN-based platform, the proposed approach will be discussed based on VMs which are connected to the cloud data centers with the help of the SDN framework through Openflow switches. Openflow switches of SDN interconnect the controller and the different VMs in the network which defines the rules of dynamic architecture of networking needs to be updated into the physical cloud network hardware layer in order to change the configuration periodically. The SDN controller manages a global insight of the underneath network [25]. However, failure of the SDN controller makes the entire network impassive and the attacker can easily penetrate into the network through vulnerable points by spoofing the controller and modifying the network table which elevates bandwidth and creates huge anonymity. As cloud data centers provide hosting services of server-centric equipment to consumers, it should maintain a strong security paradigm in order to engage tenants without exposing potential information through some data leakage. The proposed approach has been implemented in the cloud data centers by aiming to resolve the above-stated issue. This prevention approach has been employed with the help of one of the significant machine learning modules named GBDT by preceding the dimensionality reduction for the extraction of essential features from the dataset using PCA. The machine learning framework consists of two significant modules: The first one is the training module which is trained to distinguish the DDoS attack from the dataset that occurs in individual VM in advance. The testing module, which is considered the second module, is updated to introduce the training model. The testing modules use a monitored dataset as a training sample to modify the parameters of the trained module. Based on the framework of SDN architecture controller policy has been kept separate from the machine learning module. The entire system has been built up at the end of cloud data centers. Statistical information derived from the network can arrive at the cloud data centers through a network router which can be hosted in the gateway of the cloud platform.

However, the network packets coming from the respective client's machine, are considered to be normal in nature. The proposed approach cannot break the privacy of the inbound and outbound traffic. In the statistical information collection phase, the required details have been fetched and stored in the VM. The most significant statistical information is the port address, incoming IP address, and control signals for our methodology. The block diagram presents the architectural details of the proposed framework in Figure 15.3.

15.4.1 Workflow of the Proposed Framework

Disrupting the availability of the services of SDN- based cloud data centers and bottlenecking the network traffic are considered the prime objective of DDoS attacks. DNS services, HTTP services, and FTP services are some of the significant network services that can be easily tampered with by DDoS attacks significantly.

As all the VMs are connected to the SDN-based controller residing in the cloud data center by OpenFlow switches, the DDoS attacks occurring in VM, in turn, can maliciously infect the SDN-based controller when they are connected via OpenFlow switches. These DDoS attacks can rattle the entire cloud network and create congestion in the network bandwidth of cloud data centers. In cloud architecture, the DDoS attack can be generated by gaining complete control of the entire VM or utilizing a particular VM as a bot to generate a DDoS attack. The identity is kept secret by the attacker while penetrating the VM. As the attacker cannot have the physical acquisition of the cloud data centers, remote penetration is used at the time of a DDoS attack.

The central SDN controller monitors the status of the connected VMs and evaluates the statistical information coming through the router in a cloud platform. As per the

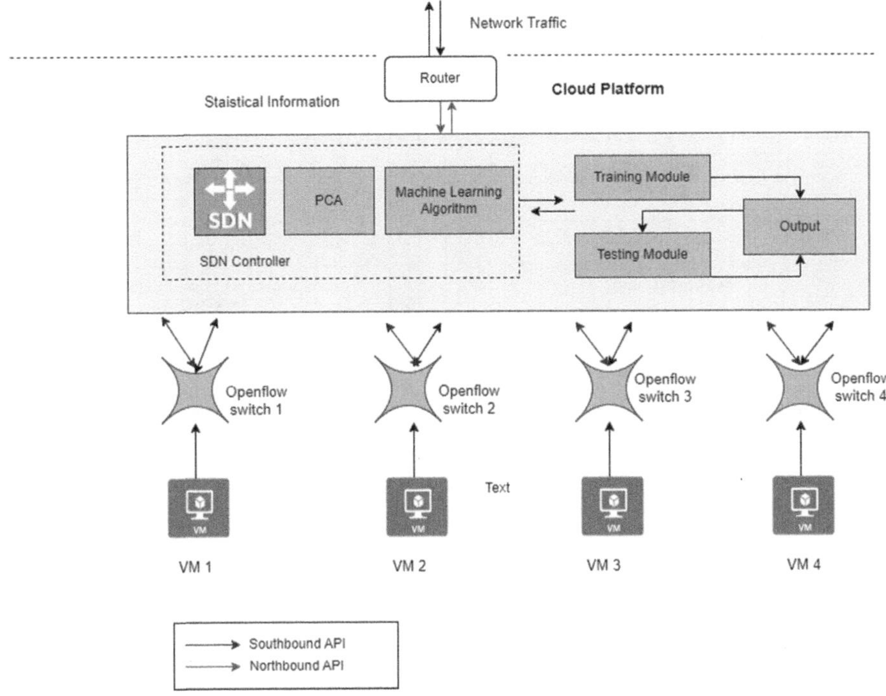

FIGURE 15.3 Proposed framework.

proposed method, all the statistical information collected from the cloud platform must be fed into the proposed approach. The proposed framework is divided into two parts. The first unit is dedicated to the filterization of the dataset using PCA, which is responsible for the dimensionality reduction of the incoming network traffic from the dataset. The classified dataset is further redirected to the next unit, that is, machine learning unit. This unit is responsible for the detection of suspicious network traffic which comes from the VM. GBDT is used by this approach to find malicious traffic in the cloud environment. If it is found that the DDoS attack is limited only to a specific VM, then as per the proposed framework, the specific VM is terminated. The DDoS attack is almost certainly underway if the suspicious behavior is detected across multiple VMs. As a result, the network connection should be suspended to prevent the DDoS attack as per the proposed framework.

Different phases of the proposed methodology have been illustrated below. The entire methodology has been described based on the dataset "Dataset_SDN.csv." All the computations based on the dataset have been performed in the Databrick community edition platform. Figure 15.4 exhibits the step-by-step workflow of the proposed approach.

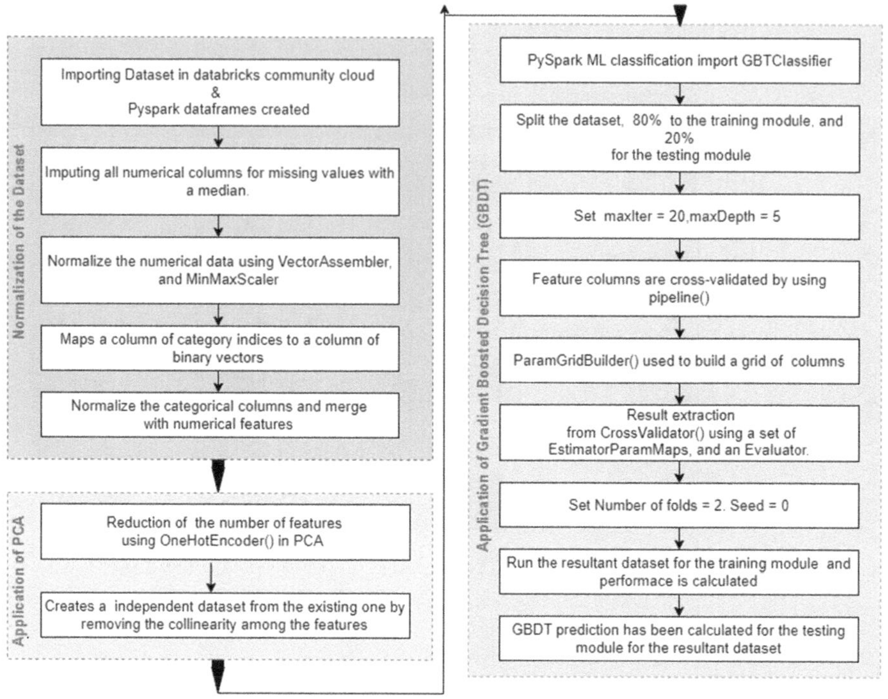

FIGURE 15.4 Workflow of the proposed approach.

Phase I: Normalization of the Dataset

1. Uploading of the *dataset_SDN.csv* file and retrieving all the features from the .CSV file and *PySpark* data frames have been created to access those features as per the corresponding columns.
2. Importing of *Pyspark* library to databricks community cloud edition and returning the total number of rows in the .CSV file.
3. Imputing all numerical columns for missing values with a median.
4. Normalizing the numerical data using *VectorAssembler* and *MinMaxScaler*. *VectorAssembler* helps to transform the list of columns into one single vector column. Scaling of all the datasets into 0 to 1 has been done using the *MinMaxScaler* function by preserving the shape of the original distribution.
5. Mapping a column of category indices to a column of binary vectors and merging multiple columns into a vector column.
6. Normalizing the categorical columns.
7. Combining categorical and numerical features.

Phase II: Application of PCA over Normalized Dataset

1. PCA library has been imported from PySpark machine learning features.

2. It will reduce the number of features by using *OneHotEncoder* method.
3. This phase creates a completely independent dataset from the existing one by removing the collinearity among the features if any exist. Independent features are one of the underlying assumptions for the Linear Regression Algorithm.

Phase III: Application of Gradient Boosted Decision Tree over the Reduced Dataset

1. From PySpark machine learning classification, import *GBTClassifier*.
2. Split the dataset where **80%** data is assigned to the training module, and **20%** data is reserved for the testing module.
3. Initialize the resulting feature column after applying PCA using *GBTClassifier* where *maxIter* = **20** and *maxDepth* = **5** is set.
4. Feature columns are cross-validated over several steps using *pipeline*().
5. *ParamGridBuilder*() is used by GBDT to build a grid of featured columns for grid-search-based model selection.
6. Extraction of the result from *CrossValidator()* by using a set of *Estimator ParamMaps*, and an *Evaluator*.
7. Number of folds for the *CrossValidate* has been taken as **2**, and the *seed* is calculated as **0**.
8. Run the resultant dataset for the training module first.
9. Resulting performances have been derived from the training module.
10. In a similar way, GBDT prediction has been calculated for the testing module for the resultant dataset.
11. Calculate the accuracy and *AOC_ROC* based on the test data.

15.5 RESULT AND DISCUSSION

Databricks community cloud has been used by this proposed approach as a platform for implementation along with the Spark application interface for development. For the proposed approach we use Intel 2.30 GHz i3 processor with 8 GB RAM along with 1TB HDD as a basic configuration. Databricks cloud edition is a free version of Databrick's cloud-based Big Data platform for the enterprise. It helps to create a micro-cluster which is based on the configurable versions of Apache Spark. Databricks community cloud also helps to create, manage, and execute Spark code with more efficiency free of charge. This community edition has been designed for developers, data scientists, and data engineers to perform Spark applications.

Dataset_SDN [26] has been taken as a dataset by this approach, which has a total of 104,345 different DDoS attack entries. This dataset has ICMP flooding attacks, SSH brute-force attacks, TCP SYN attacks, and DNS reflection attacks as divergent types of DDoS attack examples. These datasets have been filtered and normalized by using PCA for further use by the GBDT-based machine learning model. In this machine learning module, twofold GBDT is used as a machine learning algorithm in this chapter. We have taken 80% data for training and 20% for the testing module

TABLE 15.1
List of columns presented in Dataset_SDN

Column Name	Description
dt	Date and Time
switch	ID in the topology
src	IP of Source
dst	Destination IP
pktcount	No. of packets sent in a transmission
bytecount	No. of bytes sent in a transmission
dur	Time duration of the flow in the switch
dur_nsec	Duration in nanoseconds
tot_dur	Summation of dur_sec along with dur_nsec
flows	Inflow of packets between source IP and destination IP
packetins	No. of packet_in transmitted to the controller
pktperflow	No. of total packets in a flow
byteperflow	Counting of packets in bytes in a flow
pktrate	Packets sent per second
Protocol	Protocol associated with the traffic
port_no	Port address of the switch
tx_bytes	No. of bytes sent on a specified port
rx_bytes	No. of bytes transmitted through a port
tx_kbps	Kilobytes transferred per second
rx_kbps	Kilobytes received per second
tot_kbps	Total bandwidth in kilobytes of a port
label	Identification of benign and malicious flow

in the process of a twofold approach. Significant dataset columns are presented in Table 15.1.

By maintaining the above-stated protocols for this approach, GBDT takes 1.30 minutes to train the above dataset and 0.27 seconds to test the dataset. The evaluation of the area under the curve-receiver operating characteristic (AUC-ROC) curve shows 98.884% accuracy for this dataset.

Different machine learning approaches have been compared with our proposed approach based on the above dataset for analysis of the accuracy and performance. Logistic Regression, SVM with linear kernel, and Random Forest methods have been chosen by rigorous analysis of the literature survey to compare the efficiency of our proposed approach. All the divergent approaches have been experimented on the normalized dataset which has been transformed after PCA. To maintain the similarity with our proposed approach, these algorithms have been compared based on the training and testing modules. Trial and error methods have been followed while calculating the value of the maximum iteration, the number of trees, and the maximum depth.

Table 15.2 illustrates the comparative analysis results of different machine learning approaches. Performance and accuracy have been calculated and compared based on training time and testing time. Based on that AUC-ROC evaluation has been calculated and compared.

TABLE 15.2
Comparative results of different machine learning algorithms

Algorithm	Training Time	Testng Time	AUC-ROC Evaluation
GBDT	1.30 min	0.27 sec	98.884%
Random Forest	59.54 sec	0.21 sec	97.679%
Logistic Regression	2.73 min	6.46 sec	82.947%
SVM kernel	1.18 min	7.81 sec	75.780%

It is evident from the table that among the four best-suited machine learning algorithms, GBDT outperforms all. Though the random forest model takes less time for training and testing, the accuracy is still better for GBDT, 98.884%. Random forest is considered the best suitable machine learning algorithm, producing 97.679% accuracy. However, the logistic regression and SVM kernel model do not perform well in this case, producing an overall accuracy of 89.947% and 75.780%, respectively. The time taken for training and testing is also relatively high than GBDT and Random Forest.

The comparative accuracy analysis is presented in Figure 15.5. Receiver Operating Characteristic (ROC) curve is used here to exhibit the accuracy analysis shown in Figure 15.6. Here it computes the true positive rate (detection rate) against the false positive rate (FPR) of the determined values from the methods. It is observed that our proposed method has a high detection rate compared to other methods.

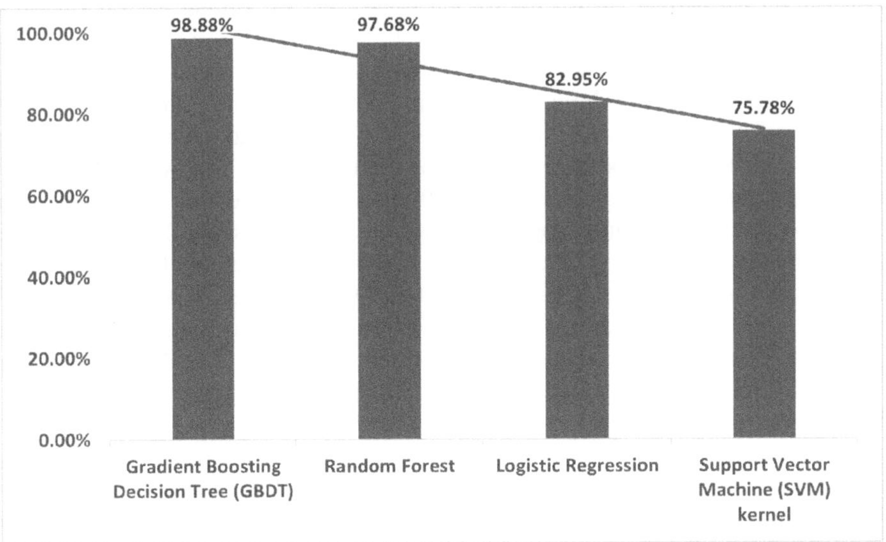

FIGURE 15.5 Comparative analysis of overall accuracy.

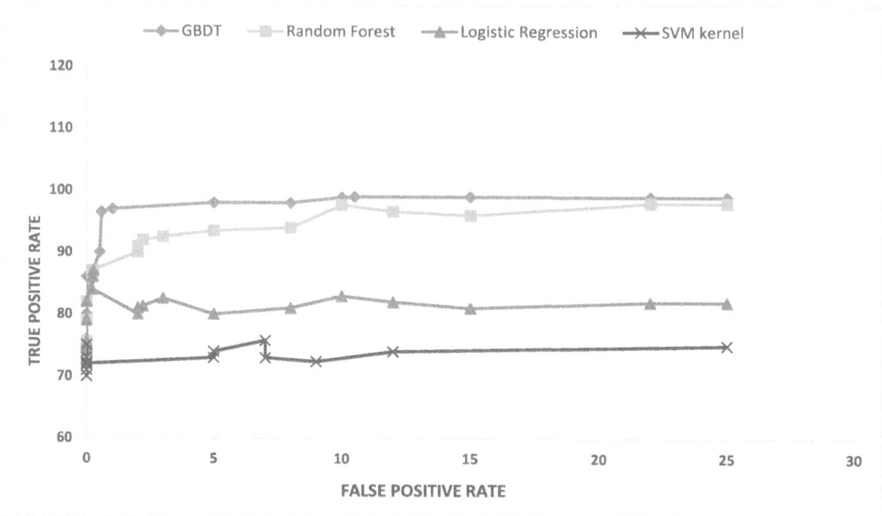

FIGURE 15.6   Comparative ROC curve analysis.

15.6   CONCLUSION

The DDoS attack has been considered as one of the significant attacks in cyberse-curity that violates the cyber-physical infrastructure of cloud computing, and these attacks threaten the whole world. Considerable research work has been done to prevent DDoS attacks in the client's machine's application layer. However, the cloud infrastructures residing on the server side are equally violated by these attacks, which in turn barred the availability of the cloud services coming from the servers. Creating a prevention mechanism based on the destination is considered a passive defense that cannot detect the attacks that have already penetrated the cloud server and the malicious services provided to them [27]. So creating a mechanism based on the cloud VM is essential. Most cloud server infrastructure has been shifted to the SDN-based framework from the traditional foundation. It follows the concept of a separated data plane and control plane. By using the concept of SDN, individual cloud VM has been designed nowadays. This chapter proposed a DDoS attack prevention methodology based on this SDN-based VM in the cloud. GBDT has been used as the latest machine learning algorithm. PCA has been applied here for the filterization of essential features. The proposed framework has been implemented in Databricks community cloud, a distributed cloud framework. The Pysperk interface is used here for the implementation of the machine learning algorithm. Dataset_SDN has been used, which has a 104345 DDoS attack dataset. The dataset comprises divergent DDoS attack instances such as TCP SYN attacks, ICMP flooding attacks, SSH brute-force attacks, and DNS reflection attacks. Our proposed method successfully detected malicious traffic among them. Other algorithms have been compared to prove the efficiency of our proposed method. Our approach proves to have a high accuracy (98.884%) and outperforms all.

REFERENCES

1. Wen-Kuei Hsieh, Wen-Hsu Hsieh, Jiann-Liang Chen, Feng-Yi Chou, and Yung-Sheng Lee. Load balancing virtual machines deployment mechanism in SDN open cloud platform. In *2015 17th International Conference on Advanced Communication Technology (ICACT)*, pages 329–335, Phoenix Park, PyeongChang, South Korea, July 2015. IEEE.

2. H.M. Anitha and P. Jayarekha. SDN based secure virtual machine migration in cloud environment. In *2018 International Conference on Advances in Computing, Communications and Informatics (ICACCI)*, pages 2270–2275, Bangalore, September 2018. IEEE.

3. Stefanos Kiourkoulis. DDoS datasets: Use of machine learning to analyse intrusion detection performance. page 81, 2020. [Dissertation]. Available from: https://urn.kb.se/resolve?urn=urn:nbn:se:ltu:diva-78980

4. Thamarai Selvi Somasundaram and Kannan Govindarajan. Virtual machine placement optimization in SDN-aware federated clouds. In *2015 IEEE International Conference on Electro/Information Technology (EIT)*, pages 379–385, Dekalb, IL, USA, May 2015. IEEE.

5. Galura Muhammad Suranegara, Dimas Agil Marenda, Rifqy Hakimi, Aris Cahyadi Risdianto, and Eueung Mulyana. Design and implementation of VM migration application on SDN-based network. In *2018 4th International Conference on Wireless and Telematics (ICWT)*, pages 1–6, Nusa Dua, July 2018. IEEE.

6. Kannan Govindarajan and Vivekanandan Suresh Kumar. An intelligent load balancer for software defined networking (SDN) based cloud infrastructure. In *2017 Second International Conference on Electrical, Computer and Communication Technologies (ICECCT)*, pages 1–6, Coimbatore, February 2017. IEEE.

7. Jehad Ali, Byeong-hee Roh, Byungkyu Lee, Jimyung Oh, and Muhammad Adil. A machine learning framework for prevention of software-defined networking controller from DDoS attacks and dimensionality reduction of big data. In *2020 International Conference on Information and Communication Technology Convergence (ICTC)*, pages 515–519, Jeju, Korea (South), October 2020. IEEE.

8. Myo Myint Oo, Sinchai Kamolphiwong, Thossaporn Kamolphiwong, and Sangsuree Vasupongayya. Advanced support vector machine- (ASVM-) based detection for distributed denial of service (DDoS) attack on software defined networking (SDN). *Journal of Computer Networks and Communications*, 2019:1–12, March 2019.

9. Jan Fransens and Frank Van Reeth. Hierarchical PCA decomposition of point clouds. In *Third International Symposium on 3D Data Processing, Visualization, and Transmission (3DPVT'06)*, pages 591–598, June 2006. IEEE.

10. Tom Howley, Michael G. Madden, Marie-Louise O'Connell, and Alan G. Ryder. The effect of principal component analysis on machine learning accuracy with high-dimensional spectral data. *Knowledge-Based Systems*, 19(5):363–370, September 2006.

11. Khaled Labib and V. Rao Vemuri. An application of principal component analysis to the detection and visualization of computer network attacks. *Annals of Telecommunications*, 61, 218–234 (2006). https://doi.org/10.1007/BF03219975

12. Tianqi Chen and Carlos Guestrin. XGBoost: A scalable tree boosting system. In *Proceedings of the 22nd ACM SIGKDD International Conference on Knowledge Discovery and Data Mining*, pages 785–794, San Francisco California USA, August 2016. ACM.

13. Alexey Natekin and Alois Knoll. Gradient boosting machines, a tutorial. *Frontiers in Neurorobotics*, 7:21, December 2013.

14. Ping Xuan, Chang Sun, Tiangang Zhang, Yilin Ye, Tonghui Shen, and Yihua Dong. Gradient boosting decision tree-based method for predicting interactions between target genes and drugs. *Frontiers in Genetics*, 10:459, 2019.

15. Soner Yıldırım. Understanding the LightGBM, September 2020. https://towards datascience.com/understanding-the-lightgbm-772ca08aabfa

16. Zecheng He, Tianwei Zhang, and Ruby B. Lee. Machine learning based DDoS attack detection from source side in cloud. In *2017 IEEE 4th International Conference on Cyber Security and Cloud Computing (CSCloud)*, pages 114–120, New York, NY, USA, June 2017. IEEE.

17. DARPA Intrusion Detection Evaluation Dataset | MIT Lincoln Laboratory, 2023.

18. Riyan Hakak and Manzoor Ahmad. Automatic defense against distributed denial of service using anomaly based method in machine Learning. In *2021 Third International Conference on Intelligent Communication Technologies and Virtual Mobile Networks (ICICV)*, pages 804–811, Tirunelveli, India, February 2021. IEEE.

19. KDD Cup 1999 Data | Kaggle, 2023.

20. Omar Abdel Wahab, Jamal Bentahar, Hadi Otrok, and Azzam Mourad. Optimal load distribution for the detection of VM-based DDoS attacks in the cloud. *IEEE Transactions on Services Computing*, 13(1):114–129, January 2020.

21. Minh Nguyen, Amitangshu Pal, and Saptarshi Debroy. Whack-a-Mole: Software-defined networking driven multi-level DDoS defense for cloud environments. In *2018 IEEE 43rd Conference on Local Computer Networks (LCN)*, pages 493–501, Chicago, IL, USA, October 2018. IEEE.

22. Joel Anderson and Jin-Hee Cho. Software defined network based virtual machine placement in cloud systems. In *MILCOM 2017 - 2017 IEEE Military Communications Conference (MILCOM)*, pages 876–881, Baltimore, MD, October 2017. IEEE.

23. Sayed Qaiser Ali Shah, Farrukh Zeeshan Khan, and Muneer Ahmad. Mitigating TCP SYN flooding based EDOS attack in cloud computing environment using binomial distribution in SDN. *Computer Communications*, 182:198–211, January 2022.

24. Anchal Ahalawat, Korra Sathya Babu, Ashok Kumar Turuk, and Sanjeev Patel. A low-rate DDoS detection and mitigation for SDN using Renyi Entropy with Packet Drop. *Journal of Information Security and Applications*, 68:103212, August 2022.

25. Richard Cziva, Fung Po Tso, and Dimitrios P Pezaros. SDN-based virtual machine management for cloud data centers. *IEEE Transactions on Network and Service Management*, 13(2), 212–225, June 2016.

26. Nisha Ahuja, Gaurav Singal, Debajyoti Mukhopadhyay, and Neeraj Kumar. Automated DDOS attack detection in software defined networking. *Journal of Network and Computer Applications*, 187:103108, August 2021.

27. Sudakshina Mandal, Danish Ali Khan, and Sarika Jain. Cloud-based zero trust access control policy: An approach to support work-from-home driven by COVID-19 pandemic. *New Generation Computing*, 39(3–4):599–622, November 2021.

16 Cloud-Based Decentralized Freelancing Marketplace

Harsh Rai, Vamshi Adouth, Deepika Saxena, and Jitendra Kumar

16.1 INTRODUCTION

In recent years, the decentralized cloud market has emerged as a response to the high costs and lack of security offered by traditional public cloud service providers. By leveraging blockchain technology, these newer providers aim to offer cheaper and more secure alternatives that are based on pay-as-you-go business models. The worldwide infrastructure as a service (IaaS) market grew 41.4% in 2021, to total $90.9 billion, up from $64.3 billion in 2020, according to Gartner Inc.[1] Inc. Amazon retained the number 1 position in the IaaS market in 2021, followed by Microsoft, Alibaba, Google, and Huawei. As a result, consumers can become locked into a particular provider's ecosystem, limiting their optionsnd opportunities. In the future, we can expect to see the emergence of a more open, fair, and trustworthy cloud resource trading market powered by blockchain, which will provide more choices and opportunities for both providers and consumers alike [1, 2].

One area where the emergence of a decentralized cloud market could have a significant impact is in the realm of freelancing. The rise of platforms such as Upwork[2] and Fiverr[3] have made it easier than ever for freelancers to find work, but these platforms can often be opaque and unfair, with high fees and arbitrary rules. In a decentralized market powered by blockchain, freelancers would have more control over their own work and be able to access a wider range of opportunities. Moreover, the transparent and secure nature of blockchain would ensure that all parties are treated fairly and that payments are processed efficiently. In short, the decentralized cloud market has the potential to revolutionize the freelancing industry and create a more open, fair, and equitable environment for all participants.

Blockchain technology offers a potential solution through the creation of decentralized applications with no single point of ownership. Decentralized applications can provide fair and protected execution of tasks, as shown in Figure 16.1 and discussed in various studies [3, 4, 5, 6, 7, 8, 9, 10, 11, 12, 13, 14, 15]. This new paradigm uses smart contracts to enforce rules and store data on the blockchain, allowing for automated transactions and rules that can only be changed with the agreement of both parties. Decentralization also eliminates third-party fees, with the only additional cost being transaction fees paid to miners for changing data on the blockchain. Some platforms [16] also use private authenticated chains to deploy and interact

DOI: 10.1201/9781003390954-16

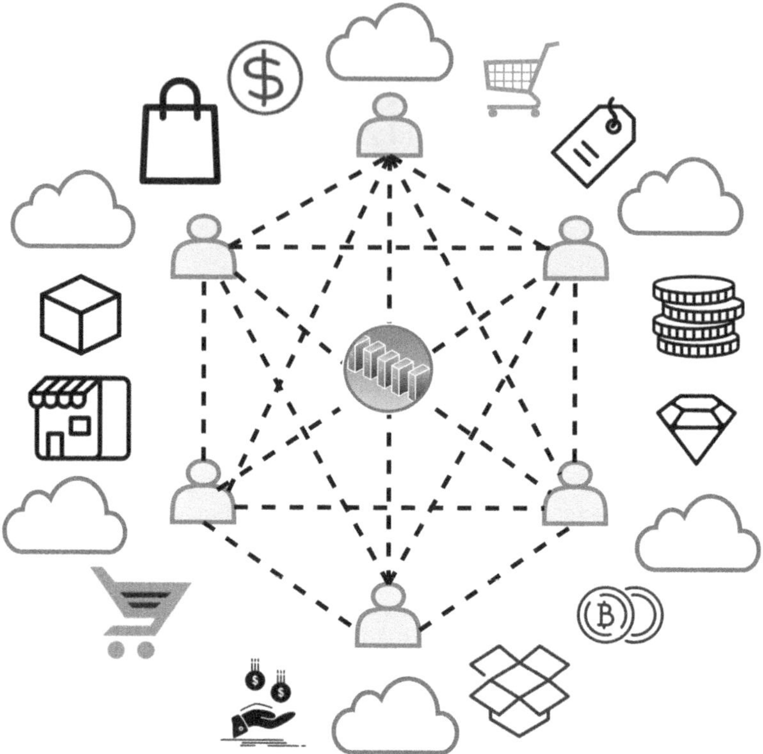

FIGURE 16.1   Blockchain technology: powering the future of cross-industry business.

with freelancing smart contracts, creating a more secure and transparent freelancing marketplace. However, a common challenge with decentralized applications is high transaction fees for on-chain storage, which can be addressed by using a centralized database and storing only the hash of the data on the blockchain, or using an Interplanetary File System [17] for peer-to-peer storage with the hash of the data on the blockchain.

16.2 CONTRIBUTIONS

In this chapter the main contributions are as follows:

1. We provide a comprehensive review of existing decentralized architectures proposed for the freelancing and crowdsourcing marketplace, highlighting their limitations and identifying areas for improvement. Our review offers novel insights into scalability, privacy, user experience, and socioeconomic implications, distinguishing it from prior reviews in the field.
2. We propose a decentralized application that addresses the lack of transparency and trust between parties due to a mediator. It also makes it easier for new

freelancers to enter the market with little or no upfront cost and ensures equal opportunities without a central server that is biased toward a subset of users.

3. In the proposed application, the employer must deposit the project budget during the assignment of the project to an employee, providing collateral and increasing overall trust.

4. The project is divided into milestones with predefined reward values, representing the fees or salary for each completed stage. This ensures transparent payments within the smart contract and enables seamless project continuity by allowing the employer to assign a new employee at the last pending milestone if needed.

5. Our decentralized freelancing marketplace eliminates high commission rates charged by centralized platforms. Only gas fees are required for users to complete a transaction.

16.3 RELATED WORK

In 2017, Levy [18] proposed the IsraTo network, an abstraction layer built on top of the Ethereum blockchain for decentralized fundraising and freelancing. The paper introduces the IRT token (which implements the standard ERC20 specifications) for freelancing and fundraising markets. IRT allows users to earn value and spend it on services/products that are internal to the IsraTo platform. This platform helps employers to raise funds and find freelancers for the project. Curators are present to do an audit on the employer and their project to increase the authenticity of the employer and the work. In 2019 Pallam and Gore [16] presented a blockchain-based freelance paradigm on Hyperledger Fabric. Hyperledger Fabric is an open-source blockchain framework hosted by Linux Foundation. Hyperledger fabrics are authorized networks where all participating member's identities are authenticated beforehand, which is different from public networks like Ethereum or Bitcoin, where anyone can participate. The proposed model was named Boomerang.

Another similar idea was implemented by Deshmukh et al. [19] in 2020, where they used Solidity programming language for writing smart contracts. The proposed paradigm used Ethereum virtual machine (EVM) for contract deployment. In 2021 a freelancing platform prototype was proposed by Afrianto et al. [20] on a public Ethereum network. The proposed architecture used hosted database for storing relevant user and project data on a central server and used smart contracts to automate transactions between the employer and the employee. Web3.js is used for interacting with on-chain smart contact functions. IPFS stores work files created by an employee, which are shared with the employer. An initial payment percentage of the overall project budget is deposited into the smart contract to increase trust between freelancers and employers. In 2021 a blockchain-based crowdsourcing solution TFCrowd was proposed by Li et al. [21], which mainly focused on the problem of reasonably rewarding workers with incentives upon completion of work. The framework calculates the reputation of both employers and workers. Upon completion of work, if an employer rejects the work and the worker disagrees with the employer's comments, they can raise a conflict which a committee of miners resolves. According to the miner's evaluation, reputation of workers and employers is updated. In 2022 Kodjiku et al. [22] reported a similar work on a decentralized crowdsourcing framework with a decision tree technique to analyze workers' performance. This framework is for crowdsourcing, where a worker is automatically selected from a pool of workers. Existing methods use a reputation-based selection of workers, but this framework introduces a decision tree model to assess workers' performance according to existing assigned work and its status. In 2022 Batool and Byun [23] proposed a biometric-based blockchain solution to freelancing, where a private hyperledger blockchain is used for the deployment of smart contract between the employer and the employee, which is signed by biometric data of both parties, which increased the authenticity of agreement, and blockchain's immutability behavior guarantees fairness to both parties. This solution also used MySQL Database for storage of user and project details. Only the hash of the stored data is recorded in the blockchain, so it has some centralized behavior.

16.4 SYSTEM WORKFLOW

16.4.1 LOGIN WORKFLOW

When a user first visits the platform, they must register as either a freelancer or an employer with their relevant details stored in a database server. Personal information is securely stored for privacy purposes. The steps for the login and registration process are shown in Figure 16.2:

1. The user connects their crypto wallet, such as metamask, to the platform.
2. After the platform fetches account details from the wallet, the Registration component collects user role and other relevant information, which is sent to the moralis-backed central server.
3. The server stores the data in a cloud-hosted MongoDB cluster.
4. If a registered user returns to the platform, they are prompted with a moralis data signature metamask prompt. The data is generated by the moralis server and signed by the user's private key from their crypto wallet.
5. After successful authentication, a JSON Web token is generated and sent as a cookie to remember the user's session.
6. This completes the registration and login process, redirecting the user to a different page based on their role.

16.4.2 PROJECT LIFE CYCLE

After registering, an employer needs to create a project so that other entities in the platform can interact with the project data in the smart contract. Figure 16.3 Figure 16.2 illustrates the processes involved in the life cycle of a project, from its creation to completion.

1. An employer creates a new project, providing relevant details such as the project title, description, and duration.

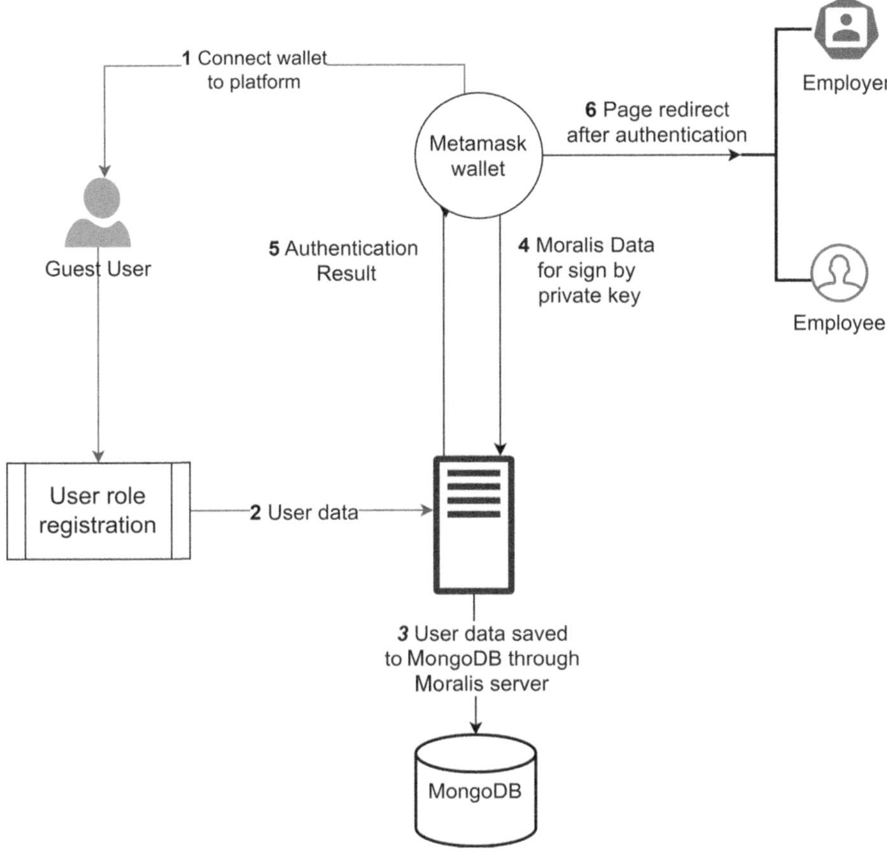

FIGURE 16.2 Login and registration workflow.

2. A freelancer retrieves the list of available projects from all employers.
3. The freelancer applies for a project.
4. The employer selects the project id and retrieves the list of all applicants who have applied for that project.
5. The employer then chooses a freelancer for the job and assigns the project to them. The employer also deposits the project budget value in the smart contract as collateral.
6. Upon completion of the work, the freelancer submits the work-related data, which is then verified by the employer. This action automatically triggers the payment of the reward value to the freelancer's account.

16.5 SMART CONTRACT DESIGN

16.5.1 Smart Contract Design

The smart contract for the decentralized freelancing marketplace consists of several functions and variables that enable the interaction between clients and freelancers.

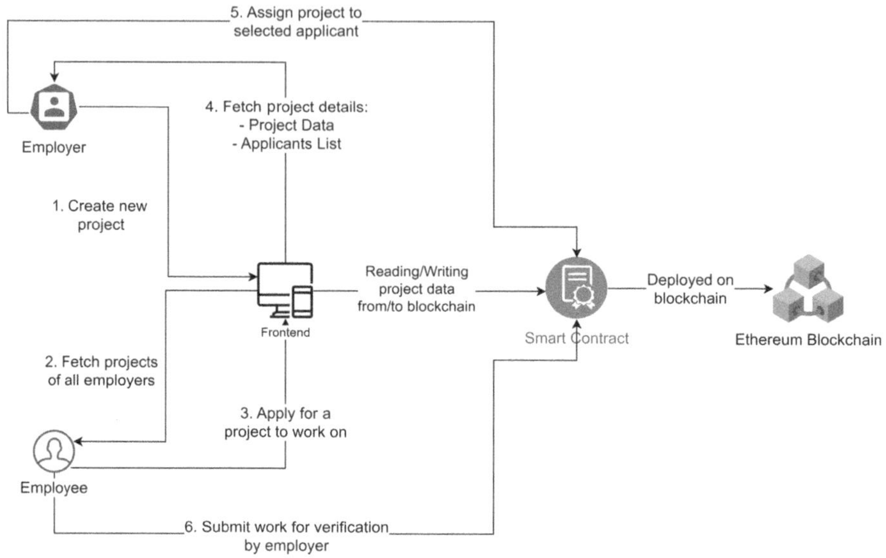

FIGURE 16.3 Project life cycle.

The main data structure used in the contract is a structure called "Project" which contains the details of a project such as the project name, description, employer, freelancers, and payment.

The **"addProject"** function allows an employer to create a new project by providing the necessary details and making a payment. This function adds the new project to an array of "Project" structs, which is stored in the contract's storage.

The **"applyForProject"** function allows a freelancer to apply for a project by providing their address and the ID of the project they want to apply for. This function adds the freelancer's address to the list of addresses for the selected project, and it also updates the 'applicants' field of the "Project" struct with the new list of addresses.

The **"cancelApplication"** function allows a freelancer to cancel their application for a project by providing the ID of the project they want to cancel their application for. This function removes the freelancer's address from the list of addresses for the selected project, and it also updates the "applicants" field of the "Project" struct with the new list of addresses.

The **"assign"** function allows an employer to assign a project to a freelancer by providing the ID of the project and the address of the freelancer. This function sets the "assignedTo" field of the "Project" struct to the freelancer's address and updates the "status" field to "assigned,"

The **"unassign"** function allows an employer to remove an assigned freelancer from a project by providing the ID of the project and the address of the freelancer. This function sets the assignedTo field of the Project struct to zero and updates the status field to "unassigned."

The **"setCheckpointLink"** function allows a freelancer to submit their work to the employer for verification by providing the ID of the project and a link to their work. This function updates the checkpointLink field of the Project struct with the provided link.

The **"verifyCheckpoint"** function allows an employer to verify a freelancer's work by providing the ID of the project and a boolean value indicating whether the work is accepted or rejected. This function updates the status field of the Project struct based on the provided Boolean value. If the work is accepted, the status is set to "completed" and the payment is released to the freelancer. If the work is rejected, the status is set to "in progress" and the freelancer is given the opportunity to resubmit their work.

Overall, the smart contract provides a transparent and fair way for clients and freelancers to interact and complete projects on the decentralized marketplace.

16.6 CONSTRUCTION DIAGRAM

In this section, we present a series of construction diagrams that illustrate the various components and their relationships in our system.

16.6.1 FREELANCER

Figure 16.4 shows the various actions that a freelancer can perform in our system. A freelancer can interact with functions in the deployed smart contract to fetch all projects, fetch details of a single project, apply for a project, cancel an application, or submit work. A freelancer can perform the following actions:

1. Fetch all projects created by employers on the blockchain.
2. Fetch all details of a single project by providing the project id as a parameter.
3. Apply for an existing project, which appends the freelancer's account number to the list of all applicants for a project. Refer to Table 16.1 for transaction details of the method called *applyForProject*.
4. Cancel an application for a project that the freelancer has already applied for, which removes the freelancer's account number from the list of all applicants for a project. Refer to Table 16.2 for transaction details of the method called *cancelApplyForProject*.

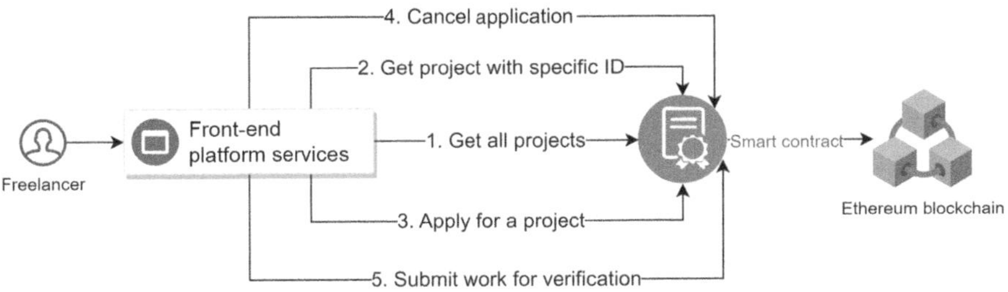

FIGURE 16.4 Actions by freelancer.

TABLE 16.1
Project creation by employer

Contract Method Name	Transaction Fee	Gas Fee
addProject Method interaction link	0.004030323306 ETH	0.000000010256 ETH

Project creation adds the details of the project to an array of struct type Project in smart contract. This operation changes the state of the blockchain and therefore is a write operation, requiring gas fees to be paid by the employer.

TABLE 16.2
Applying for a project

Contract Method Name	Transaction Fee	Gas Fee
applyForProject Method interaction link	0.004030323306 ETH	0.000000010256 ETH

An employee/freelancer applies for a project. This operation adds the function caller's address to a list of addresses, making it a write operation. In this case, the gas fees for the application of the project need to be paid by the freelancer.

TABLE 16.3
Cancel application for a project

Contract Method Name	Transaction Fee	Gas Fee
cancelApplyForProject Method interaction link	0.003477941784 ETH	0.000000110075 ETH

An employee/freelancer can also cancel an existing pending project application. This operation deletes the function caller's address from the list of applicants' addresses, making it a writable operation. In this case, the freelancer must pay the gas fees for application cancellation.

5. Submit work to the employer (only a freelancer whom the employer has assigned the project can perform this operation), as illustrated in Algorithm 16.1. Refer to Table 16.3 for details regarding the transaction for the method called *submitWork*.

16.6.2 Employer

In this section we describe the various actions performed by an employer as shown in Figure 16.5. The employer is able to interact with the smart contract by performing several actions. The employer can create a new project, fetch all of their owned projects, fetch details for a single project by providing the project ID, assign a freelancer to a project, and verify a freelancer's work for a given milestone. These actions

---

Algorithm 16.1 Store checkpoint work's data

Can only be called if:

1. The given project id exists.
2. The project is assigned to someone.
3. The function caller is the assignee of the project.

function SET CHECKPOINT LINK(projectId, checkpointIndex, checkpointData)
    if The given checkpoint index is out of range **then**
        Raise Error **"Invalid checkpoint index"**
    end if
    if The checkpoint at the given index is already completed **then**
        Raise Error **"Checkpoint already completed"**
    end if
    $project \leftarrow projects[projectId]$
    Set checkpoint data at the given index
    $project.checkpointLinks[checkpointIndex] \leftarrow checkpointData$
end function

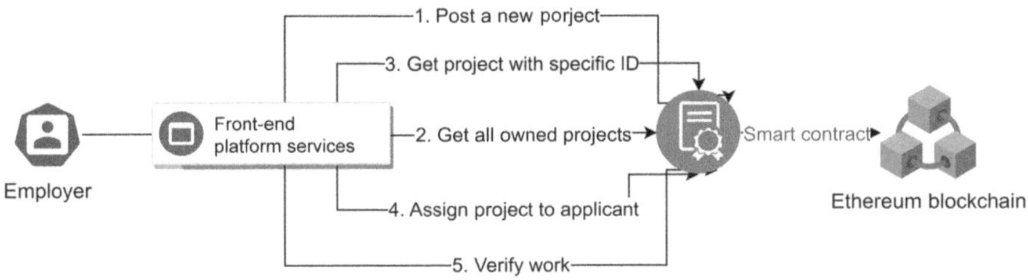

FIGURE 16.5   Actions by employer.

are implemented as functions in the smart contract and are called by the employer. The algorithms for each function are provided, along with tables showing the transaction details for each function. An employer can interact with functions in the deployed smart contract which provides the following functionalities:

1. Creation of new project as illustrated in Algorithm 16.2 (see Table 16.4 for transaction details).
2. Fetch all projects owned by him/her.
3. Fetch all details of a single project by providing the project id as a parameter.
4. Select a freelancer from the list of applicants and assign the project to him/her. During the assignment of the project, total project value should be deposited into the smart contract by the employer as illustrated in Algorithm 16.3 (see Table 16.5 for transaction details).
5. Verify the freelancer's work for a given milestone, after which the reward value is transferred automatically from the smart contract to the freelancer's

---

Algorithm 16.2 Add new project

---

function ADD PROJECT(ProjectData)
    $id \leftarrow (projectIndex + 1)$
    if project already exists **then**
        Raise Error **"Project Already Exists"**
    end if
    Push ProjectData to end of projects array
    $projects[id] \leftarrow ProjectData$
    Move project pointer to newly added project
    $projectIndex \leftarrow (projectIndex + 1)$
end function

TABLE 16.4
Assigning project to a freelancer

Contract Method Name	Transaction Fee	Gas Fee
assign Method interaction link	0.005602403439 ETH	0.000000115795 ETH

Employee selection and project assignment are done by the employer. This operation changes the assignee address in the Project block of the specified project. This write operation needs **gas fees + the total project amount to be paid by the employer.** The project value is sent with the function call with the in-built property **msg.value**.

TABLE 16.5
Submitting work for verification

Contract Method Name	Transaction Fee	Gas Fee
setCheckpointLink Method interaction link	0.005602403439 ETH	0.000000101675 ETH

Assigned employee can send the project work link to the employer, which is stored in an array inside specified Project data block in the smart contract. This operation is also a write operation; hence, the employee must pay a transaction fee.

---

account, as illustrated in Algorithm 16.4 (refer to Table 16.6 for transaction details).

6. Remove the assigned freelancer from the project. This operation can only be performed if there is no pending milestone verification, as illustrated in Algorithm 16.5 (see Table 16.7 for details regarding transaction for the method called *unassign-project*).

Algorithm 16.3 Assigning project to freelancer

Can only be called if:
(a)   Given project id exists
(b)   Function caller is the owner of the project

   function Assign Project(ProjectData)
      if project is already assigned **then**
         Raise Error **"Project already assigned"**
      end if
      Calculate total project reward
      $totalReward \leftarrow \mathbf{0}$
      $project \leftarrow projects[projectId]$
      for every $checkpoint$ in $project.checkpoints$ **do**
         if $checkpoint$ is not completed **then**
            $totalReward \leftarrow (totalReward + checkpoint.reward)$
         end if
      end for
      if $function\ call\ value$ is less than $Reward$ **then**
         Raise Error "Not enough fund"
      elseAssign project
         $project.assignee \leftarrow assigneeAddress$
      end if
   end function

TABLE 16.6
Verification of work by employer

Contract Method Name	Transaction Fee	Gas Fee
verifyCheckpoint Method interaction link	0.005867077791 ETH	0.000000087135 ETH

Employer needs to verify the work sent by the employee so that the employee can be rewarded with the specified amount of the milestone. After verification, the milestone status is updated, making this operation writable; in this case, the employer pays the transaction fee.

16.7   EXISITNG APPLICATIONS

There are decentralized applications deployed in the main-net blockchain which pay users in crypto coins. Following is the list of some applications running on different blockchains and providing various features to freelancers and employers:

1.   Argon[4] is based on the Binance blockchain network. It has its crypto token for transactions known as argon token. Argon does not charge any commission fees on transactions. The smart contracts written and deployed by the argon team are supervized by CertiK and protected by 7/24 Oracle Skynet technology. Argon platform needs third-party arbiters, also known as approvers.

Algorithm 16.4 Checkpoint work verification and transfer of reward

---

Can only be called if:
 (a)   Given project id exists.
 (b)   Project is assigned to someone.
 (c)   Function caller is the owner of the project.

> **function** SET CHECKPOINT LINK(projectId, checkpointIndex)
>   **if** Given checkpoint index is out of range **then**
>   Raise Error **"Invalid checkpoint index"**
>   **end if**
>   **if** Checkpoint at given index is already completed **then**
>   Raise Error **"Checkpoint already completed"**
>   **end if**
> $project \leftarrow projects[projectId]$
>   **Set checkpoint completed status at given index to true**
> $project.checkpointsCompleted[checkpointIndex] \leftarrow true$
>   **Transfer   checkpoint   reward   to   assignee's   wallet**
> $project.assignee.transfer(project.checkpointRewards[index])$
>   **end function**

TABLE 16.7
Unassigning project

Contract Method Name	Transaction Fee	Gas Fee
unassign Method interaction link	0.005602403439 ETH	0.000000101675 ETH

This operation clears out the assignee field in the specified Project data block in the blockchain. Transaction fees for changing this data need to be paid by the employer. This operation is impossible if there is a milestone's work whose verification is pending by the employer.

Approvers are rewarded for assessing jobs and determining whether they are completed. A consensus among approvers also resolves disputes between employer and employee.

2. CryptoTask[5] freelance market is a marketplace where employers and freelancers can find each other. The marketplace is decentralized, meaning that there is no central authority controlling the marketplace. This reduces the fees that are typically charged by centralized platforms. Jobs are secured by putting money up front during job assignments. Additionally, the reputation of free is stored on the blockchain, so there is no potential for censorship or hidden tampering.

3. Orbi Network[6] has developed two protocols based on non-fungible tokens (NFTs) for professional ID + reputation, and Intellectual Property, based on the ERC 721 and 725 standards and the Ethereum blockchain. This platform works on an Ethereum-based ERC20 Token called Orbicoin. Orbicoin is

---

Algorithm 16.5 Remove assigned freelancer from a project

Can only be called if:
(a)   Given project id exists.
(b)   Project is assigned to someone.
(c)   Function caller is either the owner or assignee of the project.
(d)   There are no checkpoints with work verification pending.

   function UNASSIGN PROJECT(projectId)
      project ← projects[projectId]
      if project is not assigned to someone yet **then**
         Raise Error **"Project not yet assigned"**
      end if
      Check for any pending checkpoint verification
      for every *checkpoint* in *checkpoints* **do**
         if *checkpoint* is not completed and *checkpoint.work* is not empty **then**
            Raise Error **"Cannot unassign project. Checkpoint work not empty"**
         end if
      end for
      Calculate refund reward value for employer
      totalReward ← **0**
      for every *checkpoint* in *checkpoints* **do**
         if *checkpoint* is not completed **then**
            totalReward ← (totalReward + checkpoint.reward)
         end if
      end for
      Unassign project, set assignee to 0
      project.assignee ← **0**
      Transfer pending amount value back to employer
   project.client.transfer(totalReward)
   end function

---

    designed for various purposes, including facilitating transactions, paying for gas, depositing as an escrow by the employer, rewarding nodes, and decentralising governance. For resolving disputes, users of the platform can appeal to the community to resolve potential conflicts between parties. Juries will see the validity of a request, whether or not it is approved. For these tasks, approvers will be paid in OrbiCoin, which is taken from the loser.

4. EthLance[7], as the name suggests, this decentralized freelancing platform works on the Ethereum main net and uses ether as its currency for transactions. EthLance offers a platform for creating and managing listings, including jobs, sponsorships, and other public goods. Payments are made in ether, and the platform does not hold any funds. Listings are stored on Ethereum and IPFS so that anyone can access them.

5. Coinlancer[8] is the only platform that charges a 3% commission on transactions. The platform is controlled by smart contracts that are deployed on the blockchain. Ethereum makes the entire process open and transparent. The platform will require users to provide unique identities to author contracts and receive payments. The identities will be based on addresses in the blockchain, making it difficult to forge. When a contract is awarded, the client must deposit Coinlancer tokens to sign it. Coinlancer also uses the same methodology of having a community called the Freelancers tribunal to resolve disputes between the parties and incentivise them from the losers end.

16.8 RESULTS

16.8.1 EXPERIMANTAL SETUP

The proposed decentralized freelancing marketplace was built on top of the Goerli testnet, a test network for the Ethereum blockchain. The smart contract, written in Solidity version 0.7.3, was deployed to the Goerli testnet using the Hardhat framework version 2.6.1. The front-end client was built using NextJS version 11.0.1, a React-based front-end framework, and Moralis version 1.0.0, a JavaScript-based library for interacting with smart contracts.

The experiments focused on testing the key features of the proposed solution, including project creation, application submission, and project cancellation. These tests were conducted using various inputs and observed the outputs produced by the smart contract. The results of these experiments are presented in the following section.

16.8.2 EXPERIMENTAL RESULTS

In this section, we present the results of our experiments with the smart contract implementation of our proposed solution. We provide details of the transaction and gas fees for each method in the smart contract, along with links to the specific transactions on the blockchain and the method of interactions on the contract code. These results demonstrate the feasibility and efficiency of our proposed solution for enabling decentralized freelancing on the blockchain.

16.9 CONCLUSION

The rapidly growing popularity of freelancing among people across the globe makes it an attractive option for many. However, the centralized freelancing system has many deficiencies, such as fairness, high transaction fees, lack of transparency, and the possibility of fraud. Our research shows that a decentralized system is a better alternative. This blockchain-based method ensures privacy, fairness, and transparency and lowers transaction fees, ensuring trust and transparency among freelancers and employers.

NOTES

1 Gartner says worldwide IaaS public cloud services market grew 41.4%.
2 www.upwork.com/
3 www.fiverr.com/
4 https://argon.run/
5 www.cryptotask.org/en
6 https://orbi.network/
7 https://ethlance.com/
8 www.coinlancer.com/home

REFERENCES

1. Suhan Jiang and Jie Wu. A blockchain-powered data market for multi-user cooperative search. *IEEE Transactions on Network and Service Management*, 19(1):203–215, 2022. doi: **10.1109/TNSM.2021.3125604**

2. Mallikarjun Reddy Dorsala, VN Sastry, and Sudhakar Chapram. Blockchain-based solutions for cloud computing: A survey. *Journal of Network and Computer Applications*, 196:103246, 2021.

3. Collin Meese, Hang Chen, Syed Ali Asif, Wanxin Li, Chien-Chung Shen, and Mark Nejad. BFRT: Blockchained federated learning for real-time traffic flow prediction. In *2022 22nd IEEE International Symposium on Cluster, Cloud and Internet Computing (CCGrid)*, pages 317–326, 2022. doi: **10.1109/CCGrid54584.2022.00041**

4. Beiji Zou, Fanbo Nie, Ling Xiao, Tao Zhang, and Chengzhang Zhu. Meda: Using blockchain for patient-controlled medical data auditing in institutions. In *2022 22nd IEEE International Symposium on Cluster, Cloud and Internet Computing (CCGrid)*, pages 297–306, 2022. doi: **10.1109/CCGrid54584.2022.00039**

5. Guangcheng Li, Qinglin Zhao, Yu Wang, Tie Qiu, Kan Xie, and Li Feng. A blockchain-based decentralized framework for fair data processing. *IEEE Transactions on Network Science and Engineering*, 8(3):2301–2315, 2021. doi: **10.1109/TNSE.2021.3086332**

6. Chao Li, Balaji Palanisamy, Runhua Xu, Jian Wang, and Jiqiang Liu. Nf-crowd: Nearly-free blockchain-based crowdsourcing. In *2020 International Symposium on Reliable Distributed Systems (SRDS)*, pages 41–50, 2020. doi: **10.1109/SRDS51746.2020.00012**

7. Arvind Panwar, Vishal Bhatnagar, et al. Blockchain-based web 4.0: Decentralized web for decentralized cloud computing. In *Cloud IoT*, pages 219–233. Chapman and Hall/CRC.

8. Varun Gupta, Jose Maria Fernandez-Crehuet, Chetna Gupta, and Thomas Hanne. Freelancing models for fostering innovation and problem solving in software startups: An empirical comparative study. *Sustainability*, 12(23):10106, 2020.

9. Yinghui Zhang, Robert H. Deng, Ximeng Liu, and Dong Zheng. Outsourcing service fair payment based on blockchain and its applications in cloud computing. *IEEE Transactions on Services Computing*, 14(4):1152–1166, 2021. doi: **10.1109/TSC.2018.2864191**

10. Vamshi Adouth, Eswari Rajagopal, and Syam Kumar Pasupuleti. EB-CSPA: Efficient blockchain-based certificateless short signature public auditing scheme for cloud-based cyber-physical systems. *Journal of Electronic Imaging*, 32(2):023002, 2023.

11. T. Tharun, A. Vamshi, and R. Eswari. NFT application for music industry using blockchain smart contracts. In *2023 4th International Conference on Innovative Trends in Information Technology (ICITIIT)*, pages 1–6, 2023. doi: **10.1109/ICITIIT57246.2023.10068684**

12. Vinay Raj and Sadam Ravichandra. A service graph based extraction of microservices from monolith services of service-oriented architecture. *Software: Practice and Experience*, 52(7):1661–1678, 2022.

13. Niharika Singh, Jitendra Kumar, Ashutosh Kumar Singh, and Anand Mohan. Privacy-preserving multi-keyword hybrid search over encrypted data in cloud. *Journal of Ambient Intelligence and Humanized Computing*, pages 1–14, 2022.

14. Jitendra Kumar, Ashutosh Kumar Singh, and Anand Mohan. Resource-efficient load-balancing framework for cloud data center networks. *ETRI Journal*, 43(1):53–63, 2021.

15. Vamshi Adouth and Eswari Rajagopal. Blockchain-based certificateless public auditing with privacy-preserving for cloud-based cyber-physical systems. *Concurrency and Computation: Practice and Experience*, 35(12):e7690, 2023.

16. Babu Pallam and MM Gore. Boomerang: Blockchain-based freelance paradigm on hyperledger. In *2019 10th International Conference on Computing, Communication and Networking Technologies (ICCCNT)*, pages 1–6. IEEE, 2019.

17. Juan Benet. IPFS-content addressed, versioned, p2p file system. *arXiv preprint arXiv:1407.3561*, 2014.

18. Adam Levy. A decentralized fundraising and freelancing network. *Whitepaper Revision*, pages 1–31, 2017.

19. Prathmesh Deshmukh, Shreyas Kalwaghe, Ajinkya Appa, and Aprupa Pawar. Decentralised freelancing using ethereum blockchain. In *2020 International Conference on Communication and Signal Processing (ICCSP)*, pages 881–883. IEEE, 2020.

20. Irawan Afrianto, Christover Ramanda Moa, Sufa Atin, Iding Rosyidin, et al. Prototype blockchain based smart contract for freelance marketplace system. In *2021 Sixth International Conference on Informatics and Computing (ICIC)*, pages 1–8. IEEE, 2021.

21. Chunxiao Li, Xidi Qu, and Yu Guo. Tfcrowd: A blockchain-based crowdsourcing framework with enhanced trustworthiness and fairness. *EURASIP Journal on Wireless Communications and Networking*, 2021(1):1–20, 2021.

22. Seth Larweh Kodjiku, Yili Fang, Tao Han, Kwame Omono Asamoah, Esther Stacy EB Aggrey, Collins Sey, Evans Aidoo, Victor Nonso Ejianya, and Xun Wang. Excrowd: A blockchain framework for exploration-based crowdsourcing. *Applied Sciences*, 12(13): 6732, 2022.

23. Amreen Batool and Yungcheol Byun. Reduction of online fraudulent activities in freelancing sites using blockchain and biometric. *Electronics*, 11(5):789, 2022.

17 Poisson-Based Energy Hole Prediction (PBEHP) for Voronoi Wireless Multimedia Sensor Network

Suseela Sellamuthu, Eswari Rajagopal, and Nickolas Savarimuthu

17.1 INTRODUCTION

Routing heavy traffic through an ephemeral network with varying channel conditions is a significantly challenging problem in WMSN are energy constraint, bandwidth, energy hole, node deployment, mobile node information, and scalability. WMSN and its presence are inevitable for the modern communication world. WMSN consists of a large number of low-power, low-cost, multifunctional sensor nodes. Sensor nodes are energized by the low-power battery in lots of applications and these power units cannot be substituted. When the sensor nodes deplete their energy level and are exhausted, and the overall network functionality is stopped. Thus, an energy-efficient algorithm is very much needed to improve the network lifetime and handle the battery efficiently. The performance of WMSN is affected by the locality of sensor nodes and the occurrence of energy holes. These holes cause the failure of the data delivery in the path. Operational lifetime, network coverage, and energy consumption are directly proportional to the distance between sensor nodes and the energy level of the sensor nodes. The prediction of energy deficiency in sensor nodes becomes an overhead in WMSN. Hence, the development of energy-efficient algorithms is needed to place the sensor in closer distance and to predict the energy holes of the WMSN in advance to increase the network performance. This motivates us to develop energy-efficient routing algorithms for WMSN.

To develop a Poisson Based Energy Holes Prediction for Voronoi WMSN (PBEHP) to bypass the energy holes using the Poisson-based method and find the alternate path and shifting the path for the routing of packets with the objectives of increasing network lifetime, increasing packet delivery ratio, and minimizing delivery delay, this research focuses on improving the performance and longevity of WMSN using energy efficient routing algorithms. It also proposes poison-based energy hole prediction to predict the energy hole WMSN. A Poisson process (Kumar et al., 2013) is a model for a sequence of discrete events where the regular time between events is known, but the exact timing of events is arbitrary. In this work,

DOI: 10.1201/9781003390954-17

prediction of energy hole is Poisson process. Discrete probability distribution predicts the number of energy holes at a given time. It counts the occurrences of energy holes at fixed regular intervals of time. The mean energy hole rate (MEHR – which is defined as the number of energy holes per unit of time) is assumed to be constant over time. In this research work, PBEHP is proposed to predict the energy holes based on possibilities of occurrence for Voronoi Wireless Multimedia Sensor Network.

17.2 LITERATURE REVIEW

Numerous tiny sensing components in WMSNs carry out the role of uninterrupted and distributed ambience contact and transfer of obtained data to central infrastructure. Because multimedia data uses a lot of bandwidth, there is a very high energy utilization. The main criteria for WMSNs in terms of communication are the timely transmission of essential information and careful adherence to Quality of Service (QoS).

Routing in WMSNs is challenging due to the inherent characteristics of mobile ad hoc networks or cellular networks. Routing discovers the appropriate path to transmit the multimedia data and preserves the routes in the network. Path selection plays an important role in conveying multimedia data over WMSN. Due to the energy gap created by power shortage and traffic congestion (Feng et al., 2020), single path selection results in routing failure. The packets are dropped at any hop of the route from the source node to the destination node due to the energy shortage of a certain node in the routing path. In contrast to single route routing, the multipath routing algorithm selects multiple paths to speed up packet delivery. In order to reduce the actual power consumption of the sensor nodes, routing must choose the shortest path possible.

17.2.1 MULTIPATH ROUTING

Since there are many paths to travel from one place to another, multipath routing in WMSN is preferable to single-path routing (Macit et al., 2014). However, there is a processing overhead issue in multipath routing that will impact the network performance.

As a result, many researchers concentrate on creating a reactive multipath algorithm to enhance network performance by raising the packet delivery ratio (PDR), throughput, and reducing delay. Mohammed Zaki Hasan et al. (2017) proposed a Lagrangian relaxation method to prioritize the sensor nodes. The authors used only the link quality parameter to calculate the prioritization. The authors avoided talking about delays and sensor node energy levels. An adaptive power control approach for effective routing was introduced by Vijay Ukani et al. (2014). Multipath routing was discovered by the authors using simply energy constraints. Incebacak et al. (2013) developed a method to decrease the energy demands of wireless sensor networks which used mixed integer programming and only took into account two channels from multipath networks. In order to enhance performance, Kai Lin et al. (2011) suggested a QoS trust estimation model that takes advantage of cluster structure and cellular topology. They considered only energy constraints for multiple pathfinding and not other factors such as distance and hop count. A probabilistic modeling-based multipath routing technique for WMSN was suggested by Dubey et al. (2013). However, their technique requires more redundancy because it permits identical packets to go over numerous paths and updates the routing table using local data. A randomized forwarding system was suggested by Deb et al. (2003) to reduce errors and overheads. But it makes unnecessary consideration for every data forwarding route. The energy constrained multipath routing (ECMP) algorithm was developed by Bagula et al. (2008) to maximize bandwidth utilization while using the least amount of energy. However, the ECMP algorithm mainly focused on energy level, disregarding delay and network quality. A reliable discontinuous and braided multipath routing for sensor networks was proposed by Yang et al. (2010). By preserving local routing information, they used hop-by-hop multipath routing to lower routing broadcast and energy usage. However, for multipath detection, the authors did not prioritize the high energetic path.

17.2.2 MULTIPATH ROUTING PROTOCOLS

Ad-hoc On-Demand Distance Vector (AODV) Protocol: It is a multipath routing protocol (Hasan et al., 2017) that supports the mobility of the sensor nodes, link failures, and packet losses. In AODV, nodes discover the next neighborhood based on the route request sent to the sensor nodes. AODV is best suited for multipath routing, but it is unable to foresee the energy hole information beforehand.

Destination Sequenced Distance Vector (DSDV) Protocol: This technique is used for proactive multipath routing (Kushwaha et al., 2015). At each sensor node, the routing table is updated periodically and supports dynamic changes of network topology. Regular updates of the routing table consume more energy and it is supported only for small networks. DSDV is not suitable for highly dynamic networks.

Optimized Link State Routing Protocol (OLSR) Protocol: It is a proactive multipath routing protocol (Clausen & Jacquet, 2003) that is well suited for a large optimized network. A sensor node continuously updates routes for all destinations in the network and it is suited for high random and periodic patterns. OLSR is suffering from massive overhead due to the usage of multipoint relays for forwarding the data to the destination.

Dynamic Source Routing (DSR) Protocol: It is deployed in mobile and static node multi-hop wireless ad hoc networks (Al-Ariki et al., 2017). The simple and efficient DSR allows self-organizing and self-configuring of network infrastructure. DSR uses source routing or path addressing for forwarding the data from source to destination instead of updating the routing table in each sensor node. The performance of the sensor networks reduces rapidly with increasing mobility. Routing overhead is involved which is directly proportional to the length of the path.

17.2.3 ENERGY HOLE PREDICTION

Energy holes (Robinson et al., 2021) are created in WMSN when the sensor nodes could not select their neighborhood because of depletion of their energy level and when the neighborhood nodes are not in the transmission range. The routes' energy level will shortly run out, and as a result, the routing paths will become broken. Network performance is significantly impacted by energy depletion. By cutting off the sensor nodes, energy holes have an impact on the network topology. Therefore, a hole must be found before creating the multipath from the source to the sink. To enhance network performance, researchers focused on the development of energy hole prediction algorithms (Sharmin et al., 2020). In order to transmit data from source to destination, Parvin et al. (2015) reviewed a number of void management approaches, including bound holes, perimeter routing, and Anchored Geodesic Packet Forwarding (AGPF). The relocation of a node is constructed based on hole and distance from the source to destination. The AGPF these technique did not provide an optimal path and guaranteed delivery. Their technique does not support scalability but has high overhead and complexity. Fang et al. (2006) proposed a BOUNDHOLE algorithm to identify the energy hole in the whole boundary of the void region. But processing overhead occurs because their proposed algorithm analyzes the entire boundary of the void region.

Chen et al. (2017) proposed a preventive void handling technique, AGPF, to prevent the energy hole in the network where each source node maintains anchors (list of temporary destinations). The data packet is forwarded using the greedy method and cannot move in backward. The main drawback of their proposed method is that the sender needs to update and acquire the position of anchors periodically. It is very difficult in a large network. Leong et al. (2006) identified the void node using the planar graph by utilizing the Relative neighborhood graph (RNG) planarization algorithm and Group-based grouping planner (GGP) algorithm. In this perimeter routing, the right-hand rule is used to identify the void node (Chaaf et al., 2021), and the greedy method is used to forward the data. The traversal path of perimeter routing with stuck nodes did not provide the optimal path and yields high cost.

Energy-Aware Multipath Geographic Routing (EMGR): It (Huang, Junbao et al., 2017) is a multipath attractive routing that supports desirable scalability and simplicity. EMGR uses geographic information to select the next neighborhood. It uses a dynamic anchor list to change the routing path for load balance. Routing overhead is high in EMGR and it consides only energy cost for next-hop selection.

Energy-Aware Dual-Path Geographic Routing (EDGR): EDGR (Huang et al., 2017) is a multipath routing algorithm that uses local information to route the data. EDGR suffers from energy holes which lead to routing failures. Only two metrics – the location information and residual energy are used to make routing decisions. Routing holes are identified only using forwarding direction and not backward direction.

Energy-Efficient Beaconless Geographic Routing (EBGR): It is proposed by Zhang et al. (2009) which supports scalability and efficiency but does not support highly dynamic scenarios. EBGR selects a neighborhood using the handshaking mechanism RTS/CTS. There are no packet loss and no link failures but consumption of more energy.

17.3　POISSON-BASED ENERGY HOLES PREDICTION ALGORITHM FOR VWMSN

The proposed PBEHP algorithm concentrates on multipath routing to avoid energy holes and find the shortest path from source node to destination node with the objectives of maximizing energy efficiency, network lifetime, and success rate, minimizing energy consumption, latency. It consists of three modules: Construction of Voronoi WMSN, Poisson-Based Energy Holes Prediction, and Path Shifting.

17.3.1　CONSTRUCTION OF VORONOI WMSN

In the Euclidean plane, Sk (S1, S2, S3,..., Sn) is a finite set of sensor nodes that together form VWMSN. The network is set up using the Voronoi cells (VC) and Voronoi edges (VE). The width w and height h of the VC are formed at (-wvc,-hvc), (2wvc,-hvc), and (2wvc, 2hvc), correspondingly (-wvc,2hvc). The VC with coordinates Sk (1=k=n) is made up of any sensor nodes in the plane whose proximity from Sk is smaller than their distance from each other node Sk+1. Each polygon in the Voronoi WMSN is created by intersecting half-spaces with the sensor Sk's bisector perpendicular. The sensors Sk in the plane that are uniformly spaced from the two closest cells make up the Voronoi WMSN. Figure 17.1 and Algorithm 17.1 show the construction of VWMSN.

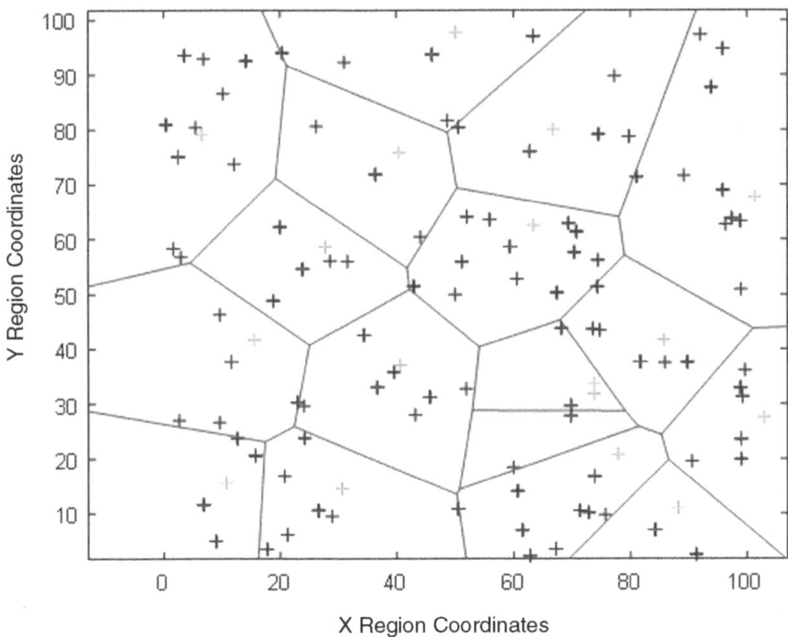

FIGURE 17.1　Voronoi WMSN.

Algorithm 17.1: Poisson_Based_Energy_Holes_Prediction()

Input: Voronoi WMS Networks
Output: Energy Holes
1:**begin**
2: set n, e_l,e_{lth}// no. of nodes,energy level, energy level threshold
3:    **for** each node i in n
7:      **if**(e_l<= e_{lth})
8:          Set $P_{EH}[i]$ = True
9:          Find Probability of Prediction of Energy Hole (P_{EH}) using eqn. 17.3
10:          Find Probability Density Function of inter energy hole using eqn. 17.4
11:          Find Average number of energy holes(L_n) using eqn. 17.5
12:          Find Service Time (ST) using eqn. 17.8
13:    **end if**
14: **end for**
16:**end**

If the residual energy of the sensor nodes is below the energy threshold level, VWMSN detects the energy holes. The average number of energy holes and the time required to inform the neighborhoods about the occurrence of the energy holes are calculated. Energy hole occurrences in VWMSN are also updated in each sensor node. The time required to find the energy holes by the sink nodes is also calculated. The calculated values are used to find alternate paths.

17.3.2 Poisson-Based Energy Holes Prediction

Initially all sensor nodes are deployed randomly in VWMSN. The prediction of an energy hole is a Poisson process. Energy hole occurrences are counted at predetermined, regular periods of time and tend to cluster or be dispersed in some way. A PBEHP is a discrete probability distribution which predicts the number of energy holes in a given time.

The MEHR is assumed to be constant over time. The probability of arrival time (PAT) of energy holes between time t and t + dt is calculated using equation 17.1

$$P_{AT} = MEH_R \times dt$$

$$PH_0 = \cfrac{1}{\sum_{n=0}^{\rho-1} \frac{\left(\frac{\lambda}{\hbar}\right)n}{n!} + \frac{\left(\frac{\lambda}{\hbar}\right)n}{\rho!} \times \frac{(\rho\hbar)}{\rho\hbar-\lambda}}$$

$$PH_0 = \cfrac{1}{\sum_{n=0}^{\rho-1} \frac{\left(\frac{\lambda}{\hbar}\right)n}{n!} + \frac{\left(\frac{\lambda}{\hbar}\right)n}{\rho!} \times \frac{(\rho\hbar)}{\rho\hbar-\lambda}} \tag{17.1}$$

Initial probability of energy hole occurrence is calculated using equation 17.2

$$PH_0 = \cfrac{1}{\sum\limits_{n=0}^{\rho-1} \cfrac{\left(\frac{\lambda}{\hbar}\right)^n}{n!} + \cfrac{\left(\frac{\lambda}{\hbar}\right)^n}{\rho!} \times \cfrac{(\rho\hbar)}{\rho\hbar-\lambda}} \tag{17.2}$$

where λ is the searching time of energy holes in path ρ; $\hbar$ is the servicing time of energy Holes in path ρ; ρ is the number of paths; and n is the number of sensor nodes. The probability of prediction of energy holes at time t is calculated using equation 17.3.

$$P_{EH} = (MEH_R \times t)n \times e^{-MEH_R} \times t/n! \tag{17.3}$$

The probability density function (PDF) of inter energy hole at time t (time interval between two consecutive holes) is calculated using equation 17.4.

$$PDF_{EH} = MEH_R \times e^{-MEH_R} \times tPDF_{EH} = MEH_R \times e^{-MEH_R} \times t \tag{17.4}$$

The average number of energy holes processing in the queue is calculated using equation 17.5.

$$L_n = \frac{\lambda\hbar \left(\frac{\lambda}{\hbar}\right)^\rho}{(\rho-1)!(\rho\hbar-\lambda)^2} \times P_{EH} + \frac{\lambda}{\hbar}$$

$$L_n = \frac{\lambda\hbar \left(\frac{\lambda}{\hbar}\right)^\rho}{(\rho-1)!(\rho\hbar-\lambda)^2} \times P_{EH} + \frac{\lambda}{\hbar} \tag{17.5}$$

The average number of energy holes processing in the queue is calculated using equation 17.6.

$$L_q = L_n + \frac{\lambda}{\hbar}L_q = L_n + \frac{\lambda}{\hbar} \tag{17.6}$$

Energy hole service time (EHSt) is the time required to inform the neighborhoods about the occurrence of the energy holes in the region. The Mean Energy Hole Service Rate (MEHSR) is the number of neighbors receiving information about occurrences of the energy holes in the region. Probability of a service time completed between t and dt is calculated using equation 17.7.

$$P_{ST} = MEHS_R \times dt \tag{17.7}$$

The service time to update nodes about energy holes' occurrences is calculated using equation 17.8.

$$ST = (MEHS_R \times t)n \times e^{-MEHS_R} \times t/n! \tag{17.8}$$

$$ST = (MEHS_R \times t)n \times e^{-MEHS_R} \times t/n!$$

SL_x, SL_y	nb_{tr}	e_l	nb_m	P_{EH}	L_n

FIGURE 17.2   Log table of path shifting.

The average time a sink node spends to find the energy holes in the network is calculated using equation 17.9.

$$\omega = \frac{L_n}{\lambda} \tag{17.9}$$

In VWMSN, the energy holes are identified if the residual energy of the sensor nodes is lesser than the energy threshold level. The average number of energy holes and the time required to inform the neighborhoods about the occurrence of the energy holes are calculated. Energy hole occurrences in VWMSN are also updated in each sensor node. The time required to find the energy holes by the sink nodes is also calculated. These calculated values are used to find alternate paths.

17.3.3 PATH SHIFTING

In WMS networks, if an energy hole (Parvin et al., 2015) occurs, a path shifting algorithm is explored to identify a more effective detour. The steps are given in Algorithm 17.2. The path shifting algorithm is iterative, asynchronous, distributed, and self-terminated. Each node sensor node has a log table that the path shifting algorithm updates on a regular basis. It finds alternate paths if an energy hole occurs. The log table is presented in Figure 17.2.

In the figure SLx, SLy: Location of a sensor node; nbtr: Transmission range of neighbors; nbm: Minimum distance of Neighbors; Ln: Average number of energy holes; e_l: Energy level; P_{EH} : Probability of energy hole occurrence.

The path shifting is done using Algorithm 17.1. Multiple paths are identified using Algorithm 17.2. If an energy hole exists in a path, then it finds the alternate path by checking the neighboring nodes. The nearby node that is both within the transmission range and is closer to the current node will be taken into account. The connection will be established between the selected neighboring node and the current node. This path shifting helps to bypass the energy hole and find alternate energized paths. The path shifting is done using Algorithm 17.2. Multiple paths are identified using Algorithm 17.3.

17.4 EVALUATION METRICS

Olayinka et al. (2017) state that the evaluation metrics provided below are used to compare the performance of new and existing algorithms.

Algorithm 17.2: Path_Shifting()

Input: VoronoiWMSN
Output: Shifted path
1:**begin**
2:    Multipath_Construction()
3:    **for** each node n in *path p*
4:    **if**(P_{EH}=True)
5:        **for** each neighbor nb of n
6:            **if** $nb_{tr} <= T_r$
7:                d_{nb} = find_ dist(nb, n)   //Find distance of neighbors
8:            **endif**
9:        **endfor**
10   :nb_m = findmindist(d_{nb})
11:        establish connection between n-1 to nb_m
12:    **endif**
13:**endfor**
14:**end**

Algorithm 17.3: Neighborhood_Selection()

Input: WMSN
Output: Neighboring nodes
1: **begin**
2: set Number of nodes (n); Transmission Range (T_r), Energy Level (e_l);
   Energy Level Threshold (el_{th})
3:    **repeat**
4:      **foreach** hop in the network
5:          **foreach** neighborhood node
6:            A disjoint node sends R_{req} message to neighborhood
7:            **if**(e_h==False && Ni_{el}>el_{th})
8:                Ni_{dist} = min(Ni_{dist})
9:                Send reply R_{rep} message to the disjoint node
10   **endif**
11:               **endfor**
12:       **endfor**
13:    **until** all neighboring nodes are visited
14: **end**

Packet Delivery Ratio

The ratio of total packets delivered to destinations to total packets sent from sources
is known as the packet delivery ratio.

$$\chi = \frac{Total\ Packets\ delivered\ to\ destination}{total\ Packet\ sent\ by\ the\ source} \qquad (17.10)$$

Delivery Delay

The amount of time it takes for a bit to get from a sender to a recipient is known as the delivery delay (∂). It is the ratio of distance between the sources to destination to transmission speed and is calculated using equation 17.11

$$\partial = \frac{Dist(Source, Destination)}{Transmission\,speed} \tag{17.11}$$

Network Lifetime

It is referred to as the node's operating time, during which it is capable of carrying out the assigned task.

$$N_{lt} = \frac{I_c - W_P}{E_{con} + S_t \times R_{erel}} \tag{17.12}$$

where I_c is the initial energy; is the W_e wasted energy; E_{con} continuous power consumption of the network; S_t is the average sensing reporting time; and R_{erel} is the estimated reporting energy level.

Energy Consumption

$$E_{con.tx} = \begin{cases} L \times \alpha_t + L \times \alpha_{fs} \times D^2 D < d_0 \\ L \times \alpha_t + L \times \alpha_{mp} \times D^4 D \geq d_0 \end{cases} \tag{17.13}$$

where α_t is the energy dissipated in transmitter electronics per bit; α_r is the energy dissipated in receiver electronics per bit; α_{amp} is the energy dissipated in transmitter amplifier; α_{fs} is the energy spent by the transmitter amplifier in the free space; α_{mp} is the energy needed by the transmitter amplifier in multipath model; and d_0 is the threshold value.

17.5 EXPERIMENTAL SETUP

The NS2.34 simulator is used to simulate the performance of PBEHP. Two hundred sensor nodes are distributed in a region of more than 1000 m by 1000 m in line with IEEE 802.11.

The performance of the VWMSNs is calculated based on the network densities. The range of TCP flows and Constant Bit Rate (CBR) is 1 Mbps and 0.5 Mbps, respectively. The packet sizes are considered as 128, 256, and 512 bytes (Torrieri et al., 2015). The transmission range (Tr) of each sensor node is 40 m at maximum. The transmission speed is 54 Mbps at 2.4 GHz. All nodes' initial transmission power is 7.5 dBm. The delay ranges from 0 to 200 seconds, with intervals of 1 second. The experimental parameters are displayed in Table 17.1.

17.6 RESULTS AND DISCUSSIONS

The new PBEHP algorithm is tested with the preexisting EDGR, EBGR, and EMGR, in terms of energy usage, network lifetime, packet ratio, and delivery delays. Significantly the energy consumption of PBEHP is decreased than the other routing method

TABLE 17.1
Parameters for VWMSN

Parameters	Values
Simulated space	1000 m * 1000m
Packet size	512 bytes
Initial energy node	10 J
Initial location	(0,0)
Transmission range (T_R)	35 m, 2.5 GHz, 54 Mbps
Transmission power	7.5 dBm
Sensor nodes	200
Simulation seconds	500 s

as shown in Figure 17.3 As the network size grows, a considerable energy decrease occurs. If the network density increases it will help to find a more energy efficient path. In terms of energy consumption, the proposed PBEHP outperforms EBGR, EMGR, and EDGR by 18.9%.

Figure 17.3 shows the comparison results of the algorithms in terms of network density versus network lifetime. Compared with EBGR, EMGR, and EDGR, the proposed PBEHP increases the lifetime of the sensor network by 4%. Figure 17.4 shows the PDR with various densities. The simulation results indicate that the existing methods have low Packet Delivery Ratio (PDR) under a higher density area than PBEHP.

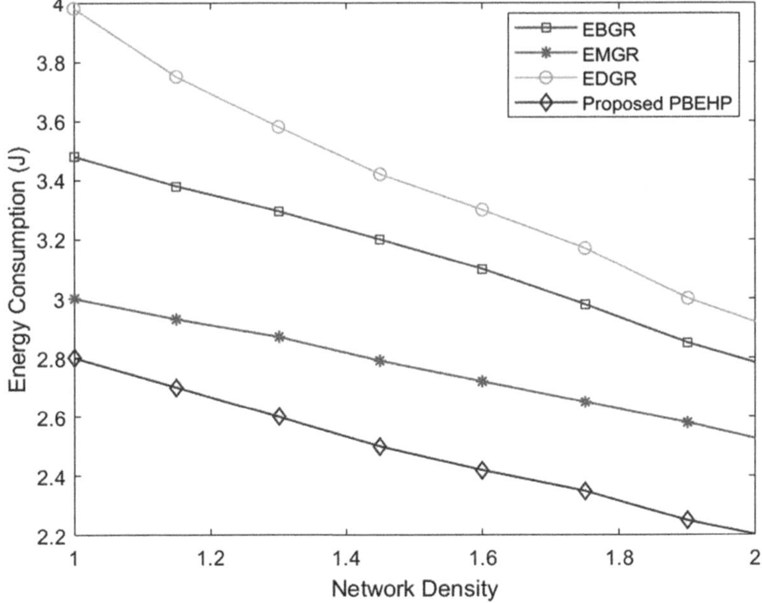

FIGURE 17.3   Energy consumption.

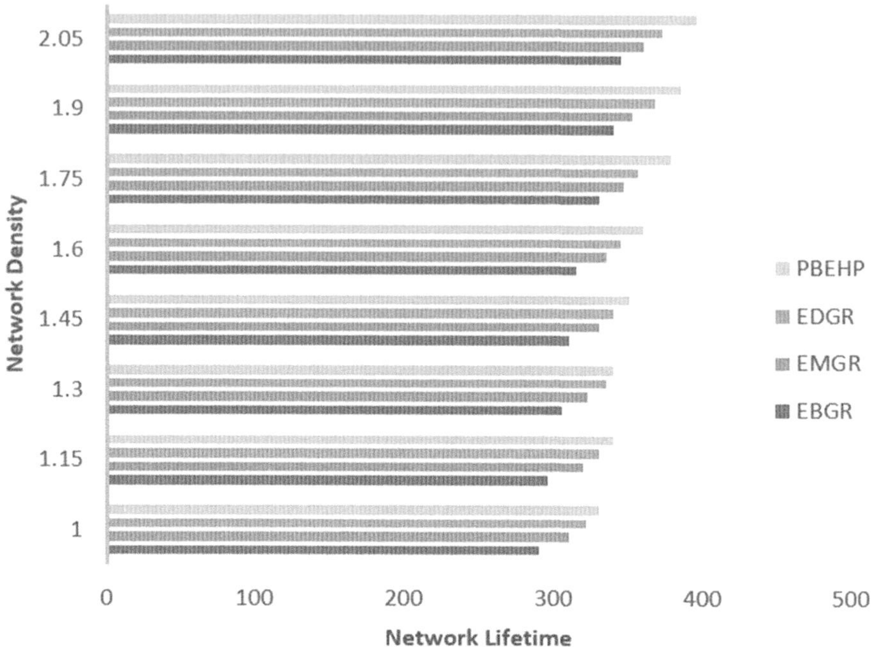

FIGURE 17.4   Network lifetime.

Through the energy hole prediction in advance and shifting a path based on that hole occurrence, the PDR is increased by 4% by PBEHP when compared to other algorithms.

Figure 17.5 shows the delivery delay of all routing algorithms. In the proposed algorithm, the network density increases and the delivery delay decreases. The result shows that the delivery delay of the proposed PBEHP is reduced to 2% compared to other algorithms.

17.7   CONCLUSION

WMSN is affected by energy holes due to the inability of sensor nodes to select neighbors for data delivery. The proposed PBEHP algorithm provides efficient routing of packets for VWMSNs. It predicts the energy holes in each path based on Poisson process and then alters the path to improve the VWMS network's performance. The PBEHP also predicts the average number of energy holes, average time for finding energy holes, and efficiency of the system. The findings of the simulation show that the proposed PBEHP algorithm increases PDR by 4%, increases network lifetime by 4%, reduces delivery ratio by 2%, and decreases energy consumption by 18.9 % than other existing EBGR, EMGR, and EDGR algorithms. In PBEHP algorithm, the delay decreases as the network density increases. A is proposed (PBEHP) is proposed which predicts energy holes based on the Poisson process and shifts the

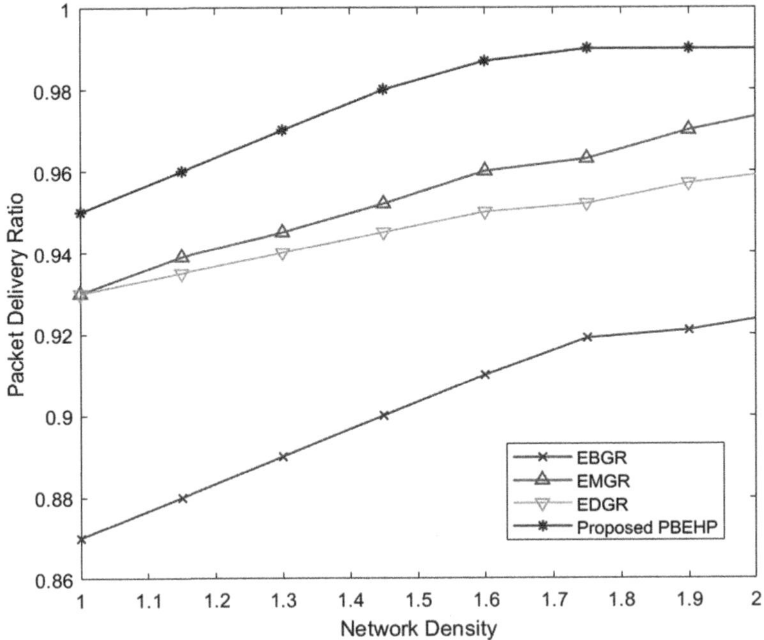

FIGURE 17.5   Packet delivery ratio.

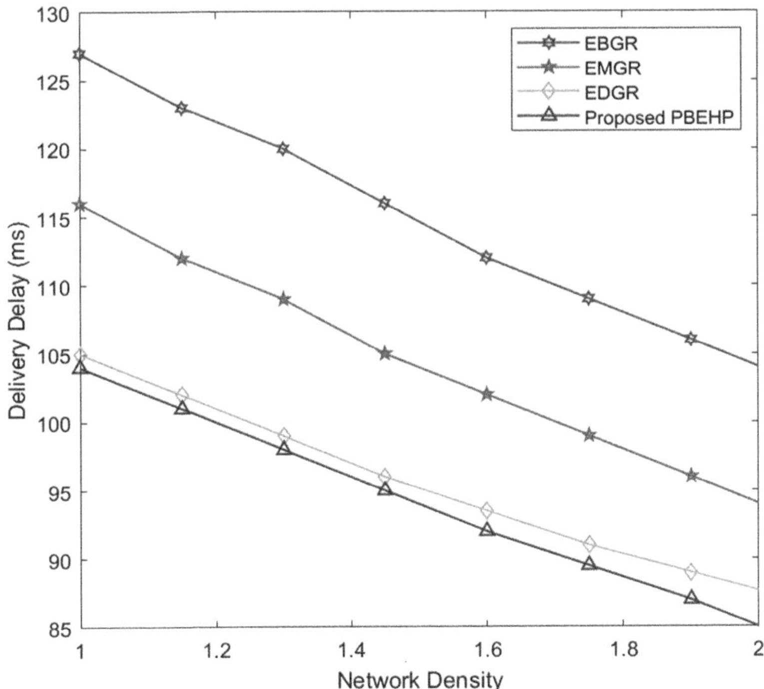

FIGURE 17.6   Delivery delay.

routing path. The proposed method will be implemented in is proposed based applications in the agriculture field, Intra Vehicular communication, weather monitoring, and sports/military personnel.

REFERENCES

1. Ab Aziz, Nor Azlina Bt, Ammar W. Mohemmed, and BS Daya Sagar (2007). "Particle swarm optimization and Voronoi diagram for wireless sensor networks coverage optimization." In *2007 International Conference on Intelligent and Advanced Systems*, pp. 961–965. IEEE.
2. Al-Ariki, Hasib DaowdEsmail, and MN Shanmukha Swamy. (2017). "A survey and analysis of multipath routing protocols in wireless multimedia sensor networks." *Wireless Networks*, 23(6), 1823–1835.
3. Bagula, Antoine B., and Kuzamunu G. Mazandu. (2008). "Energy constrained multipath routing in wireless sensor networks." In *Ubiquitous Intelligence and Computing: 5th International Conference, UIC 2008, Oslo, Norway, June 23-25, 2008 Proceedings 5*, pp. 453–467. Springer Berlin Heidelberg.
4. Chaaf, Amir, Mohammed Saleh Ali Muthanna, Ammar Muthanna, Soha Alhelaly, Ibrahim A. Elgendy, Abdullah M. Iliyasu, and Ahmed A. Abd El-Latif. (2021). "Energy-efficient relay-based void hole prevention and repair in clustered multi-AUV underwater wireless sensor network." *Security and Communication Networks* 2021, 1–20.
5. Chen, Dazhi, and Pramod K. Varshney. (2017). "A survey of void handling techniques for geographic routing in wireless networks." *IEEE Communications Surveys & Tutorials* 9, no. 1, 50–67.
6. Clausen, Thomas, Philippe Jacquet, Cédric Adjih, Anis Laouiti, Pascale Minet, Paul Muhlethaler, Amir Qayyum, and Laurent Viennot. (2003). "Optimized link state routing protocol (OLSR)." inria-00471712f, 1–53.
7. Deb, Budhaditya, Sudeept Bhatnagar, and Badri Nath. (2003). "ReInForM: Reliable information forwarding using multiple paths in sensor networks." In *28th Annual IEEE International Conference on Local Computer Networks, 2003. LCN'03. Proceedings*, pp. 406–415. IEEE.
8. Dubey, Shivendra, and Chetan Agrawal. (2013). "A survey of data collection techniques in wireless sensor network." *International Journal of Advances in Engineering & Technology* 6, no. 4, 1664.
9. Fang, Qing, Jie Gao, and Leonidas J. Guibas. (2006). "Locating and bypassing holes in sensor networks." *Mobile Networks and Applications* 11, 187–200.
10. Feng, Jie, Hongbin Chen, Xianjun Deng, Laurence Tianruo Yang, and Fangqing Tan. (2020). "Confident information coverage hole prediction and repairing for healthcare big data collection in large-scale hybrid wireless sensor networks." *IEEE Internet of Things Journal* 8, no. 23, 16801–16813.
11. Huang, Haojun, Hao Yin, Geyong Min, Junbao Zhang, Yulei Wu, and Xu Zhang. (2017). "Energy-aware dual-path geographic routing to bypass routing holes in wireless sensor networks." *IEEE Transactions on Mobile Computing* 17, no. 6, 1339–1352.
12. Huang, Haojun, Junbao Zhang, Xu Zhang, Benshun Yi, Qilin Fan, and Feng Li. (2017). "EMGR: Energy-efficient multicast geographic routing in wireless sensor networks." *Computer Networks* 129, 51–63.
13. D. Incebacak, B. Tavli, K. Bicakci, and A. Altın-Kayhan. (2013). "Optimal number of srouting paths in multi-path routing to minimize energy consumption in wireless sensor networks." *EURASIP Journal on Wireless Communications and Networking*, 1, 252–261.

14. Kai Lin, Joel J. P. C. Rodrigues, Hongwei Ge, NaixueXiong, and Xuedong Liang. (2011). "Energy efficiency QoS assurance routing in wireless multimedia sensor networks." *IEEE Systems Journal*, 5(4), 495–505.

15. Kolahdouzan, M., and Shahabi, C. (2004). "Voronoi-based k nearest neighbor search for spatial network databases." In *Proceedings of the Thirtieth International Conference on Very large Data Bases*, 30, 840–851.

16. Kumar, C., and B. Nair. (2013). "On stuttering hyper-Poisson distribution and its properties." *Sri Lankan Journal of Applied Statistics* 14, no. 1.

17. Kushwaha, Uday Singh, P. K. Gupta, and S. P. Ghrera. (2015). "Performance evaluation of AOMDV routing algorithm with local repair for wireless mesh networks." *CSI transactions on ICT* 2, no. 4, 253–260.

18. Leong, Ben, Barbara Liskov, and Robert Tappan Morris. (2006). "Geographic routing without planarization." In *NSDI*, vol. 6, p. 25.

19. Liu, Y., Han, Y., Yang, Z., and Wu, H. (2014). "Efficient data query in intermittently-connected mobile ad hoc social networks." *IEEE Transactions on Parallel and Distributed Systems*, 26(5), 1301–1312.

20. Macit, Muhammet, V. CagriGungor, and Gurkan Tuna. (2014). "Comparison of QoS-aware single-path vs. multi-path routing protocols for image transmission in wireless multimedia sensor networks." *Ad Hoc Networks*, 19, 132–141.

21. Mohammed Zaki Hasan, Fadi Al-Turjman, and Hussain Al-Rizzo. (2017). "Optimized multi-constrained quality-of-service multipath routing approach for multimedia sensor networks." *IEEE Sensors Journal*, 17(7), 2298–2309.

22. Mohammed Zaki Hasan, Hussain AI-Rizzo, and Fadi AI-Turjman. (2017). "A survey on multipath routing protocols for QoS assurances in real time wireless multimedia sensor networks." *IEEE Communications Surveys & Tutorials*, 19(3), 1424–1456.

23. Olayinka O. Ogundile, and Attahiru S. Alfa. (2017). "A survey on an energy-efficient and energy-balanced routing protocol for wireless sensor networks." *Sensors J.*, 1–51.

24. Parvin, Sanaz, Mehdi Agha Sarram, Ghasem Mirjalily, and Fazlollah Adibnia. (2015). "A survey on void handling techniques for geographic routing in VANET network." *International Journal of Grid & Distributed Computing* 8, no. 2, 101–114.

25. Robinson, Y. Harold, T. Samraj Lawrence, E. Golden Julie, Raghvendra Kumar, Pham Huy Thong, and Le Hoang Son. (2021). "Enhanced border and hole detection for energy utilization in wireless sensor networks." *Arabian Journal for Science and Engineering*, 1-13.

26. Sharmin, Nusrat, Amit Karmaker, William Luke Lambert, Mohammad Shah Alam, and MST Shamim Ara Shawkat. (2020). "Minimizing the energy hole problem in wireless sensor networks: A wedge merging approach." *Sensors* 20, no. 1, 277.

27. Torrieri, Don, Salvatore Talarico, and Matthew C. Valenti. (2015). "Performance comparisons of geographic routing protocols in mobile ad hoc networks." *IEEE Transactions on Communications* 63, no. 11, 4276–4286.

28. Vijay Ukani, Ankit Kothari, and Tanish Zaveri. (2014). "An energy efficient routing protocol for wireless multimedia sensor network." In *2014 International Conference on Devices, Circuits and Communications (ICDCCom)*, pp. 1–6. IEEE.

29. Yang, Jing, Mai Xu, Wei Zhao, and Baoguo Xu. (2010). "A multipath routing protocol based on clustering and ant colony optimization for wireless sensor networks." *Sensors* 10, no. 5, 4521–4540.

18 A Viral Decoy Environment for Ransomware Defense

Kuldeep Mohanty, Ghanshyam S. Bopche, and Kshira Sagar Sahoo

18.1 INTRODUCTION

The convergence of the Internet of Things (IoT) and cloud-based technologies presents a fresh approach to delivering conventional information and communications technology (ICT) services to organizations and governments. By integrating platforms, operating systems, storage components, databases, and other ICT equipment, these Cloud of Things (CoT) solutions are progressively finding applications in residential and professional settings, leading to improved service delivery and enhanced productivity. Various industries including healthcare [19], retail [15], and manufacturing such as Industrial IoT [42], transportation [6], defense (such as Internet of Battle Things [24]) have eagerly adopted IoT and cloud technologies. Because of such convergence, things have changed in a way no one could have predicted. IoT devices produce massive amounts of data, and the cloud offers a scalable and effective platform for handling, storing, and analyzing such data. IoT devices may offload computing work to cloud services and use machine learning algorithms for sophisticated analytics and just-in-time decision-making. While CoT infrastructure offers numerous benefits to organizations, governments, and end users, it does come with its unique set of challenges concerning security and privacy. IoT device upgrades, configurations, and monitoring are made more accessible by the cloud's centralized management and control. As a result, the attack surface throughout the cyberspace has increased, luring attackers to breach cloud services and steal data.

The healthcare industry is one of the most targeted one since it generates a vast amount of data in electronic health records (EHRs) over the cloud. EHR may contain sensitive information such as patient medical history, biographical details, contact information, real-time diagnostics reports, the unique identification number of implanted devices like pacemakers, etc. Figure 18.1 shows a typical healthcare management system that uses the CoT for enhanced service delivery. Numerous Internet of Medical Things (IoMT) may be attached to the patient's body, including remote patient monitoring machines, infusion pumps, biosensors (wearable and implants), and sensors that track medication orders. Over an IoMT edge network, the recorded health parameters of the patients will be sent periodically to the cloud for further processing, such as predictive analytics. Various stakeholders, such as hospitals, pharmacies, labs, and patients, can access the recorded data in the cloud. In

DOI: 10.1201/9781003390954-18

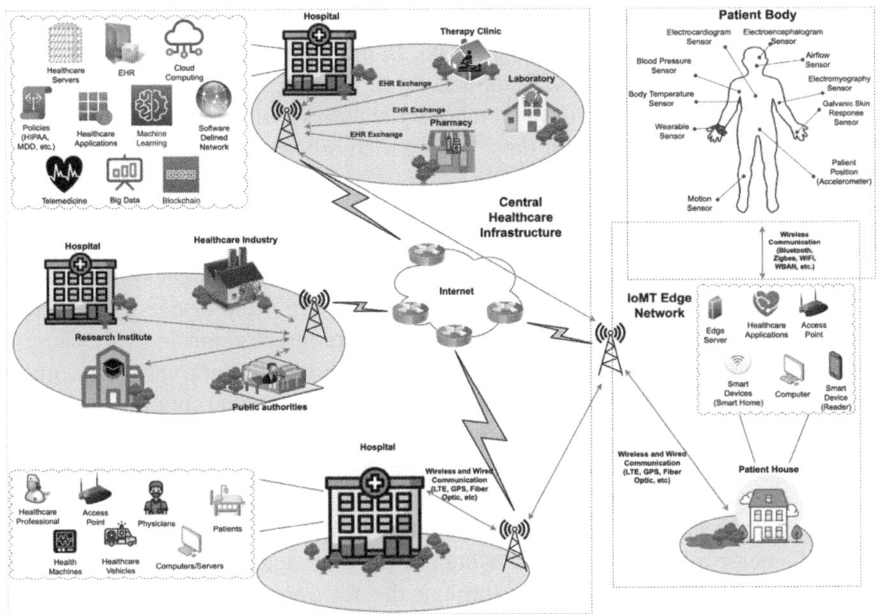

FIGURE 18.1 Healthcare overall scenario. (Adopted from [29].)

such an intelligent environment, vital information from IoMT sensors can be made available to patients and doctors, as it is crucial for real-time health monitoring. These data collectively contribute to the EHR. These data are helpful for follow-up therapies and, most crucially, healthcare research projects since they provide information on a patient's entire course of treatment. Think about the Covid19 epidemic, where the mutated virus affected the world's population. The diverse symptoms and responses to the treatments set off a chain reaction and affected the whole healthcare sector. The EHR findings must have significantly contributed to understanding the SARS-CoV-2 virus, which led to the development of vaccines to combat it. What if a cyberattacker could have encrypted or deleted the patient's data or EHR?

There are potential risks of cyberattacks at every level in the cloud-based healthcare management system mentioned above. For instance, steganography-based attacks may occur when data in the form of X-ray or Magnetic Resonance Imaging scan images are shared between a lab and hospitals, meaning that computer malware may be embedded in the digital photos and infect the whole CoT-based healthcare infrastructure. But what if the malware encrypts the cloud-based electronic health records? Millions of patients' complete medical history and EHR records will vanish, putting all hospital tasks on hold. Think about being in the middle of a major surgical procedure, then bang! Imagine vital IoMT devices are compromised, and all medical reports and real-time patient data, including those from the IoMT sensors, have vanished. How will the doctors carry out the procedure? Sounds weird, huh? It illustrates how seriously malware, especially ransomware, may damage the CoT ecosystem.

In the ocean of malware, ransomware is one of the most well-known and inevitable threats to mission and business-critical data. In the latest ransomware attack on the healthcare industry (WannaCry, SamSam, Ryuk, Conti), a prominent hospital network was targeted, resulting in the temporary shutdown of critical systems and the compromise of sensitive patient data, posing a significant risk to both patient care and data security. Ransomware is one of the malware that constantly changes its modus operandi and attack vectors for infection. By breaching firewalls and bypassing the other standard defense systems like intrusion detection and prevention system (IDPS), ransomware can enter the system and gain access to the target system without authorization. In essence, ransomware locks down the system or encrypts essential information in exchange for a Bitcoin ransom. Cryptographic ransomware encrypts the target's business or mission-critical files. It is more common in PCs, workstations, servers, and cloud storage and demands a ransom for decryption keys without guaranteeing release. The locker ransomware, in contrast, locks the target machine and prevents the user from accessing it until the ransom demand is fulfilled. It is more common on mobile, mainly Android systems. The increased use of ICT in various domains, such as finance, healthcare, etc., led to a surge in sophisticated ransomware attacks. Cybercriminals have begun using ransomware attacks as a primary weapon and a means of generating money through the Ransomware-as-a-Service (RaaS) model. The harm done by earlier variants of ransomware strains was repairable. However, the damage caused by existing sophisticated ransomware families cannot be fixed without a suitable data backup. We need an active detection technique that can identify ransomware early.

There are several strategies and solutions in place to combat ransomware. Some researchers recommend using machine learning algorithms to detect ransomware at the earliest stages of intrusion using a static or behavioral analysis approach. In contrast, other researchers think that data backup is one of the most outstanding solutions to deal with data loss caused by ransomware. Even said, some methods for dealing with ransomware rely on self-healing techniques or self-recovery of data from the network traffic. The solutions don't just stop at this; they also change as the ransomware does. Despite the abundance of ransomware solutions that are currently available, the frequency and severity of ransomware attacks have remained relatively high. As security administrators and cybersecurity analysts develop new machine learning strategies and other techniques to combat ransomware, ransomware creators develop more sophisticated strategies to fight them, circumventing the updated security measures and rendering the existing security countermeasures ineffective. Many additional reasons contribute to ransomware's escalating effectiveness; therefore, this is hardly the sole cause. Human mistake or ignorance is among these most crucial factors. Defense-in-depth and proactive security strategies considerably improve an organization's security posture but do not provide 100% assurance against ransomware attacks. While existing security measures are essential to a robust security plan, it's vital to recognize that determined attackers can still discover methods to overcome deployed defenses. It necessitates the development of a proactive technique for ransomware detection. In this chapter we have proposed a conceptual framework that not only detects and prevents ransomware but also inflicts damage to the adversary. We propose using a viral decoy environment to defend against ransomware. Furthermore, we discuss the benefits, limitations, and ethical difficulties in implementing such a framework.

The rest of the chapter is organized as follows: Section 18.2 discusses different phases in the typical ransomware attack and RaaS business model. Section 18.3 explores existing security controls for securing business or mission-critical data from the ransomware attack and the need for a new deception-based viral framework to defend against the ransomware attack. We present the framework based on the viral decoy environment (VDE) in Section 18.4. Section 18.5 discusses the issues, challenges, and ethical concerns in implementing the VDE-based framework. Section 18.6 concludes the chapter with future work.

18.2 RANSOMWARE-AS-A-SERVICE MODEL

Ransomware performs lateral movement to spread throughout the network for valuable files. Ransomware follows a five-step strategy to compromise any network, as shown in Figure 18.2. The first and most crucial stage of a ransomware attack is infection. It begins with the selection of an attack vector. Trapping the target through a typical cyberattack is essential to inject the ransomware into the victim's system/network. Although many attack vectors may exist, social engineering-based attacks are the most prevalent. Phishing is one of the most often used social engineering attacks. These are systematic methods through which malware, including different types of ransomware, enters the system. The genuinely looking emails may contain malicious links or attachments. The associated malware is downloaded into the system and executed upon clicking the link or downloading the email attachment.

The second stage of a ransomware attack involves communicating with the attacker's remote Command and Control (C&C) server, where the encryption keys necessary to encrypt the target's system data are stored. There are two methods to establish C&C: Hard-coded IP or Domain Generation Algorithms (DGAs). Some ransomware families may use the same IP address or domain to establish communication with the C&C server. It offers a dependable connection between the target machine and C&C server throughout the onslaught. Since the IP address is static, the firewalls can quickly identify and stop the attack. Hard-coded IPs render the ransomware useless, which results in a botched attack attempt. To establish a dynamic

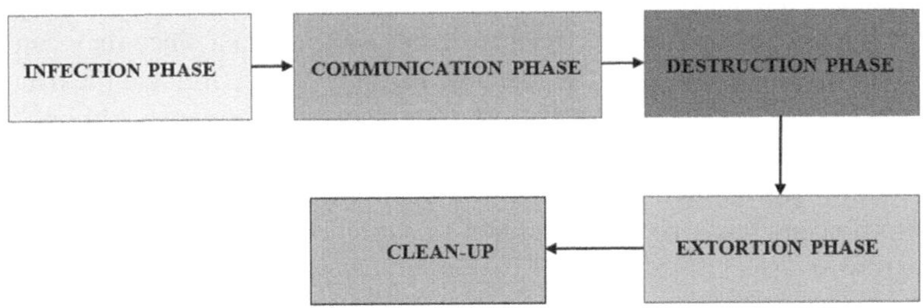

FIGURE 18.2 Phases in the typical ransomware attack.

connection with the C&C, the attackers employ DGAs. Rapidly fluxing or changing the domain names gives the server a different domain name for each communication. The most significant benefit of domain creation methods over hard-coded IP is that firewalls cannot readily identify them, providing cybercriminals with an environment conducive to successful attacks.

Unauthorized data exfiltration from the targeted system is the focus of the third phase, also called the "destruction phase," which encrypts the system with robust encryption algorithms that render the victim incapable of decrypting the data. The victim is forced to choose between paying a ransom without guarantee of receiving the decryption keys or inadvertently losing the data. It occurs in the case of a family of cryptographic ransomware, but what might the flow be in the case of a family of locker ransomware? In this stage, the crucial turning point for most ransomware families is to lock the victim's system, preventing the victim from accessing the system or services. The locking mechanism could be categorized into three types as follows:

- Screen Locking: When the system Graphical User Interface (GUI) is locked, and a ransom is requested to unlock it, this is called screen locking. While some families, like LockerPin, set the specific parameters to Android System APIs to make the Android screen persistent, mobile ransomware families do this to lock the mobile device.
- Browser Locking: Web browser locking occurs when ransomware seizes control of the victim's browser and demands payment. Attackers lock the victims' browsers by sending victims to a website containing malicious JavaScript code.
- Master Boot Record (MBR) Locking: The system's MBR, which stores data needed to start the OS, is the target of MBR locking. It either substitutes a fake MBR for the genuine one or encrypts the genuine MBR to stop the system from loading the boot code. The ransomware family PETYA is an illustration of one that locks the MBR.

The attacker's true motive during the ransomware attack is to obtain the ransom from the victim at the attack's forth and most crucial stage. The victim must exclusively pay the desired ransom in Bitcoin. The apparent explanation is that Bitcoin ransom retains the attacker's identity, making it impossible to track them. Bitcoin is decentralized, uncontrolled, and not governed by law enforcement agencies. But does the extortion stage always fruitful? A few factors may influence the victim's decision to pay the ransom, such as (1) the victim may be unfamiliar with cryptocurrency payments, (2) the victim may prefer to lose their data since they can't meet the substantial ransom demand, (3) the attacker may have exfiltrated the data before encrypting it, which they might sell to make money on the darknet. Although it may appear that the attacker has lost if the victim refuses to pay the demanded ransom, the attacker still retains ultimate control of the situation. During the ransomware attack, data extortion can take one or two of the following forms:

- Single Extortion: A ransom is requested from the victim in exchange for the decryption keys to his valuable encrypted file in single-party extortion, the most general extortion approach used by attackers. This approach could be more typical and only applicable to end users.

- Double Extortion: An additional extortion technique in which the attacker exfiltrates all the data before encrypting the organization's business-critical data. This attack is more prevalent in end-user and organizational attacks since it may cause a company's expulsion from the commercial world and always involves two parties. Let's say there is a corporation called "X" with "n" essential clients. The firm is under attack, and the attacker has exfiltrated the data (e.g., client information, source codes, or business-critical files) before encrypting it. The attacker demands a ransom from a victim organization in exchange for the stolen data. The attacker may leak stolen data into the dark web if the firm declines to pay the ransom. There is no guarantee that the data won't be disclosed in public forums or the dark net, even if the ransom is paid. Egregor, the Maze, and Darkside ransomware are some examples of using the double extortion technique.
- Triple Extortion: In this category, the extortion takes place through three steps: (1) data exfiltration, (2) data encryption, and (3) the initiation of a distributed denial-of-service (DDoS) attack on the victim organization's business-critical services. The attacker exfiltrates the data, encrypts it to make it inaccessible to the victim, and initiates a DDoS attack. The business organization is most vulnerable to this kind of attack since it can ruin and wipe the organization off the market for the reasons listed below. Such an attack could harm the business-customer relationship: customer data may become public, and DDoS attacks knock down critical business services. Suncrypt and RaagnarLocker are examples of ransomware that uses triple extortion techniques.
- Quadruple Extortion: It is one of the worst forms of extortion because in addition to data exfiltration, encryption, and DDoS attacks, the ransomware also directly informs the business clients through emails or text messages that their data is under seizure and can be made public if the company refuses to pay the demanded ransom. An attack technique like this has negative socio-psychological impacts on the victim and impacts the company, and pushes it out of the mainstream. It causes a problem with trust between the victim, the customer, and the market. It causes a recurring problem and permanent exclusion of the organization from the market. Because it costs the victim their job and damages their reputation in the marketplace, this extortion also negatively affects their psychological well-being. One of the finest strategies for deliberate ransomware attacks for commercial rivals may be quadruple extortion.

The fifth and last step of the attack is known as the concluding step, where the attacker may or may not provide the decryption keys depending on whether the ransom demand is fulfilled. During this stage, the attackers erase their digital traces (to complicate digital forensics) and leave the network. There are now several ransomware families, not just one, so no longer a lone wolf. They have planted their roots all over the dark web through a well-known business model called RaaS, as shown in Figure 18.3. The business model involves an agreement between operators and affiliates. The operators are the ransomware developers who create and provide

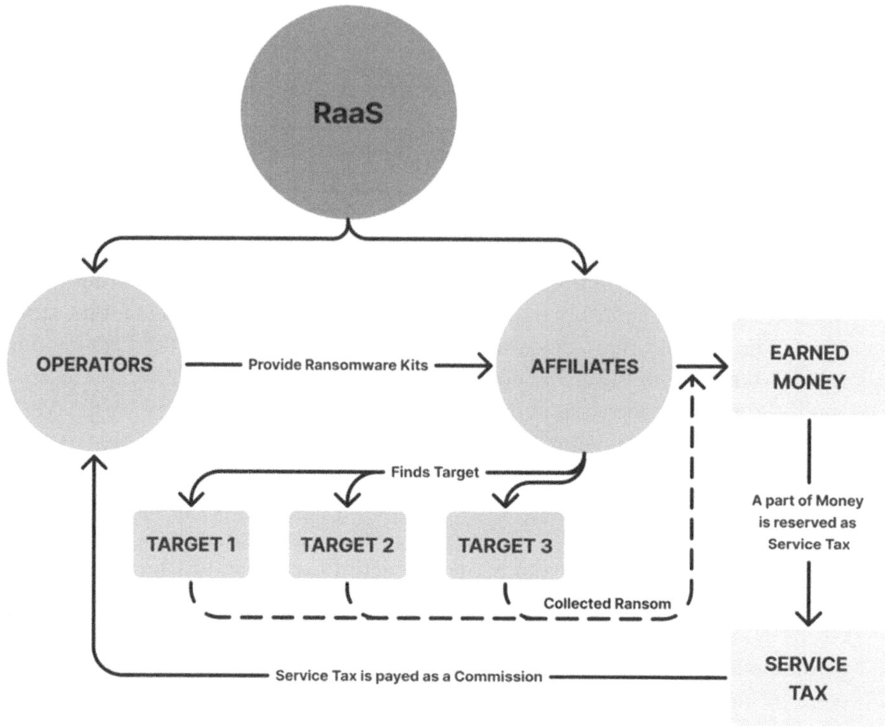

FIGURE 18.3   Ransomware-as-a-Service (RaaS) model.

ransomware development kits on the dark web, and the affiliates are the ones that purchase these ransomware development kits, create their ransomware, and carry out attacks without difficulty, regardless of their level of expertise. Essentially, operators search forums for affiliates and then put up a unique control and command dashboard so that the affiliates can track the shipment. Furthermore, they establish a victim payment portal to aid affiliates in ransom negotiations with victims. Affiliates pay the operators to employ ransomware in exchange for a service charge for each obtained ransom. When affiliates have chosen a victim, launched a ransomware attack, and demanded payment, they have the decryption keys.

RaaS is now developing into a significant dark web industry. It has been relatively effective and strengthened as a revenue model for malware producers due to the ongoing connection between operators and affiliates. When we compare the two, there are significant differences between the RaaS and non-RaaS models, as shown in Figure 18.4. Compared to a non-RaaS model, the well-distributed chain of the RaaS model makes it possible to raise more money for malware creators. RaaS allows anybody, regardless of technical proficiency, to join the chain, create their ransomware using ransomware development kits, and quickly launch an attack. Because the ransomware team is more effective, launching massive strikes on several organizations takes less time. In RaaS, a variety of revenue models are available. Affiliates may opt

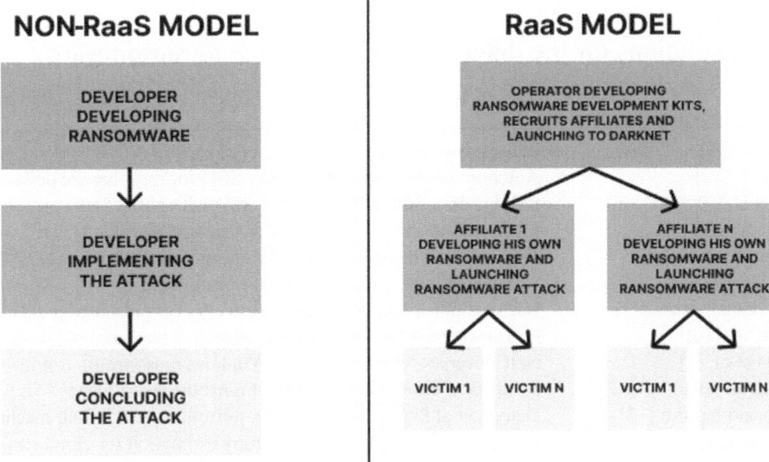

FIGURE 18.4 Non-RaaS vs. RaaS model.

for a monthly subscription model, where they must pay a set amount each month for a subscription. An affiliate program revenue model is very similar to a subscription model, except that affiliates must pay ransomware developers a portion of their earnings. In a single-time license model, affiliates must pay a large sum once. In profit sharing, affiliates may get little to nothing, but the ransomware developers receive the full benefit. In cyberspace, RaaS has become a greater menace and presented a new problem for cyber attribution. Ransomware represents a significant barrier to adopting technologies like the IoT, Industrial Internet of Things (IIoT), Cloud Computing, 5G, etc.

18.3 EXISTING SOLUTIONS TO DEFEND FROM THE RANSOMWARE

Ransomware has created a severe menace in cyberspace. Many academicians and researchers have been working around the clock to solve the ransomware problem and deal with the issues of cyberattack attribution. Despite the enormous research into developing anti-ransomware systems, none of the existing solutions is sufficient to detect ransomware proactively. Backup-based solutions are considered the first line of defense against ransomware. The other potential solutions are self-healing techniques to recover files after a ransomware attack, access control techniques to protect systems from ransomware, use of sandboxing environments to stop the spread of ransomware, machine learning techniques to proactively detect ransomware, static and dynamic analysis of ransomware, blacklisting, and whitelisting-based solutions, etc. Table 18.1 shows the category of existing solutions available in the literature for the detection and prevention of ransomware.

TABLE 18.1

Existing solutions for the detection and prevention of ransomware

Work	Description
Lee et al. [25]	Ransomware prevention using secure backup of encryption keys
Min et al. [30]	Autonomous backup and recovery SSD for ransomware defense
Baek et al. [7]	SSDInsider – Internal defense of SSD with perfect data recovery
Paik et al. [34]	Ransomware-aware storage buffer management policy
Park et al. [35]	Time out-based backup, a low overhead ransomware proof SSD
Chen and Bridges [11]	Automated behavioral analysis of ransomware
Vinayakumar et al. [39]	Use of shallow and deep neural networks for ransomware defense
Chen et al. [10]	Ransomware detection using Generative Adversarial Network
Alhawi et al. [4]	NetConverse – classification of Windows ransomware traffic
Daku et al. [14]	Behavioral-based classification of ransomware variants
Cohen and Nissim [12]	Detection of known and unknown ransomware in virtual machines
Lee et al. [26]	Ransomware detection using entropy estimation for cloud services
Continella et al. [13]	ShieldFS – Identification of ransomware behavioral pattern
Al-Dwairi et al. [3]	SH-VARRi – Protection of XML documents from ransomware
Berrueta et al. [9]	Restoration of ransomware encrypted files from network traffic
Kim and Lee [22]	Access control mechanism to restrict unauthorized operations
Kim et al. [21]	An improvised whitelisting-based approach to identify ransomware
Azmoodeh et al. [5]	Detection of ransomware in IoT by analyzing energy consumption
Lee et al. [28]	Using MTD to tackle ransomware by altering specific file extension
Monge et al. [31]	MTD-based approach to defend against cryptographic ransomware
Moore [32]	Use of fake directories to fight against ransomware
Wang, et al. [41]	RansomTracer – trapping and gathering clues about the adversary
Gómez-Hernández et al. [16]	R-Locker – preventing ransomware using a unique file-based trap
Gómez-Hernández et al. [17]	Technique for preventing ransomware compatible to Windows OS
Saleh et al. [37]	SentryFS – Use of NLP to make honey files more believable
Pascariu and Barbu [36]	Low cost, low power device to host network shares with decoy files
Wang et al. [40]	Tracing back source of RDP-based ransomware attack

18.3.1 Backup-Based Solutions to Tackle Ransomware

Lee et al. [25] proposed a method to keep the encryption keys in a secure location for data recovery in case of a ransomware attack. Min et al. [30] proposed "Amoeba," an autonomous backup and recovery Solid State Drive (SSD) for ransomware defense. Amoeba specifically embeds specialized Ransomware Attack Risk Indicator (RARI) compute hardware on the Direct Memory Access (DMA) module for quick ransomware detection and carries out independent backup and recovery within the device. Another study by Baek et al. [7] suggests SSD-Insider, a solid-state drive's internal defense against ransomware that offers a flawless data recovery technique from the infected files utilizing the solid-state drive's delayed deletion capability. Even though ransomware incidents have risen daily, crypto-ransomware has been particularly vicious. Paik et al. [34] proposed a ransomware-aware buffer management policy and an access pattern-based detector. Another study by Park et al. [35] aims to reduce storage overheads of the delayed deletion, thus using RBlocker or RansomBlocker, a time-out-based backup strategy. A typical study conducted to improve storage mechanisms for ransomware protection by Ahn et al. [2] proposed a Key-SSD-based solution that is a disk drive using an access control mechanism that claims it is the last barrier to data protection. There has been much more research on backup-based methods for ransomware prevention. However, these solutions still need to guarantee total security against ransomware attacks because there are still ways for cybercriminals to operate. Backup-based solutions are only effective in single-extortion ransomware cases. However, when it comes to double, triple, or quadruple extortion, the backup-based solutions fall short because even though data retrieval may be possible after the attack, the exfiltrated data could be made available on the dark web by the adversary. Adopting backup systems may also result in service downtime, which hinders a firm's operation.

18.3.2 Machine Learning-Based Solution to Tackle Ransomware

Researchers have developed additional methods, including machine learning algorithms, after realizing that backup-based solutions aren't adequate to combat ransomware. One such study deals with automating malware analysis based on a case study of WannaCry ransomware conducted by Chen and Bridges [11]. The authors proposed a methodology to automatically extract malware features from the host logs. Another study conducted by Vinayakumar et al. [39] aims to evaluate the efficacy of shallow and deep neural networks for detecting and classifying ransomware. The study shows that deep neural networks perform better than a traditional shallow network, where the Multi-Layer Perceptron (MLP) accounted for better and more accurate results. Chen et al. [10] proposed a robust methodology for identifying ransomware. The authors proposed a framework in which dynamic ransomware samples are created using the Generative Adversarial Network (GAN) and a set of adversarial quality measures to gauge how dangerous the GAN-created ransomware samples are. Baldwin and Dehghantanha [8] proposed using Support Vector Mchine (SVM) for Opcode density-based cryptographic ransomware detection. The study illustrated that the analysis of opcodes, i.e., the CPU instructions, can distinguish between cryptographic ransomware and goodwares with higher precision. Alhawi et al. [4] conducted yet another research and proposed a substantial NetConverse model, which uses machine learning to analyze network data for Windows ransomware and obtain a high detection rate. The Bayes Network (BN), Logistic Model Tree, K-Nearest Neighbor (KNN), MLP, Decision Tree (J48), and Random Forest (RF) were just a few of the machine learning classifiers that were evaluated using the NetConverse model. Daku et al. [14] worked on identifying and classifying ransomware variants using machine learning approaches. The purpose of this study was to classify ransomware based on their behaviors. Three machine learning algorithms, i.e., J48 Decision Tree, Naive Bayes, and KNN, were considered, among which the J48 Decision Tree accounted for a maximum classification accuracy of 78%.

18.3.3 Cloud-Based Solutions to Tackle Ransomware

Due to the growing amount of data generated every day, organizations have begun keeping their data in cloud-based storage, allowing cloud computing on a global scale. Ransomware and other cyberattacks have been drawn to the cloud, thus making cloud security imperative. Cohen and Nissim [12] proposed a volatility framework for detecting known and unknown ransomware in virtual machines on private clouds by analyzing volatile memory dumps (RAM). Real-world ransomware families were considered for experimentation, and the Random Forest classifier provided the best detection results in all experiments. Another work by Lee et al. [26] examines the use of machine learning-based file entropy analysis to identify ransomware. The authors demonstrated that entropy-based methods correctly detected infected files in most machine learning models and had extremely low false positivity and negativity rates. A different work by Hirano and Kobayashi [18] aims detection of ransomware using machine learning and storage access patterns obtained from a live forensic hypervisor called WaybackVisor. The authors used machine learning based algo-rithms like KNN, SVM, and Random Forest for conducting experiments. Although the studies based on using machine learning to identify ransomware are not confined to those described here, it is undoubtedly restricted to a few circumstances, which keeps ransomware a threat even in the face of such diligent research. The availability of datasets and the changing behavior of evolving ransomware families limit these machine learning models to a certain extent since most of them only make predictions based on static and behavioral analysis, thus demanding more accurate ransomware detection models.

18.3.4 Self-Healing Mechanism for Data Recovery Post Ransomware Attack

Machine learning and backup-based solutions are not the only solutions proposed by the researchers; there is even a lot to them. Some researchers proposed self-healing mechanisms to restore files post-ransomware attacks. An approach to making operating systems more resistant to ransomware and other encryption-based attacks was suggested by Continella et al. [13]. ShieldFs was released to identify ransomware-based behavioral patterns and undo any alterations or harm brought on by ransomware. ShieldFS maintains a database of recently modified files by closely monitoring the file system, in contrast to standard backup systems that keep track of long-term file changes. In this way, ShieldFS ensures better security and a mechanism to self-repair the files after an attack. A similar study by Al-Dwairi et al. [3] proposed a resilient methodology, i.e., Self Healing Version Aware Ransomware Recovery, to protect XML documents from ransomware attacks. The system is comprised of two modules: the first module is a decentralized control system that creates regular backup versions of each file and keeps the newly modified ones, while the second module is an access control module that performs particular actions executed with administrator privileges to protect the backup versions of the file. Berrueta et al. [9] suggested a software application that may restore files encrypted by ransomware in a network shared volume situation. The program examines user activity on a file recorded in network traffic. It incorporates any user edits to rebuild the file's content. It makes it possible to retrieve data from network traffic more straightforwardly. Self-healing techniques are an excellent and surefire way to recover data after a ransomware attack, making them appropriate for circumstances of single extortion. These self-healing processes typically need to catch up in large-scale attacks on companies and multinational companies (MNCs) when double, triple, or quadruple extortion is the norm. It is because data is still exfiltrated and made available on the dark web, raising privacy issues. In addition, self-healing mechanisms provide a technique to retrieve data but do not stop an attack from happening when it first enters the network. Moreover, the system is burdened more by regular data backups. Considering the limitations of the available solutions, many academicians and researchers have also proposed solutions based on blacklisting and whitelisting techniques.

18.3.5 WHITELISTING-BASED APPROACH TO TACKLE RANSOMWARE

Kim and Lee [22] contrasted the standard blacklisting approach with the whitelisting-based method for identifying and preventing ransomware. The typical blacklisting-based approach uses known ransomware code signatures. However, the authors suggested an entirely novel strategy based on whitelisting, which examines how an operating system handles files, i.e., file operation procedures, and integrates access control using a whitelist as part of the file operation mechanism, allowing only authorized actions and restricting all unauthorized operations. In comprehensive research, Kim et al. [21] suggested a whitelist-based approach to identify ransomware in a real-time environment without requiring updates from a reliable third party, thereby preventing the spread of new and well-known ransomware variants. Lee et al. [27] suggested a brand-new anti-ransomware whitelist-based method dubbed "Alohomora." It uses an application's I/O activity as a useful tool to manage I/O whitelisting. When the software application or service submits write request to an SSD, Alohomora provides a program context value associated with the application along with the request. A host CPU register can support this program context value. The SSD then decides if the write request has been pre-approved based on the program context value. This technique prevents ransomware from encrypting or changing files stored on the SSD by guaranteeing that only authorized programs may modify those contents. Despite offering a potential defense against ransomware attacks, whitelisting-based solutions have certain limitations. Privilege escalation attacks make it simple for ransomware attackers to get around whitelist-based access control mechanisms, getting unapproved access and rendering whitelist-based solutions useless. In addition, whitelist-based solutions may not be effective against zero-day attacks, need adequate upkeep and monitoring, and may result in a poor user experience owing to possible false-positives.

18.3.6 IoT-BASED SOLUTION FOR RANSOMWARE

Another area that has drawn cybercriminals to gain access is the IoT. Whether it's a simple smartwatch, a smart TV, a smart house, or a smart city, we are all linked to

IoT devices and use them in our day-to-day lives. It necessitates an IoT ecosystem that is truly safe and secure for computing. Azmoodeh et al. [5] proposed a framework to detect ransomware in IoT networks by analyzing their energy consumption. The authors propose distinguishing ransomware from non-malicious apps based on power usage patterns. The process entails breaking up the sequence of energy various applications use into smaller chunks known as subsamples. Then, these subsamples are categorized to determine if they are ransomware or non-malicious software. The method produced a detection rate of 95.65% and an accuracy rate 89.19% during a series of tests. Since there haven't been significant ransomware attacks on IoT networks, a limited number of solutions are available for the ransomware problem.

18.3.7 Moving Target Defense-Based Approach for Defense Against Ransomware

Using the idea of moving target defense (MTD), Lee et al. [28] proposed a method for dealing with ransomware by altering the file extensions that are more susceptible to encryption. Essentially, ransomware employs a whitelisting-based scheme to encrypt files with specific extensions. Monge et al. [31] proposed a strategy for defending against cryptographic ransomware that uses self-organizing techniques and cutting-edge communication technology to coordinate various defense measures depending on the circumstances and ransomware risk level. However, these solutions don't ensure a suitable ransomware solution because the majority of studies are based on assumptions, and using an MTD-based defense could affect system performance by constantly changing system attributes, giving false positives and false negatives, which challenge its accuracy and also give rise to operational overhead. Additionally, MTD increases complexity, as a result of which attackers may create countermeasures to detect and exploit patterns in MTD-based systems.

18.3.8 Cyber Deception-Based Solution to Tackle Ransomware

Since attackers have been deceiving end users and organizations for years to access their networks, security professionals have also developed the notion of cyber deception to deceive attackers. Figure 18.5 shows the conventional cyber deception environment for a typical enterprise network. Moore [32] proposed a decoy-based cyber deception system that creates fake directories to lure and trap the ransomware attacker, thus restricting access to the original system or network. Nevertheless, the success of using particular files to entice the malware was restricted since it is difficult to get the malware to interact with those files. Another study by Wang et al. [41] proposed RansomTracer – a cyber deception-based strategy to trace ransomware. Their research demonstrates that RansomTracer successfully sets traps in the decoy environment and gathers as many clues as possible. Then, it examines any clues that can point to the attacker. However, the likelihood of ransomware falling into the deception trap and dealing with zero-day exploits could be RansomTracer's largest limitation. In addition, ransomware may employ cutting-edge strategies like obfuscation or an anti-analysis mechanism to circumvent the decoy environment.

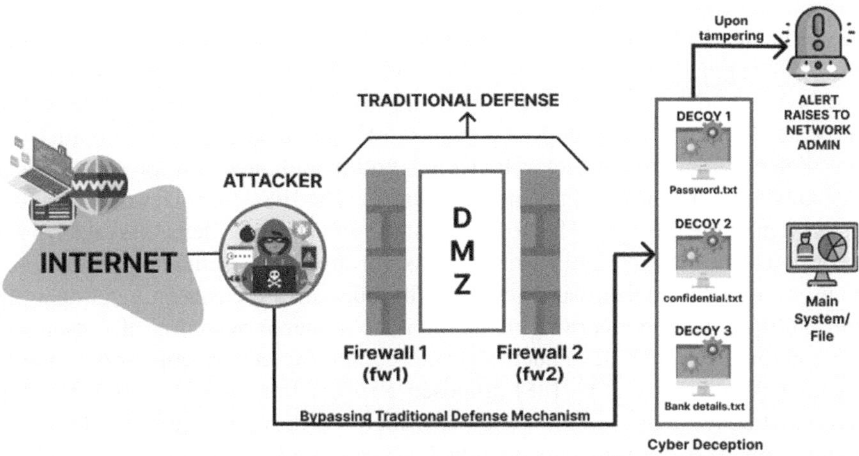

FIGURE 18.5 Traditional cyber deception model.

Gómez-Hernández et al. [16] presented R-Locker, a technique for preventing ransomware by using a unique file-based trap. When ransomware tries to access such a file, it gets blocked, preventing the encryption of the other files. This methodology was used with the Linux operating system and showed effective outcomes. However, the technique wasn't tested on other operating systems like Windows, MacOS, or others because it was developed for the Linux operating system, so it doesn't guarantee 100% protection when the operating systems' diversity is considered. Apart from that, zero-day exploits continue to be a big issue. Gómez-Hernández et al. [17] developed and released an enhanced version of R-Locker that is compatible with the commonly used Windows operating system to solve the issue of compatibility to a certain operating system exclusively. A novel method for distributing and managing honey files is proposed in the study, and the use of lists to automatically discern between safe processes and ransomware activities was also taken into consideration. Despite positive outcomes, the experiment failed to identify Cerber ransomware. Saleh et al. [37] presented SentryFS, a model to make honey files more believable to lure the ransomware more effectively to it. This was done to solve the uncertainty of ransomware going into the trap. These honey files are created to appear more appealing to intelligent ransomware that picks out its victim files with care. Natural Language Processing (NLP) is used to create the content of these honey files, and their metadata is often changed to increase their attraction to ransomware. SentryFS links to a web service for anti-ransomware that offers the most recent information on new ransomware techniques, allowing the honeyfiles to be updated appropriately. By creating file clones that are identical copies of the original files, SentryFS offers an extra degree of protection. If a ransomware attack is successful and remains unnoticed, it encrypts copies of the files rather than the originals.

More research was available on a honeypot-based solution since it appeared more promising than others. To trick the ransomware into falling into the trap, Pascariu and

Barbu [36] researched various honeypot-based solutions and offered a low-power, low-cost embedded device that the host network shares with decoy files. The primary goal of this approach was to redesign the existing honeypot concept so that the software layer could be implemented on routers and other types of consumer networking hardware. Ng et al. [20] proposed a framework that consisted of six modules (IPS, static detector, dynamic detector, honeypot, notification, and gateway) for detecting and analyzing ransomware in computer systems. The IPS recognizes downloads and sends them to the gateway. The honeynet, static and dynamic detectors analyze samples to identify the kind and family of ransomware. The notification module alerts the user about any ransomware attempt. The study also focuses on a voting system with multiple machine learning models for the dynamic detection of ransomware. The emergence of IoT demanded a secure computing ecosystem. Addressing this issue, Chakkaravarthy et al. [38] proposed a robust Intrusion Detection Honeypot (IDH) that consists of (1) a Honeyfolder, which is a decoy directory modeled using Social Leopard Algorithm (SoLA), which acts as a bait for ransomware attacks and serves as an early warning system, thus alerting the user when any suspicious activity occurs in the file system; (2) an Audit Watch which acts as an entropy module that examines the randomness level or predictability in files and folders; and (3) the Complex Event Processing engine that aggregates data from different security systems to identify ransomware behavior and respond promptly. Essentially, IDH is the successor to IDPS. Although there were many other solutions by different academics, researchers, and security specialists, the methods have several drawbacks that still need to be solved today, including the issue of cyber attribution. We witness ransomware attacks, attempt to thwart them, and recover data after an attack, but we can never determine who is behind them. The face behind an attack remains a question mark. Wang et al. [40], in their study, attempted to trace back the source of the Remote Desktop Protocol-based ransomware attack. They proposed a method that involves luring attackers to a deceptive environment that collects identifiable clues left by attackers, analyzes them using NLP and machine learning based techniques, and finally traces them back.

18.4 TIT-FOR-TAT MODEL FOR DEFENSE AGAINST RANSOMWARE

This section presents a deception-based VDE to deal with ransomware. In any network environment, there could be two scenarios, i.e., authorized user access or unauthorized user access. When an unauthorized user penetrates the network and bypasses the traditional defense system, it will face the VDE, as shown in Figure 18.6.

18.4.1 TRAP FOR THE UNAUTHORIZED USERS

A VDE is a network of dummy systems that appear genuine but a fake and tempting version of the system that lures the malicious user to access or operate on those systems or files. Once the files or systems are tampered with, the VDE invokes the

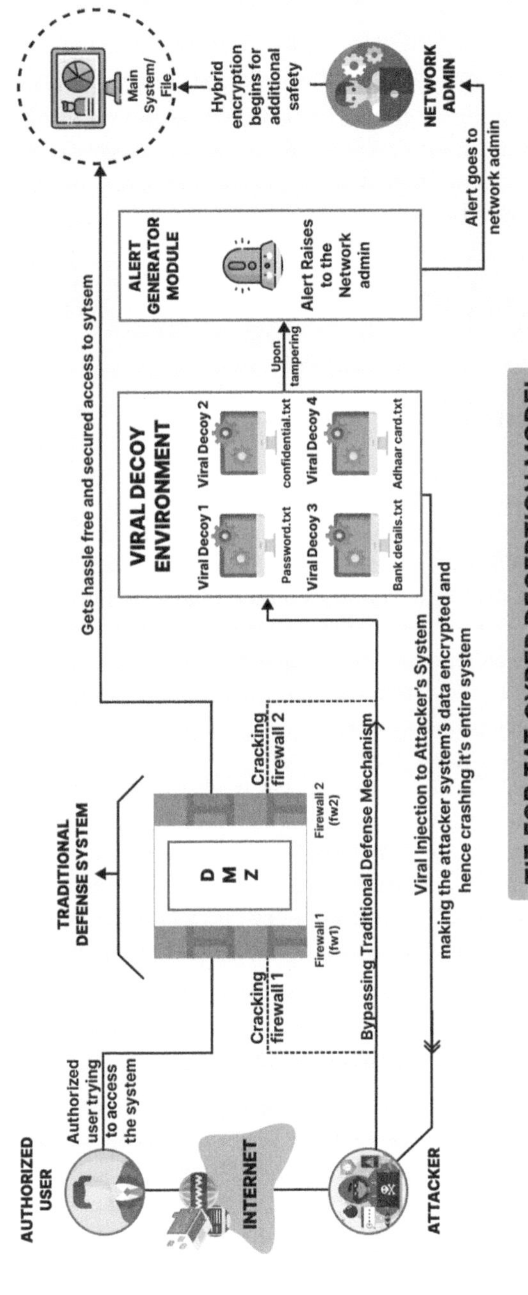

FIGURE 18.6   Tit-for-Tat cyber deception model.

alert generator module that generates two alerts to the security administrator, one indicating that someone has entered the VDE and another signal on files or systems being tampered with in that network. It will help the network administrator identify the particular IP address and block its access to the system. Consequently, the proposed trap mechanism will block real users/IPs from accessing the resources. Things are not like an open and shut case here, but there is more than it appears. It provides us with a methodology not only to tackle the double extortion techniques [33] followed by the ransomware families to steal data before encryption and sell it to the dark net but also to gather information about the attacker's system and destroy all the files and credentials stored by the attacker. What if the attacker follows techniques like IP address mutation to change their IP address every moment, thus protecting themselves from getting tracked easily? Our deception strategy also has a method to tackle this, which is elaborated further in the following subsection.

18.4.2　Viral Injection to the Attacker

Consequently, this is where the actual game starts! Our "Tit-for-Tat" deception system injects a piece of malware, often ransomware created by the victim organization, onto the attacker system during data exfiltration, in contrast to conventional deception systems that just entice the attacker into a trap and prevent them. As described in Section 18.2, most ransomware families target a large organization and use double extortion or even more extreme forms to steal their data and sell it on the dark web. Therefore, when they attempt to take the data, they steal it from the VDE while seeming to have access to the actual database. In reality, however, they steal a ransomware file that the network administrator on the victim end controls. Using just one malware, the victim organization can compromise the attacker's system, wipe out the stolen data to ensure data privacy, and track the attacker for further threat intelligence. The injected malware employs the same operational procedures as a conventional ransomware strain. The tables have now finally turned! It makes it an aggressive defense technique against the ransomware family since the attacker is now on the victim's end, and the victim turns out to be an attacker. C&C paved a route to the attacker's system, much like the attacker attempted to do with our system.

18.4.3　Worst Cases Are handled as well!

Although finding ways to go around or trick the deception system and effectively get access to the target is highly challenging, it is not impossible. In such a situation, we also have a proactive policy to guard against data breaches. Even in the unlikely event that the attacker manages to evade the VDE, there is still a very high likelihood of misdirecting the attacker to the VDE because the attacker was not a legitimate member of the target organization. The alert generator module will notify the network administrators of the security breach by triggering the first warning, indicating the presence of an unauthorized user in the VDE. But what if the second warning intended to reveal the altered files is not triggered? Such a situation may indicate that the attacker is aware of the trap and working to find a means to access the system. It might prompt the network administrators to launch file encryption techniques on the mission or business-critical systems. The network administrators might devise some clever methods the attacker uses to encrypt massive data quickly, given that the amount of data is sufficiently significant to be encrypted quickly. Network administrators might use hybrid encryption, which ransomware families typically do. The data could first be encrypted using symmetric encryption using the Advanced Encryption Standard (AES) algorithm [1] to create a public key. Then the generated public key could be encrypted using asymmetric encryption using the RSA algorithm [23] to create a public and a private key, double-encrypting the data in a shorter time. In addition, the administrator can back up the encrypted data into a cloud server entirely separate from the infected network chain. Now, even if the attacker were to steal the data, it would be of little use since it would already have been encrypted, and the only person who possesses the decryption keys is the network administrator. What if the ransomware already encrypts the system's files? So why should we be concerned? Because the backup data is already available to continue business operations.

18.4.4   Green Corridors for Authorized Users

According to our approach, any successful organization may satisfy its responsibilities by being straightforward. We adhere to the abstraction principle, whereby the authorized user is constantly unaware of any complexity, preventing them from entering the VDE and granting them more straightforward access to the primary system. Such an arrangement encourages confidence inside a company and guarantees dependable functioning for authorized users. Remember that only IP addresses that forcefully enter the network and are not a part of the organization are captured by the VDE and all other unauthorized data access attempts. It would be dubious to claim that the "Tit-for-Tat" model will be the most effective, loophole-free means of preventing ransomware attacks. However, it offers a proactive approach for spotting and reducing ransomware at its earliest phases of network intrusion. Our model uses a more effective version of a cyber deception approach and reactive defense strategy to counter the ransomware families.

18.5   DISCUSSION

Ransomware has been an unavoidable menace to the cyber world and economies, despite numerous academics and cyber specialists putting out several detection models and approaches. The proposed viral decoy-based environment offers a mechanism to directly interact with the attacker, allowing us to track him down and take him out only using his tactics. The VDE-based solution not only gives authorized users a safe means to access the services but also makes it impossible for unauthorized users to access the data by forcing them to enter the VDE. Unfortunately, no tests were performed as a proof-of-concept; hence, the suggested framework's solutions for cyber attribution and ransomware are based only on theorizing. To verify the model's practical applicability and compare its accuracy with state-of-the-art solutions, we want to undertake an experiment in the future under secured circumstances with the prior permission of all the parties involved, thus abiding by the law. Meanwhile, considering the hypothetically suggested solution, if this model resolves the problem of cyber attribution, the frequency of ransomware attacks might reduce to a great extent as the attacker's anonymity is compromised. Although it dramatically enhances the idea of the conventional honeypot and honeyfile systems and offers a great answer to cyber attribution, it still has several significant flaws that must be addressed.

Furthermore, a few technological and ethical concerns continue to stand in the way of the suggested solution. One such worry is what happens if the adversary compromises a network of computers using spoofing to deploy malicious payloads from a remote computer. It will result in a partial failure of the suggested system since, even if the data is saved anyhow, the issue of cyber attribution will continue to be terrifying. The evil character of the approach that reverses attacks on the system from which the malicious payloads are originating is another worry or ethical concern. It can further sour international ties between nations.

What if the attacker employs cutting-edge tactics like spoofing to conceal its identity and remotely controls another system to launch malicious payloads on the target system? A straightforward illustration might help you to understand this. For instance, firm X wishes to destroy firm Z because of business rivalry or competition. Firm X, which has a prominent international presence, will never want its reputation to suffer, so even if it is already anonymous before an attack, it might make X even more determined to protect it by initiating a ransomware attack. A situation where firm X might deploy malicious payloads to the targeted company Z using spoofing on some company Y and get remote access to their system is possible. Can the suggested framework identify the offender, assuming that the above case scenario occurred? No, the attack was carried out remotely by X through firm Y on firm Z. Therefore, our suggested approach will identify firm Y as the attacker while the actual offender, firm X, will stay hidden. And this is where our model will fail to recognize the actual offender. However, this strategy may become popular with all RaaS affiliates that deploy malicious payloads to their target victims. Most of these affiliates come from non-technical backgrounds and use ransomware kits purchased from RaaS operators. If applied here, the proposed model is more likely to identify the affiliates who launched the attack because they are limited to launching attacks only rather than using obfuscation techniques like spoofing. This might ultimately result in the breakdown of the ransomware shadow economy by not only assisting in identifying and prosecuting cybercriminals linked to certain ransomware families and RaaS but also negatively affecting the income and profitability of the RaaS business model. Because of the removal of widespread tentacles of the RaaS model and only the developers could carry out attacks locally, a drop in ransomware occurrences may be observed. The suggested model also has other issues, but several more pressing ones need to be addressed.

Recall the example of a commercial conflict between the three companies X, Y, and Z? What if the same event happened worldwide? The model can result in strained ties between the nations and their boundaries. A straightforward illustration might help you understand this. Consider the scenario where nation A subverts and compromises country B's system to send malicious payloads to country C through country B without country B's knowledge. If the proposed "Tit-for-Tat" model is in this case, an innocent nation may be painted as an aggressor, and the real culprit will be given free reign to launch attacks. As a result, there may be tensions between the nations on the world stage, and in the worst-case scenario, there may even be cyberwar. This problem of not analyzing the targeted culprit system is called an unanalyzed revert system attack. However, the problem continues. Hacking poses a significant risk and necessitates proper planning before execution. Hacking back and getting into the attacker's system could lead to retaliation for an attack the targeted organization might not be ready to face. Being offensive here could lead to more losses rather than getting profitable.

The proposed model also poses some serious ethical concerns that should be noted. Hacking back as a part of an offensive defense approach could be illegal per the laws and regulations. The Computer Fraud and Abuse Act [43] of 1986, a bill passed by the United States legislation, clearly states that no retaliation or hack-back technique would be used as a measure of cyber protection, thus posing one of the most ethical barriers to our proposed model. Therefore, being defensive is a better option rather than getting offensive.

18.6 CONCLUSIONS AND FUTURE WORK

The ultimate objective of organizations and governments is to maintain the security posture (confidentiality, integrity, and availability) of the valuable business or mission-critical data they possess. Each sensitive or critical data item collected, stored, processed, transmitted, and accessed must be protected against potential cyberattacks. The loss of crucial or strategic data availability renders any organization or government incapable of achieving its objectives. Ransomware, a cyber criminals business model, threatens the business operations of organizations by encrypting business-critical data. Even though organizations maintain backup copies of data to restore business operations, adversaries try to extort companies by selling the exfiltrated data to business competitors or in the dark net. In this chapter, we have discussed ransomware, its modus operandi, its consequences, and existing solutions to detect and prevent such attacks. Furthermore, we presented the need for a VDE to detect and prevent ransomware attacks. Finally, we discussed the issues, challenges, and ethical concerns in implementing the proposed framework. Our future work will implement the proposed framework and verify the model's practical applicability compared to state-of-the-art ransomware defense solutions.

REFERENCES

1. Ako Muhamad Abdullah et al. Advanced encryption standard (AES) algorithm to encrypt and decrypt data. *Cryptography and Network Security*, 16(1):11, 2017.
2. Jinwoo Ahn, Donggyu Park, Chang-Gyu Lee, Donghyun Min, Junghee Lee, Sungyong Park, Qian Chen, and Youngjae Kim. KEY-SSD: Access-control drive to protect files from ransomware attacks. *arXiv preprint arXiv:1904.05012*, 2019.
3. Mahmoud Al-Dwairi, Ahmed S Shatnawi, Osama Al-Khaleel, and Basheer Al-Duwairi. Ransomware-resilient self-healing XML documents. *Future Internet*, 14(4):115, 2022.

4. Omar MK Alhawi, James Baldwin, and Ali Dehghantanha. Leveraging machine learning techniques for windows ransomware network traffic detection. *Cyber Threat Intelligence*, pages 93–106, 2018. Springer.

5. Amin Azmoodeh, Ali Dehghantanha, Mauro Conti, and Kim-Kwang Raymond Choo. Detecting crypto-ransomware in IoT networks based on energy consumption footprint. *Journal of Ambient Intelligence and Humanized Computing*, 9(4):1141–1152, Aug 2018.

6. Claudine Badue, Rânik Guidolini, Raphael Vivacqua Carneiro, Pedro Azevedo, Vinicius B Cardoso, Avelino Forechi, Luan Jesus, Rodrigo Berriel, Thiago M Paixao, Filipe Mutz, et al. Self-driving cars: A survey. *Expert Systems with Applications*, 165:113816, 2021.

7. SungHa Baek, Youngdon Jung, Aziz Mohaisen, Sungjin Lee, and DaeHun Nyang. SSD-Insider: Internal defense of solid-state drive against ransomware with perfect data recovery. In *IEEE 38th International Conference on Distributed Computing Systems (ICDCS)*, pages 875–884, 2018.

8. James Baldwin and Ali Dehghantanha. Leveraging support vector machine for opcode density based detection of crypto-ransomware. *Cyber Threat Intelligence*, pages 107–136, 2018. Springer.

9. Eduardo Berrueta, Daniel Morato, Eduardo Magaña, and Mikel Izal. Ransomware encrypted your files but you restored them from network traffic. In *2nd Cyber Security in Networking Conference (CSNet)*, pages 1–7, 2018.

10. Li Chen, Chih-Yuan Yang, Anindya Paul, and Ravi Sahita. Towards resilient machine learning for ransomware detection. *arXiv preprint arXiv:1812.09400*, 2018.

11. Qian Chen and Robert A. Bridges. Automated behavioral analysis of malware: A case study of wannacry ransomware. In 16th *IEEE International Conference on Machine Learning and Applications (ICMLA)*, pages 454–460, 2017.

12. Aviad Cohen and Nir Nissim. Trusted detection of ransomware in a private cloud using machine learning methods leveraging meta-features from volatile memory. *Expert Systems with Applications*, 102:158–178, 2018.

13. Andrea Continella, Alessandro Guagnelli, Giovanni Zingaro, Giulio De Pasquale, Alessandro Barenghi, Stefano Zanero, and Federico Maggi. ShieldFS: A self-healing, ransomware-aware filesystem. In *Proceedings of the 32nd Annual Conference on Computer Security Applications, (ACSAC'16)*, page 336–347, New York, NY, USA, 2016. Association for Computing Machinery.

14. Hajredin Daku, Pavol Zavarsky, and Yasir Malik. Behavioral-based classification and identification of ransomware variants using machine learning. In *17th IEEE International Conference on Trust, Security and Privacy in Computing and Communications/12th IEEE International Conference on Big Data Science and Engineering (TrustCom/BigDataSE)*, pages 1560–1564. IEEE, 2018.

15. Nicolaie L. Fantana, Till Riedel, Jochen Schlick, Stefan Ferber, Jürgen Hupp, Stephen Miles, Florian Michahelles, and Stefan Svensson. IoT applications—value creation for industry. In *Internet of Things*, pages 153–206. River Publishers, 2022.

16. J.A. Gómez-Hernández, L. Álvarez González, and P. García-Teodoro. R-Locker: Thwarting ransomware action through a honeyfile-based approach. *Computers & Security*, 73:389–398, 2018.

17. José Antonio Gómez-Hernández, Raúl Sánchez-Fernández, and Pedro García-Teodoro. Inhibiting crypto-ransomware on windows platforms through a honeyfile-based approach with R-Locker. *IET Information Security*, 16(1):64–74, Sep 2021.

18. Manabu Hirano and Ryotaro Kobayashi. Machine learning based ransomware detection using storage access patterns obtained from live-forensic hypervisor. In *6th International Conference on Internet of Things: Systems, Management and Security (IOTSMS)*, pages 1–6, 2019.

19. Mostafa Haghi Kashani, Mona Madanipour, Mohammad Nikravan, Parvaneh Asghari, and Ebrahim Mahdipour. A systematic review of IoT in healthcare: Applications, techniques, and trends. *Journal of Network and Computer Applications*, 192:103164, 2021.

20. Chee Keong Ng, Sutharshan Rajasegarar, Lei Pan, Frank Jiang, and Leo Yu Zhang. VoterChoice: A ransomware detection honeypot with multiple voting framework. *Concurrency and Computation: Practice and Experience*, 32(14):e5726, July 2020.

21. Dae-Youb Kim, Geun-Yeong Choi, and Ji-Hoon Lee. White list-based ransomware real-time detection and prevention for user device protection. In *IEEE International Conference on Consumer Electronics (ICCE)*, pages 1–5, 2018.

22. DaeYoub Kim and Jihoon Lee. Blacklist vs. whitelist-based ransomware solutions. *IEEE Consumer Electronics Magazine*, 9(3):22–28, 2020.

23. Çetin Kaya Koç, Funda Özdemir, Zeynep Ödemis Özger. Rivest-shamir-adleman algorithm. *Partially Homomorphic Encryption*, pages 37–41, 2021. Springer.

24. Alexander Kott. Challenges and characteristics of intelligent autonomy for internet of battle things in highly adversarial environments. *arXiv preprint arXiv:1803.11256*, 2018.

25. Kyungroul Lee, Insu Oh, and Kangbin Yim. Ransomware-prevention technique using key backup. In Jason J. Jung and Pankoo Kim, editors, *Big Data Technologies and Applications*, pages 105–114, Cham, 2017. Springer International Publishing.

26. Kyungroul Lee, Sun-Young Lee, and Kangbin Yim. Machine learning based file entropy analysis for ransomware detection in backup systems. *IEEE Access*, 7:110205–110215, 2019.

27. Sanggu Lee, Yoona Kim, Dusol Lee, Inhyuk Choi, and Jihong Kim. Alohomora: Protecting files from ransomware attacks using fine-grained I/O whitelisting. In *Proceedings of the 14th ACM Workshop on Hot Topics in Storage and File Systems*, HotStorage '22, page 113–118, New York, NY, USA, 2022. Association for Computing Machinery.

28. Suhyeon Lee, Huy Kang Kim, and Kyounggon Kim. Ransomware protection using the moving target defense perspective. *Computers & Electrical Engineering*, 78:288–299, 2019.

29. Antonio López Martínez, Manuel Gil Pérez, and Antonio Ruiz-Martínez. A comprehensive review of the state-of-the-art on security and privacy issues in healthcare. *ACM Comput. Surv.*, 55(12), 2023.

30. Donghyun Min, Donggyu Park, Jinwoo Ahn, Ryan Walker, Junghee Lee, Sungyong Park, and Youngjae Kim. Amoeba: An autonomous backup and recovery SSD for ransomware attack defense. *IEEE Computer Architecture Letters*, 17(2):245–248, 2018.

31. Marco Antonio Sotelo Monge, Jorge Maestre Vidal, and Luis Javier García Villalba. A novel self-organizing network solution towards crypto-ransomware mitigation. In *Proceedings of the 13th International Conference on Availability, Reliability and Security*, ARES 2018, New York, NY, USA, 2018. Association for Computing Machinery.

32. C. Moore. Detecting ransomware with honeypot techniques. In *Cybersecurity and Cyberforensics Conference (CCC)*, pages 77–81, Los Alamitos, CA, USA, Aug 2016. IEEE Computer Society.

33. Harun Oz, Ahmet Aris, Albert Levi, and A Selcuk Uluagac. A survey on ransomware: Evolution, taxonomy, and defense solutions. *ACM Computing Surveys (CSUR)*, 54(11s): 1–37, 2022.

34. Joon-Young Paik, Joong-Hyun Choi, Rize Jin, Jianming Wang, and Eun-Sun Cho. A storage-level detection mechanism against crypto-ransomware. In *Proceedings of the 2018 ACM SIGSAC Conf. on Computer and Communications Security*, pages 2258–2260, 2018.

35. Jisung Park, Youngdon Jung, Jonghoon Won, Minji Kang, Sungjin Lee, and Jihong Kim. RansomBlocker: A low-overhead ransomware-proof SSD. In *Proceedings of the 56th Annual Design Automation Conference 2019, DAC '19*, New York, NY, USA, 2019. Association for Computing Machinery.

36. Cristian Pascariu and Ionut-Daniel Barbu. Ransomware honeypot: Honeypot solution designed to detect a ransomware infection identify the ransomware family. In *11th International Conference on Electronics, Computers and Artificial Intelligence (ECAI)*, pages 1–4, 2019.

37. Abdul Rahim Saleh, Gihad Al-Nemera, Saif Al-Otaibi, Rashid Tahir, and Mohammed Alkhatib. Making honey files sweeter: SentryFS–a service-oriented smart ransomware solution. *arXiv preprint arXiv:2108.12792*, 2021.

38. S. Sibi Chakkaravarthy, D. Sangeetha, Meenalosini Vimal Cruz, V. Vaidehi, and Balasubramanian Raman. Design of intrusion detection honeypot using social leopard algorithm to detect iot ransomware attacks. *IEEE Access*, 8:169944–169956, 2020.

39. R. Vinayakumar, K.P. Soman, K.K. Senthil Velan, and Shaunak Ganorkar. Evaluating shallow and deep networks for ransomware detection and classification. In *International Conference on Advances in Computing, Communications and Informatics (ICACCI)*, pages 259–265, 2017.

40. ZiHan Wang, ChaoGe Liu, Jing Qiu, ZhiHong Tian, Xiang Cui, and Shen Su. Automatically traceback RDP-based targeted ransomware attacks. *Wireless Communications and Mobile Computing*, 2018, 1–13.

41. ZiHan Wang, Xu Wu, ChaoGe Liu, QiXu Liu, and JiaLai Zhang. RansomTracer: Exploiting cyber deception for ransomware tracing. In *IEEE Third International Conference on Data Science in Cyberspace (DSC)*, pages 227–234. IEEE, 2018.

42. Mohammad Zarei, Ayoub Mohammadian, and Rohollah Ghasemi. Internet of things in industries: A survey for sustainable development. *International Journal of Innovation and Sustainable Development*, 10(4):419–442, 2016.

43. S. A. Constant (2013). The Computer Fraud and Abuse Act: A Prosecutor's Dream and a Hacker's Worst Nightmare – The Case Against Aaron Swartz and the Need to Reform the CFAA. *Tulane Journal of Technology and Intellectual Property*, 16: 231–248.

19 CouPLeD

Cloud-Enabled Artificial Neural Network Framework for Plant Leaf Disease Detection

Vivek Kumar Singh

19.1 INTRODUCTION

Agricultural activities becoming more economically efficient is one of the main objectives for human agrarian. To achieve this objective many approaches have been suggested, including traditional and technological ones, but still it lacks behind the desired target due to the challenges in establishing quality and cost balance [27]. There are several factors, namely climate change, quality of grains, and plant diseases, that impact crop production and its quality. Thus, agrarian productions are heavily threatened by various plant diseases which influence the economy of agriculture and life of people associated with farming activities [16]. Therefore, disease detection in plant leaves attracts more attention to prevent the effect of the disease in geographically connected plants and the reduction of economic impacts in agriculture. These diseases originate mainly from bacteria and fungi in different plants [8]. Typically, the plant leaf disease identification approaches traditionally depend on either the plant protector's observation or the molecular examination [14]. The first approach is complicated and required well-equipped and high-precision centralized labs to conduct experiments, whereas the second approach takes a long time for detecting the plant disease, which is prone to insignificant outcomes. One of the major challenges is early stage plant leaf disease symptom identification. In recent years, plant disease identification has been a crucial issue because it plays a vital role in agricultural production and their quality.

Currently, image processing and analysis techniques are being widely employed in various interdisciplinary applications, such as image classification [12], medical imaging [13], object recognition [21], etc. Automatic recognition of plant leaf diseases in natural scenarios by using image-based technologies and machine learning is a promising approach to effectively identify plant diseases. Automation in plant disease detection reduces time consumption and overall costs required for disease detection [14]. Further, advances in computer vision are exploited in various applications such as intelligent visual processing and analysis, which provide strong evidence that plant leaf image processing and analysis can also contribute a novel direction for plant disease identification. Therefore, computer vision and machine

DOI: 10.1201/9781003390954-19

learning are widely exploited in many plant leaf disease detection methods in the literature. However, real-time plant leaf disease identification still has several noteworthy challenges, such as the severity of the disease due to the plant leaf image being captured directly from the crop field and complex background.

19.1.1 MOTIVATION

Visual identification of plant diseases by agronomists and experts is the traditional method used in various geographical locations, which is inefficient in terms of time. In contrast laboratory molecular experiments are very costly because they require a well-equipped centralized lab. Plant leaf disease identification is critical to control crop diseases over fields and economic losses in agriculture. Therefore, plant disease detection requires an efficient and robust automatic approach which detects the disease in a correct and timely manner. As delineated in Figure 19.1, the first row represents healthy or non-disease leaf images that visually appear with green color without any attractive region. ,while the second row contains unhealthy or diseased leaf images that visually appear with different colors and the regions which are affected by the disease are visually more attractive. Therefore, plant leaf disease can be detected via image processing that provides methods to analyze the plant leaf images and classify them into either disease or non-disease leaves. Motivated by this, an efficient and robust plant disease detection method is proposed in this chapter that effectively addresses the challenges of complex background in plant leaf images in a timely manner.

This chapter proposes an automatic plant disease detection method using image processing technologies and machine learning algorithms. Initially, the plant leaf images are collected from real-world agricultural environments through a digital camera and all the images are uploaded over a cloud platform for fast communication and effective usage of the computing resources. Then, these images are preprocessed

FIGURE 19.1   Example of Eggplant leaf images. (Top row) healthy leaf and (Bottom row) diseased leaf.

to capture nois-free plant leaves. In addition, image augmentation is performed over the preprocessed images which includes image flipping, cropping, rotation, and color transformation. Then, each image is partitioned into regions based on their similarity and correlations which increases the computational efficiency of the proposed framework. Afterward, Artificial Neural Network (ANN) is employed to learn a model for plant disease classification. This approach will detect diseases in time and will also reduce the overall implementation cost. The method will effectively handle the diversity present in the crop leaf visuals including complex background images in a timely manner. In order to ensure the classification accuracy of ANN, a large amount of labeled training data is required. Therefore, it is essential to generate labeled data. The proposed method develop a set of labeled plant leaf images that are annotated by various observers.

19.2 RELATED WORK

In recent years, plant leaf image-based disease detection is one of the most attractive research domains due to its significance in crop production and safety. The occurrence of plant leaf diseases impacts crop production and the reduction of economic losses in agriculture. To address plant leaf disease detection many machine learning and computer vision-based approaches have been suggested in the literature. Here, we review some recent global advancements in plant disease detection. Ashourloo et al. [4] suggested an approach which employed machine learning method for plant disease detection. Specifically, they suggested the method for wheat leaf rust disease classification. Further, the approach provides a detailed study for the evaluation of training data volume and identification of factors involved in influencing the symptoms of various diseases that effects the performance of the method. A probabilistic programming approach is employed for plant leaf disease recognition that exploits Bayesian deep learning methods with uncertainty in terms of misclassification measurement [9]. This method generates effective results due to Bayesian inference which is comparable to other counterpart methods. Simultaneously, the posterior density is exploited for the plant leaf disease recognition and analysis of uncertainty of unseen data samples during predictions.

Cap et al. [5] introduced a novel image-to-image translation architecture called LeafGAN. This architecture has an effective attention mechanism that improves recognition performance. The proposed architecture is enabled with features to transform only relevant regions of leaf images with heterogeneous and complex backgrounds that enriched a wide variety of training leaf images in the dataset. LeafGAN is utilized for data augmentation that generates different diseased leaf images through the transformation of non-diseased leaf images. This process effectively increases the significant volume of the dataset and this step helps for enhancing the recognition performance of the architecture. Ahmad et al. [3] suggested a plant leaf disease recognition method that exploited deep learning methods in natural data. The method effectively identifies plant diseases by employing memory efficient learning networks. Further, the proposed training configuration enables the method to rapidly learn patterns from the training dataset which is useful for developing industrial applications due to minimizing the learning time. Lastly, the class imbalance issue in real-world datasets is tackled with an efficient statistical technique. Liu et al. [14] systematically investigated the plant leaf disease recognition problem through computer vision and image processing jointly with the guidance of agrarian experts. Then, they built a novel plant leaf disease dataset on a large-scale consisting of 220,592 images with 271 plant disease categories. Further, a plant disease-oriented architecture is introduced for plant leaf disease classification by exploiting effective features of leaf images. Kumar et al. [11] introduced an efficient system for the recognition of a wide variety of fungal diseases, namely rust, powdery mildew, anthracnose, and root rot/leaf blight. The system employed a multi-layered perceptron framework for the predictions of various plant leaf diseases. The proposed method effectively detected the plant diseases and also increased the crop production with good quality.

Udutalapally *et al.* [26] introduced a novel method called Internet-of-Agro-Things that describes the working of an automatic plant leaf disease recognition for applications of Agriculture Cyber-Physical System. Abbas et al. [1] suggested tomato plant leaf disease classification method through deep learning technique. This approach utilized the Conditional Generative Adversarial Network for constructing non-natural images of tomato plant leaves and classified into ten categories of diseases via a DenseNet121 model that is trained on non-natural and natural images based on transfer learning. Dwivedi et al. [7] proposed an Region-based Convolutional Neural Network (R-CNN) and multitask learning architecture for grape plant leaf disease classification. This architecture is employed to classify black-rot, isariopsis, and esca diseases on the plant leaf images. Sunil et al. [24] introduced a leaf disease analysis learning architecture, U2-Net, for cardamom plant leaf images. This architecture effectively suppressed the complex and cluttered background for the leaf images to increase the ability of classification. In this architecture, EfficientNet, convolutional neural networks (CNN), and EfficientNetV2 models are trained for recognition of the plant leaf diseases. Hassan and Maji [8] presented a novel deep learning method that exploited the residual connection and inception layer to diagnose and recognize plant leaf diseases. Further, an optimization approach is used to reduce the model parameters via depthwise separable convolution. To ensure reliable performance the proposed model has been learnt over a variety of plant leaf disease datasets. The performance of the suggested method is reliable and efficient on similar and simple leaf images whereas its performance may be degraded over unseen and complex leaf images. Sharma et al. [23] introduced Deeper lightweight convolutional neural network architecture (DLMC-Net) architecture for plant leaf identification. They employed a sequence of collective blocks via the passage layer for extracting significant deep visual features. The proposed architecture has the ability to tackle the vanishing gradient problem due to utilizing deep visual features. Further, separable convolution blocks are employed to optimize the model parameters. The proposed model performs well on a wide variety of plant leaf images.

Vallabhajosyula et al. [28] analyzed the significance of several pre-trained neural networks and demonstrated the experimental analysis of a weighted ensemble of different methods for refining plant disease recognition performance. The proposed model effectively combined various pre-trained neural networks models and robustly categorized different plant diseases. Sahu and Pandey [22] suggested a novel architecture based on hybrid random forest multiclass SVM (HRF-MCSVM) that recognizes specific disease (i.e., foliar disease). In this architecture, the visual features are preprocessed and segmented by exploiting spatial Fuzzy C-means to improve the computation accuracy for the classification process. The HRF-MCSVM architecture is computationally effective due to suppressing irrelevant visual information at the initial stage. Pal *et al.* [18] introduced an Agriculture Detection model for plant disease detection. This model utilized conventional Inception-Visual Geometry Group Network (INC-VGGN) along with Kohonen-based deep learning networks for effectively classifying various types of plant leaf diseases. Initially, the input leaf image is preprocessed to remove all the destructive information from the captured plant leaf image. Then, the multivariate grabcut method is employed for meaningful partition that benefits the handling of the occlusion problem. Further, the pre-trained INC-VGGN model is utilized for accurate disease detection and recognition. They utilized weights and the features from the pre-trained model to develop a novel neural network for the specific task of plant disease detection. In addition, the proposed framework overcomes the overfitting problem by incorporating a dropout layer in the neural network. The efficacy of the proposed framework is validated via statistical analysis ensuring that the proposed framework performs well in various complex conditions.

Mattihalli *et al.* [15] introduced an approach for early recognition of plant leaf diseases. Thai *et al.* [25] introduced a transformer-based plant leaf disease recognition approach (i.e., FormerLeaf) in which two optimization techniques are employed from improving the model performance. Pandey and Jain [19] proposed a novel plant disease detection model that exploited dense learning mechanisms and merging attention of CNN for robust performance. Cheng *et al.* [6] proposed a model for the automatic recognition of eight tomato disease and pest categories by employing a chatbot controller and CNNs. Zhang *et al.* [29] suggested a novel Locally Reversible Transformer method to effectively identify plant leaf disease for grape plants. They exploited a Local Learning Bottleneck for extracting important semantic information and improve local perception in grape leaf diseases. In [17], a comprehensive study is presented for plant leaf disease recognition. This study demonstrated performance of state-of-the-art object detection and classification approaches. Based on the performance analysis, the study observed that a more effective and robust method is required to achieve the objectives of agrarian for identifying plant diseases. Hosny *et al.* [10] suggested a lightweight deep learning approach to identify and classify plant diseases. The approach fused deep features and traditional handcrafted features to recognize plant diseases. Despite the success of the abovementioned methods, several important problems of plant leaf disease identification are still unsolved, such as the visual features which play a significant role in plant leaf disease recognition. In addition, a labeled dataset is required for the implementation of the classification methods. Further, the imbalanced dataset is influenced the learning task toward specific labels. At last, an effective and robust disease detection method should be developed to achieve the goals of agrarian for plant leaf disease detection.

19.3 PROPOSED METHOD

This section presents the proposed method that employed ANN with cloud-enabled visual input for robust and effective plant leaf disease identification. The proposed method efficiently captures plant leaf from crop field via a cloud platform that provides simple yet reliable visual data capturing system and employed an ANN for learning to classify plant leaf as either diseased or non-diseased. The proposed method exploited the digital image processing algorithm and machine learning method with a cloud computing environment to detect plant leaf disease. The proposed method aims to develop a conceptual framework which effectively processes the raw plant leaf image and accurately classifies them, reducing time consumption and improving disease detection performance. In the proposed method, initially, plant leaf images are taken from the crop field through a digital camera and the captured images are sent to cloud platform for further processing. Here, cloud environment is utilized to provide efficient computing resources and services. Such ability of the cloud environment enabled the image preprocessing task very simple and reliable. Then, the image processing techniques, namely image enhancement and image cropping, are applied to capture the plant leaf and remove the other unnecessary visual information from the images taken from the crop field. Further, to increase data size and provide variation in the data image, augmentation task is also performed. Then, the processed plant leaf image is partitioned into visually meaningful segments that improve the overall computation time and provide more accurate image elements for further processing. Then, visually significant features are extracted from each region created a training dataset is created with target data. This training data is fed into ANN to learn a plant leaf disease prediction model. Lastly, the learnt model is used to predict the class of unseen plant leaf images. A conceptual overview of the proposed method is presented in Figure 19.2. The proposed method has five steps, namely data acquisition, image preprocessing on the cloud platform, image segmentation, feature extraction and ANN for disease predication. A detailed discussion of each step is provided in the following subsections.

19.3.1 DATA ACQUISITION

Initially, the proposed method captures the plant leaf images from the real environment. The images are taken by a digital camera from the crop field which provides the real challenging image samples of the particular crop for analyzing the prediction model. This process enabled the proposed method most robust in terms of various challenging scenarios of plant leaf disease detection in real-world environment. This process is under controlled conditions which limit the generalization of the process. The significance of data acquisition is to provide an appropriate and large amount of plant leaf images from real world. Such a variation in the data collection is required for deeply analyzing the challenging scenarios in plant leaf disease detection.

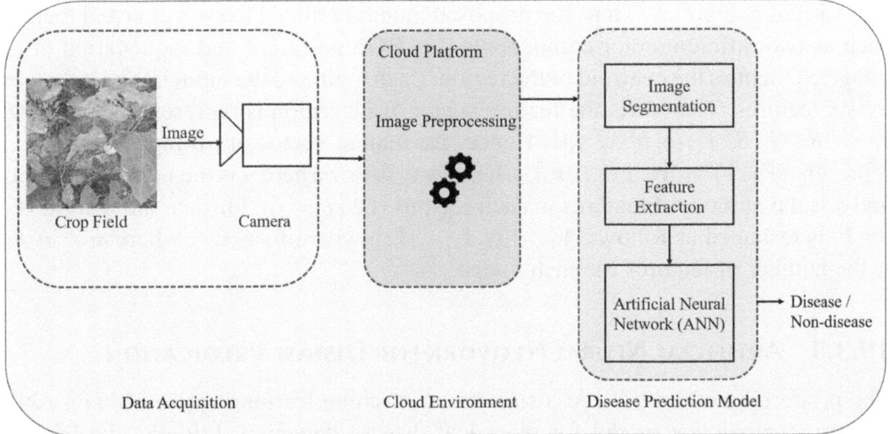

FIGURE 19.2   A conceptual overview of proposed plant leaf detection framework.

19.3.2 IMAGE PREPROCESSING

In order to provide visually significant information for further processing, the raw images taken from the crop field are required to be preprocessed before starting the analyses and learning task. This preprocessing of the image is performed in a cloud environment which provides fast and efficient computational resources and services for high-quality image processing. Specifically, the preprocessing removes background noises from the plant leaf image taken by an image capturing device such as a digital camera. For this purpose, image cropping and enhancement are performed on the raw input plant leaf image. This process supports enhancing the accuracy of the learnt model for plant leaf disease recognition and also made it computationally effective.

19.3.3 IMAGE SEGMENTATION

To facilitate the learning model with more effective visual elements, the preprocessed image is entailed to partition into visually homogeneous and meaningful regions. The proposed method employed simple linear iterative clustering (SLIC) [2] for segmenting the input image into various segments. SLIC is a region partition method that generates regular superpixels with precise boundary. By employing the SLIC algorithm, the preprocessed input image ($\mathbf{I}$) is partitioned into n regions $R = \{r_1, r_2, ..., r_n\}$ used as an image element for the learning process.

19.3.4 FEATURE EXTRACTION

Feature extraction is a very useful process for learning of predictive model. Therefore, the proposed method employed a feature extraction step after the completion of data acquisition and image segmentation. The effective feature should represent discriminated information that is used by the learning algorithm to classify input data into various categories. Thus, the proposed method utilized low-level visual features such as two different color components (i.e. RGB and Lab) and the location of the image element as the centroid of the region. Each region of the input image is denoted by six features. Therefore, the feature vector of ith region is represented as follows: $\mathbf{r}_i = [R_i, G_i, B_i, L_i, a_i, b_i, x_i, y_i]^T$. Hence, the feature vector of jth input image ($\mathbf{I}$) is represented as $\mathbf{F}_j = [\mathbf{r}_1, \mathbf{r}_2, ..., \mathbf{r}_n]$, where $\mathbf{F}_j \in \mathbb{R}^{n \times a}$, where n is the number of region and a is the number of features in each regions (i.e., $a = \mathbf{6}$). Further, the feature vector $\mathbf{F}_j$ is reshaped as follows: $\mathbf{F}_j = [\mathbf{f}_1, \mathbf{f}_2, ..., \mathbf{f}_m]$, where $\mathbf{F}_j \in \mathbb{R}^m$, where $m = n \times a$ is the number of features for each image.

19.3.5 ARTIFICIAL NEURAL NETWORK FOR DISEASE PREDICATION

The proposed method utilized a supervised machine learning approach, i.e., ANN, to learn a predictive model for plant leaf disease detection. Initially, the learning process requires a training dataset which should contain features with a target label for each image belonging to the training dataset. To construct a training dataset, the proposed method randomly chooses images from different plant leaf datasets. The training dataset is represented as $\mathbf{X} = [\mathbf{F}_1, \mathbf{F}_1, ..., \mathbf{F}_k]$, where $\mathbf{X} \in \mathbb{R}^{k \times m}$, where k is the number of training samples. The corresponding annotation is represented as $\mathbf{Y} = [y_1, y_2, ..., y_k]$, where $\mathbf{Y} \in \mathbb{R}^k$, and the value of y_i is either $\mathbf{0}$ or $\mathbf{1}$. Here, the proposed method builds an ANN that generates a robust and effective plant leaf disease predictor by using the constructed training dataset as follows:

$$\Delta : (\mathbb{R}^m | \boldsymbol{\omega}) \to \mathbb{R} \tag{19.1}$$

where $\boldsymbol{\omega}$ is the set of learnable parameters. Initially, the value of $\boldsymbol{\omega}$ is assigned zero and is gradually adjusted during the learning stage in order to reduce the class errors to develop an efficient plant leaf disease predictor.

ANN aims to learn optimal values of learnable parameters (i.e., $\boldsymbol{\omega}$) by optimizing class errors. The proposed ANN architecture consists of four layers: including input layer, first hidden layer, second hidden layer, and output layer. In this architecture, the number of neurons at input layer is determined based on the number of features that represents input image (i.e., m). Further, h_1 and h_2 are denoted as the number of neurons at the first and second hidden layers respectively. The output layer has only one neuron for binary classification.

Generally, ANN works in two phases: feed-forward and back-propagation [20]. In the feed-forward phase, a set of features is fed into ANN through the input layer neurons which are forwarded to the first hidden layer. These features are first linearly transformed with the weights $\mathbf{h}_{\omega_1}$ and biases $\mathbf{b}_1$ at the first hidden layer. Then, a non-linear transformation is applied via an activation function called sigmoid function to determine the output of the first hidden layer that is forwarded to the next subsequent hidden layer. Similarly, the second hidden layer linearly transforms the values received from the first hidden layer with the weights $\mathbf{h}_{\omega_2}$ and biases $\mathbf{b}_2$. To obtain the output values from the second hidden layer, a nonlinear transformation by sigmoid function is performed and the output values are forwarded to the output layer. Finally, the output layer processes the received values from the second hidden layer as linearly transformed and is followed by a nonlinear transformation with sigmoid function which generates the class labels in the prediction model. The ANN predicts the output as disease or non-disease for each input plant leaf image; for the ith image the predicted value is denoted by $\hat{y}_i$ and is mathematically represented as follows:

$$\hat{y}_i = \Delta(\mathbf{F}_i|\boldsymbol{\omega}) \tag{19.2}$$

In the back-propagation, an effective cost function is utilized to determine the error between each predicted value (i.e., $\hat{y}_i$) and the corresponding actual value (i.e., y_i). For this purpose, a mean square error is used as a cost function, which is denoted by $L(\mathbf{F}_i|\boldsymbol{\omega})$ and is mathematically defined as follows:

$$L(\mathbf{F}_i|\boldsymbol{\omega}) = (y_i - \Delta(\mathbf{F}_i|\boldsymbol{\omega}))^2 \tag{19.3}$$

where y_i is the target label of the ith training sample. Equation 19.3 can be represented with the help of Equation 19.2 as follows:

$$L(\mathbf{F}_i|\boldsymbol{\omega}) = (y_i - \hat{y}_i)^2 \tag{19.4}$$

Subsequently, the mean square error of the training images is denoted by $L(X|\boldsymbol{\omega})$ and is mathematically defined as follows:

$$L(\mathbf{X}|\boldsymbol{\omega}) = \frac{1}{k}\sum_{i=1}^{k}(y_i - \hat{y}_i)^2 \tag{19.5}$$

To determine the best learnable parameters $\boldsymbol{\omega}$, the objective of ANN is to minimize the error $L(\mathbf{X}|\boldsymbol{\omega})$ during iterations of back-propagation phase. Afterwards, the plant leaf disease prediction model utilizes the learnt model parameters $\hat{\boldsymbol{\omega}}$ for predicting plant leaf disease on the real-world image. For image ($\mathbf{I}$) initially, features are extracted (denoted by $\mathbf{F}_I$) and fed into the learnt ANN model to predict the image class y_I as either disease or non-disease as follows:

$$y_I = \Delta(\mathbf{F}_I|\hat{\boldsymbol{\omega}}) \tag{19.6}$$

19.4 EXPERIMENTAL RESULTS

This section presents a detailed analysis of experimental results that supports the effective performance of the proposed plant leaf disease detection. The performance of the proposed method is presented with widely used measurements such as precision, recall, and F-measure on two datasets the proposed dataset and a publicly available PlantVillage [16] dataset.

19.4.1 DATASET

This research study proposed a natural Eggplant plant leaf image dataset. This dataset consists of 200 images of which 100 images are diseased leaves and 100 images

FIGURE 19.3   Samples of proposed Eggplant leaf dataset. (Top row) diseased leaf and (Bottom row) healthy leaf.

with non-diseased leaves. All the images are captured from the crop field under a controlled environment. The data augmentation step is also applied to increase the effectiveness of the dataset. This dataset provides various challenging scenarios of plant leaf disease detection. Here, each plant leaf image taken from the crop field is annotated by various observers. The image is labeled with a target label as either disease or non-disease; for simplification the disease label and the non-disease label are denoted by 0 and 1, respectively. The sample of diseased and healthy Eggplant plant leaf images are shown in Figure 19.3. Another widely used and publicly available dataset, i.e., the PlantVillage [16] dataset, is used for performance analysis.

19.4.2   QUANTITATIVE ANALYSIS

For performance analysis, three measurements in terms of precision, recall, and F-measure are utilized for presenting the experimental results of the proposed method quantitatively. The detailed experimental results in terms of the three metrics are demonstrated in Figure 19.4 on the introduced dataset and selected plant types from the PlantVillage [16] dataset, such as Apple__Apple_scab, Apple__Black_rot, Blueberry__healthy, Grape__black_rot, Grape__healthy, Potato__healthy, and Tomato__Bacterial_spot. The proposed method achieved effective disease detection due to incorporating perceptually significant image elements as regions and extracting visually significant features for ANN learning. The ANN model is computationally more effective due to region-wise image processing and achieves better accuracy because the considered visual features are more effective.

19.5   CONCLUSION

Plant leaf disease detection is the most attentive task in agriculture due to its significant impact on crop production and its quality and safety. Recent advances in computer vision and machine learning techniques provide an effective platform to analyze plant leaf disease in a better manner. Further, benefitting from recent

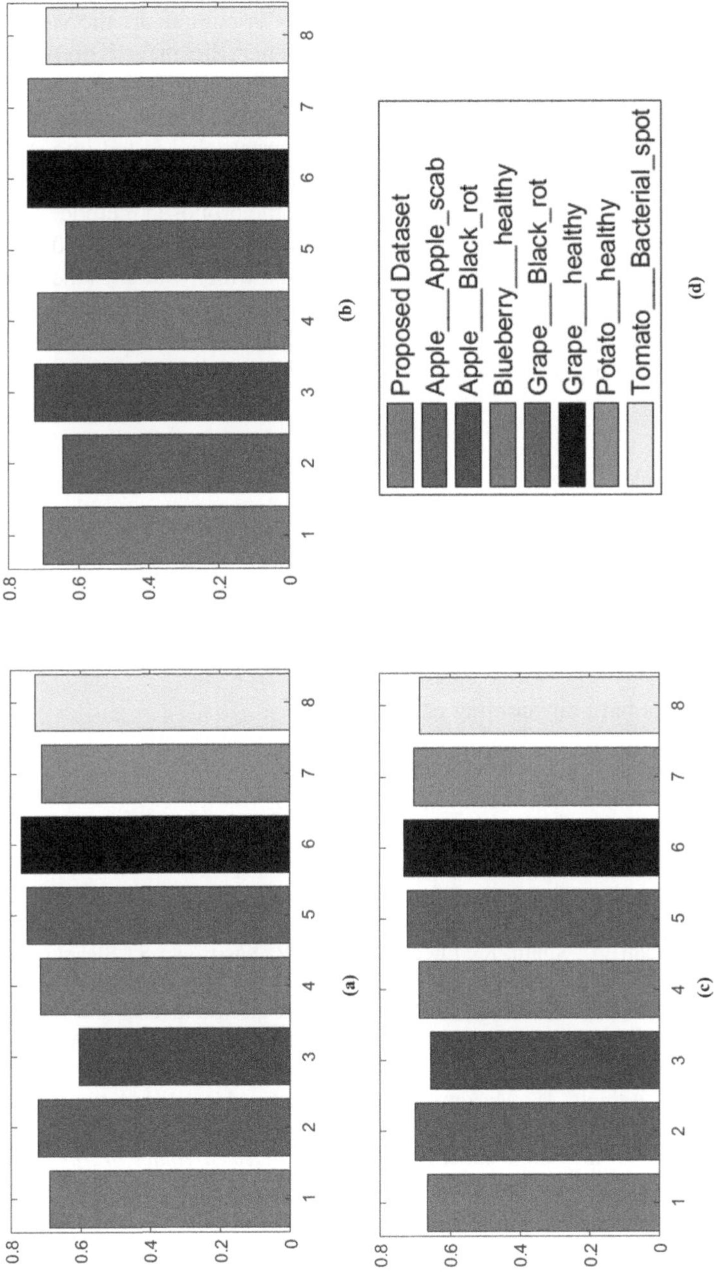

FIGURE 19.4 Quantitative performance of the proposed method on the proposed dataset and selected plant types from PlantVillage [16] dataset. (a) Precision, (b) Recall, (c) F-measure, and (d) Plant leaf dataset.

advances in cloud computing, especially effective storage and high-speed data collection from remote sensing devices, a cloud-based ANN is proposed in this chapter for plant leaf detection. Initially, the plant image is captured from the crop field and cloud supports and services are utilized to effectively communicate the visual data from the crop field to the plant leaf detection method. Then, the image is preprocessed for enhancing key visual features and segmented into meaningful regions. Visually significant features are extracted from each region to represent an image as a feature vector. Further, a set of feature vectors with a target label is fed into ANN to learn a plant leaf disease classification model. In addition, a new plant leaf dataset is introduced to provide more challenging scenarios for the proposed method. The performance of the proposed method is analyzed on the proposed dataset and publicly available PlantVillage dataset with selected plant types, which validates the accurate performance in terms of precision, recall, and F-measure. In future, we will look for more effective lightweight ANN models that could identify plant leaf disease in a more complex background efficiently.

ACKNOWLEDGMENTS

This research work is funded by the Research Development Cell, Sharda University, India, under the seed fund scheme (No. SU/SF/2022/22).

DISCLOSURE STATEMENT

The author does not have any conflict of interest.

REFERENCES

1. Abbas, A., Jain, S., Gour, M., Vankudothu, S.: Tomato plant disease detection using transfer learning with c-gan synthetic images. *Computers and Electronics in Agriculture* 187, 106279 (2021).
2. Achanta, R., Shaji, A., Smith, K., Lucchi, A., Fua, P., Süsstrunk, S.: Slic superpixels. *Technical Report 149300*, (2010).
3. Ahmad, M., Abdullah, M., Moon, H., Han, D.: Plant disease detection in imbalanced datasets using efficient convolutional neural networks with stepwise transfer learning. *IEEE Access* 9, 140565–140580 (2021).
4. Ashourloo, D., Aghighi, H., Matkan, A.A., Mobasheri, M.R., Rad, A.M.: An investigation into machine learning regression techniques for the leaf rust disease detection using hyperspectral measurement. *IEEE Journal of Selected Topics in Applied Earth Observations and Remote Sensing* 9(9), 4344–4351 (2016).
5. Cap, Q.H., Uga, H., Kagiwada, S., Iyatomi, H.: Leafgan: An effective data augmentation method for practical plant disease diagnosis. *IEEE Transactions on Automation Science and Engineering* 19(2), 1258–1267 (2020).
6. Cheng, H.H., Dai, Y.L., Lin, Y., Hsu, H.C., Lin, C.P., Huang, J.H., Chen, S.F., Kuo, Y.F.: Identifying tomato leaf diseases under real field conditions using convolutional neural networks and a chatbot. *Computers and Electronics in Agriculture* 202, 107365 (2022).

7. Dwivedi, R., Dey, S., Chakraborty, C., Tiwari, S.: Grape disease detection network based on multi-task learning and attention features. *IEEE Sensors Journal* 21(16), 17573–17580 (2021).

8. Hassan, S.M., Maji, A.K.: Plant disease identification using a novel convolutional neural network. *IEEE Access* 10, 5390–5401 (2022).

9. Hernández, S., López, J.L.: Uncertainty quantification for plant disease detection using bayesian deep learning. *Applied Soft Computing* 96, 106597 (2020).

10. Hosny, K.M., El-Hady, W.M., Samy, F.M., Vrochidou, E., Papakostas, G.A.: Multi-class classification of plant leaf diseases using feature fusion of deep convolutional neural network and local binary pattern. *IEEE Access* 11, 62307–62317 (2023).

11. Kumar, M., Kumar, A., Palaparthy, V.S.: Soil sensors-based prediction system for plant diseases using exploratory data analysis and machine learning. *IEEE Sensors Journal* 21(16), 17455–17468 (2020).

12. Li, C., Ma, Y., Mei, X., Liu, C., Ma, J.: Hyperspectral image classification with robust sparse representation. *IEEE Geoscience and Remote Sensing Letters* 13(5), 641–645 (2016).

13. Litjens, G., Kooi, T., Bejnordi, B.E., Setio, A.A.A., Ciompi, F., Ghafoorian, M., Van Der Laak, J.A., Van Ginneken, B., Sánchez, C.I.: A survey on deep learning in medical image analysis. *Medical Image Analysis* 42, 60–88 (2017).

14. Liu, X., Min, W., Mei, S., Wang, L., Jiang, S.: Plant disease recognition: A large-scale benchmark dataset and a visual region and loss reweighting approach. *IEEE Transactions on Image Processing* 30, 2003–2015 (2021).

15. Mattihalli, C., Gedefaye, E., Endalamaw, F., Necho, A.: Plant leaf diseases detection and auto-medicine. *Internet of Things* 1, 67–73 (2018).

16. Mohanty, S.P., Hughes, D.P., Salathé, M.: Using deep learning for image-based plant disease detection. *Frontiers in Plant Science* 7, 1419 (2016).

17. Moupojou, E., Tagne, A., Retraint, F., Tadonkemwa, A., Wilfried, D., Tapamo, H., Nkenlifack, M.: Fieldplant: A dataset of field plant images for plant disease detection and classification with deep learning. *IEEE Access* 11, 35398–35410 (2023).

18. Pal, A., Kumar, V.: Agridet: Plant leaf disease severity classification using agriculture detection framework. *Engineering Applications of Artificial Intelligence* 119, 105754 (2023).

19. Pandey, A., Jain, K.: A robust deep attention dense convolutional neural network for plant leaf disease identification and classification from smart phone captured real world images. *Ecological Informatics* 70, 101725 (2022).

20. Rumelhart, D.E., Hinton, G.E., Williams, R.J.: Learning representations by back-propagating errors. *Nature* 323(6088), 533–536 (1986).

21. Rutishauser, U., Walther, D., Koch, C., Perona, P.: Is bottom-up attention useful for object recognition? In: *Proceedings of the 2004 IEEE Computer Society Conference on Computer Vision and Pattern Recognition*, 2004. CVPR 2004. vol. 2, pp. II–II. IEEE (2004).

22. Sahu, S.K., Pandey, M.: An optimal hybrid multiclass SVM for plant leaf disease detection using spatial fuzzy c-means model. *Expert Systems with Applications* 214, 118989 (2023).

23. Sharma, V., Tripathi, A.K., Mittal, H.: Dlmc-net: Deeper lightweight multi-class classification model for plant leaf disease detection. *Ecological Informatics* 75, 102025 (2023).

24. Sunil, C., Jaidhar, C., Patil, N.: Cardamom plant disease detection approach using efficientnetv2. *IEEE Access* 10, 789–804 (2021).

25. Thai, H.T., Le, K.H., Nguyen, N.L.T.: Formerleaf: An efficient vision transformer for cassava leaf disease detection. *Computers and Electronics in Agriculture* 204, 107518 (2023).
26. Udutalapally, V., Mohanty, S.P., Pallagani, V., Khandelwal, V.: sCrop: A novel device for sustainable automatic disease prediction, crop selection, and irrigation in internet-of-agro-things for smart agriculture. *IEEE Sensors Journal* 21(16), 17525–17538 (2020).
27. Ünal, Z.: Smart farming becomes even smarter with deep learning—a bibliographical analysis. *IEEE Access* 8, 105587–105609 (2020).
28. Vallabhajosyula, S., Sistla, V., Kolli, V.K.K.: Transfer learning-based deep ensemble neural network for plant leaf disease detection. *Journal of Plant Diseases and Protection* 129(3), 545–558 (2022).
29. Zhang, X., Li, F., Jin, H., Mu, W.: Local reversible transformer for semantic segmentation of grape leaf diseases. *Applied Soft Computing* p. 110392 (2023).

20 Concerns, Threats, and Ethical Considerations in Using Cloud of Things

Sarita, Deepshikha Yadav, Ashutosh Yadav, and Shailu Singh

20.1 INTRODUCTION

The advent of the Cloud of Things (CoT) paradigm has introduced novel possibilities for data analytics, enabling organizations to harness the vast amounts of data generated by interconnected devices and sensors. CoT combines the capabilities of cloud computing (CC) and the Internet of Things (IoT), creating a dynamic ecosystem where devices, systems, and applications seamlessly interact and exchange data. However, as organizations strive to leverage CoT for data analytics, they face a myriad of challenges and considerations that demand careful attention and exploration.

Ethical considerations play a crucial role in the use of CoT. The interconnected nature of devices and systems raises concerns regarding privacy, security, and data ownership. Safeguarding individual privacy rights and ensuring the ethical collection, usage, and sharing of data are paramount in maintaining trust and societal acceptance. Organizations must navigate the ethical complexities inherent in CoT to avoid compromising privacy and violating ethical principles.

In parallel, data governance emerges as a critical aspect in leveraging CoT for data analytics. Robust data governance frameworks are essential to address data quality, integrity, security, and compliance challenges. Effective data governance ensures that data is accurate, consistent, and reliable across diverse sources and complies with regulations and standards. The General Data Protection Regulation (GDPR), for instance, requires organizations to implement stringent measures to protect personal data, underscoring the importance of sound data governance practices.

Furthermore, several technical challenges must be tackled to fully exploit the potential of CoT in data analytics. The enormous volume, velocity, and variety of data generated by interconnected devices necessitate scalable and efficient data processing and storage mechanisms. Real-time data processing, edge computing, and distributed analytics frameworks are vital in handling the continuous influx of data and extracting valuable insights promptly. As highlighted by Xu et al. [1], the integration of edge computing with CoT enables local data processing and reduces latency, enhancing the effectiveness of data analytics.

This chapter delves into the challenges and considerations faced in employing CoT for data analytics. By analyzing the ethical implications, data governance

DOI: 10.1201/9781003390954-20

requirements, and technical obstacles, this chapter aims to provide a comprehensive understanding of the multifaceted landscape surrounding CoT. Through a scholarly examination of these challenges and considerations, organizations can navigate the complexities of CoT and unlock its transformative power in data analytics. By establishing robust ethical frameworks, implementing sound data governance practices, and leveraging innovative technical solutions, organizations can harness the full potential of CoT to drive innovation, enhance decision-making, and gain a competitive edge in the data-driven era.

20.2 SECURITY AND PRIVACY CONCERNS IN CLOUD OF THINGS

It is not unexpected that the IoT has recently given rise to a diversity of new CoT concepts, including sensing, video surveillance, big data analytics, data, and sensor as a service, etc. IoT complexity necessitates more effective concepts and fixes. As a result, since the early 2000s, the concept CoT has been changing alongside web computing. The exponential growth in data volume necessitates the adoption of advanced storage solutions beyond conventional local or temporary storage mechanisms. As the vast amounts of data generated within the CoT paradigm continue to escalate, organizations are faced with the imperative to explore more sophisticated storage approaches. Processing and computation must adhere to the same model as the leasing of storage space, which is a rental basis, in order to provide new concepts to end users on an equitable and just basis. This must be done while taking security and privacy into account, for IoT and CC and their integration [2,3].

According to Ari et al. [4] and Narwane et al. [5], CoT refers to the integration of CC and IoT which was first used in 2016 by the German company Deutsche Telecom. The integration of ubiquitous computing has revolutionized treatment approaches, particularly with regard to the vast amount of data generated by IoT devices. The inclusion of CC as a storage and processing infrastructure has become an indispensable component due to this data influx. It opened up fresh viewpoints on issues including resource limitations, scalability, privacy, and security issues, as well as the collection and processing of big data. Regrettably, the realm of CoT still grapples with pressing security challenges, as users and IoT devices persist in sharing computing and networking resources from remote locations. In addition, the preservation of data privacy within this environment remains a critical concern. Consequently, the CoT landscape faces a constant growth in security and privacy issues, as highlighted in studies by Ari et al. [4] and Guerbouj et al. [2].

In the future, CC and IoT integration can undoubtedly aid in the accomplishment of a lot of internet objectives. But there are still some challenges in the process. The integration of ubiquitous computing presents various challenges: security, privacy, standardization, power efficiency, data storage, network control, scalability, and adaptability. Privacy and security are crucial, demanding utmost attention [4]. Threats to information security arise from flaws in the availability, confidentiality, and integrity of the method. Privacy and confidentiality are not synonymous. Denial of Service (DoS), Man-in-the-middle (MITM), and spoofing attacks are all dangers to communication. Web app vulnerabilities, control and transparency issues, unauthorized disclosure, tracking, capture, and identification without consent are some significant privacy threats in CoT [6]. Failure to ensure privacy and security in a system can lead to severe consequences. CoT devices potentially spying on and revealing our personal information, along with our true identities, are an easily understandable threat. A breach in the IoT's important applications, such as the systems that operate nuclear reactors, automobile safety systems, or medical gadgets, would make the issue considerably worse. Robust and feasible solutions must be taken into consideration in order to secure privacy in the CoT [4].

Several challenges and concerns are associated with COT services, as highlighted by Alhaidari et al. [7], Ari et al. [4], and Botta et al. [8]. Among some of them stated below, the security of data stands out as a critical challenge.

Security and Privacy: The main data is about user's personal information, so data privacy and data integrity are of prime concern. Although privacy can be protected through cryptography/encryption [7], this will help against attacks or malicious activity. This can be prevented by using multiple technologies in different segments. CoT enables data transfer from the physical world to the Cloud, yet the unresolved challenge remains in establishing proper authorization policies to restrict access to sensitive data to only authorized users. CoT integration faces challenges from provider customization, proprietary interfaces, and lack of standardized frameworks and interoperability. With around 50 billion connected devices by 2020, effectively handling the transfer, storage, processing, and access of the enormous data volume becomes paramount.

Performance: Applications for the CoT need precise quality of service and performance requirements at many different levels, including those relating to data movement, communication, computing, and storage. Because bandwidth growth does not coincide with advancements in computation and storage, achieving acceptable network performance toward the cloud is difficult. Service providers while delivering their services across nations need to ensure consideration of international regulations and societal expectations. In fact, scholars believe that the lack of standards is a major source of misunderstanding in the CoT. Numerous efforts have been made to deploy and standardize CoT, yet the need for standardizing protocols, architectures, and frameworks remains imperative. This will make it easier for heterogeneous devices to connect to one another, enabling the development of smart CoT services.

Monitoring and Scaling: Monitoring is crucial in CC and CoT for security, performance, resource management, and troubleshooting. CoT faces added challenges due to IoT's unique characteristics and vast data. It enables innovative applications with distributed devices, but large-scale behavior of systems poses new obstacles.

Fog Computing and Smart Gateway-Based, Communication: Fog computing is a model that extends traditional CC services to the network edge, catering to IoT applications with latency, mobility, and geo-distribution needs. Smart Gateways play a crucial role in preprocessing and reducing data before transmitting it to the cloud, alleviating unnecessary burden on the core network.

Energy Efficiency: The constant data exchanges in modern CoT applications, transferring data from smart devices to the cloud, significantly drain device battery life, hindering their continuous operation. Consequently, energy conservation and efficient management pose significant challenges.

20.3 THREATS IN CLOUD OF THINGS

Security and privacy are crucial concerns in CoT. Privacy, as a fundamental human right, involves protecting and appropriately utilizing customers' personal information in line with their expectations. However, the security and privacy challenges in IoT differ from those in traditional wireless networks due to unique deployment and technological aspects. IoT networks are often deployed in resource-constrained environments, limiting the use of public key encryption for device security. To address this, lightweight encryption technologies and cryptographic algorithms are necessary to ensure device and data security and privacy. Node spoofing can result in data loss, as attackers can send erroneous control messages [4]. Ferdous et al. [9] present a detailed account of various threats arising from CoT briefly discussed in this section.

20.3.1 SECURITY THREATS

This category encompasses threats that pose a risk to the system's security. Ferdous et al. [9] categorize these threats into various subcategories, including communication threats, physical threats, data threats, service provisioning threats, and other threats. Williams et al. [10] also introduce a comprehensive account of IoT security threats and solutions.

20.3.1.1 Communication Threats

The communication threats associated with a CoT system encompass various malicious activities that can exploit the system's communication channel. One such threat is the DoS attack, which can significantly impair the system's functionality by causing hardware malfunctions, resource depletion, software flaws, or the transmission of high-energy signals. CoT's device constraints make it particularly vulnerable to DoS attacks (Jensen et al. [11] as mentioned in Ferdous et al. [9]). Eavesdropping, another form of attack, involves unauthorized real-time monitoring of private conversations by gaining access to communication channels and extracting data during interactions between system elements [12,13]. Spoofing, on the other hand, occurs when a malicious entity impersonates a network device or user to target network nodes, disseminates malware, steals data, or bypasses access control systems. Different types of spoofing attacks, such as IP spoofing and Address Resolution Protocol (ARP) spoofing, can be a few examples. MITM attacks are also prevalent, enabling attackers to covertly eavesdrop on legitimate communications between senders and recipients, intercept sensitive data packets during transmission, and replace them with fake ones [14]. Lastly, replay attacks occur in unsecure networks, where attackers intercept data traffic and subsequently resend the intercepted packets to gain unauthorized access to network resources [15]. This type of attack can also lead to a minor DoS by consuming system or node resources.

20.3.1.2 Physical Threats

Physical threats pose significant risks to IoT devices and require careful consideration in terms of device and network security. One such threat is device capture, where attackers gain physical access to IoT devices and seize the data before it is securely transmitted and stored in the system [16]. To mitigate this risk, a secure architecture with restricted devices is necessary. Williams et al. [10] identify hardware trojan as yet another threat in which the attacker monitors, modifies, or disables either the data stored in the circuit or the communication of the circuit using trojan. Another physical threat is node damaging, wherein an attacker physically harms an IoT device, rendering it incapable of sensing or transmitting data. If multiple devices are damaged, it could result in a DoS attack, leading to the system's inability to provide services dependent on the affected data. Additionally, side channel attacks exploit predictable physical properties of devices, derived from knowledge obtained during system implementation, to infer sensitive information and compromise system security. These attacks leverage the relationship between logical functionalities and their corresponding physical properties, allowing attackers to make inferences that may jeopardize the overall security of the system [17–19]. Williams et al. [10] emphasize the importance of addressing physical threats in IoT systems, highlighting the need for robust security measures and secure device architectures to mitigate risks associated with device capture, node damaging, and side channel attacks.

20.3.1.3 Data Threats

Data threats in CoT systems encompass various risks and attacks that exploit the data generated by IoT devices. These threats pose significant challenges to the security and privacy of CoT systems and their users [9]. The identified threats include the potential retrieval of raw data through physical access to devices, which can be achieved through reverse engineering or micro-probing. This unauthorized access to IoT devices raises concerns regarding the security and privacy of user data. Another threat highlighted by Babar et al. [20] is data integrity and confidentiality which are critical aspects in CoT systems as data needs to be transmitted and stored securely [9]. There is a risk of data alteration during transmission (also known as eavesdropping/sniffing) [10] and storage, necessitating robust mechanisms to ensure the integrity and confidentiality of data. Ensuring device authenticity is also vital to maintain the reliability of CoT systems. Attackers may attempt to introduce unapproved devices (threat of authenticity) that generate false data, highlighting the need for proper verification and authentication of devices to mitigate this threat. In addition, the compromise of cryptographic keys (key compromization) poses a significant risk to the security of sensing data. Attackers may employ different cryptographic algorithms to compromise these keys, potentially jeopardizing the overall safety of CoT systems and the individuals utilizing them. Moreover, the introduction of false data into a CoT system (false data injection) can be achieved through the use of unauthorized IoT devices or reverse engineering authenticated devices [21]. This malicious practice compromises the accuracy and security of the system, where the reliability of sensed data is of utmost importance.

The limitations of IoT devices in terms of computing power and power sources can lead to the adoption of weak cryptographic protocols that demand fewer processing resources. However, the use of weak cryptographic techniques raises concerns as they can be easily cracked, thereby endangering the overall security of CoT systems. Data loss and leakage present additional risks, which can occur due to various factors such as service provider errors, natural disasters, inadequate disaster recovery plans, and unstable data centers. Data breaches, involving unauthorized access to data by individuals, applications, or services, can happen at different system layers and end-points [22]. These breaches may result from both malicious insiders and external attackers, highlighting the need for robust security measures.

Lastly, the sensitivity of certain types of sensor data, such as health or financial data, requires greater caution to protect user security and privacy. The handling, storage, and usage of such sensitive data demand meticulous attention to mitigate the risks associated with its invasion. Therefore, addressing these data threats in CoT systems is crucial to ensure the security, privacy, and trustworthiness of the data ecosystem.

20.3.1.4   Service Provisioning Threats

Service provisioning threats in CoT systems encompass various risks associated with the deployment and management of services and applications. These threats, highlighted by Ari et al. [4], pose challenges to ensuring the smooth operation and security of CoT systems.

One significant threat is the potential for unidentified and unauthorized access to services within the CoT system. It is crucial for these services to implement robust access control mechanisms to prevent unauthorized and unknown users from accessing sensitive data and resources. Issues in the infrastructure for permissions may result in users being granted excessive rights, thereby compromising the security of the system.

Identity theft is another pressing concern in service provisioning [23]. Attackers may gain access to important personal data of legitimate users, allowing them to impersonate these users and access services and resources that are restricted. This can lead to data loss and potentially hold the victim responsible for the actions conducted by the attacker.

Service hijacking, a specific type of network security attack, involves the attacker pretending to be one of the parties involved in a connection while taking control over it [24]. In a service hijack attack, a user intending to use a legitimate service is redirected to an unauthorized service under the control of the attacker (see [25] for a detailed discussion). This undermines the integrity and trustworthiness of the service, posing a risk to users' data and interactions.

Furthermore, insecure or compromised interfaces and application programming interfaces (APIs) are identified as significant security risks [22] and most vulnerable points of attacks [26]. Cloud providers often publish APIs that allow users to access data and utilize various services. If these interfaces are inadequately protected, attackers can exploit this vulnerability to launch fake services using authentic data, compromising the overall security and trust in the system.

Addressing these service provisioning threats requires the implementation of robust access control mechanisms, secure authentication protocols, and diligent monitoring of interfaces and APIs to prevent unauthorized access and attacks on the CoT system.

20.3.1.5 Other Security Threats

While CC offers immense processing power and benefits for businesses, it also presents opportunities for malicious activities. Some of the other security threats enlisted by Ferdous et al. [9] are as follows:

1. Malicious insiders pose a significant threat to network systems, as they launch hostile attacks while having authorized access to the system or sensitive information. These insiders can exploit their access privileges to abuse data and services, potentially causing significant harm to the organization or individuals involved.
2. Shared technology resources introduce another vulnerability. With the implementation of multi-tenancy design and virtualization, multiple users can access shared resources. However, this shared environment raises concerns about unauthorized access to other users' virtual machines (VMs). If the VM Monitor has vulnerabilities, an unauthorized user could exploit these weaknesses to gain control over someone else's simulated machine, potentially leading to unauthorized access and data compromise.
3. The ability to access significant computing power through cloud services can be misused by attackers. For instance, a single malicious actor could rent a substantial amount of computing power from another cloud service provider to launch DoS attacks, disrupting the availability and functionality of targeted systems.

Mitigating these threats requires robust security measures such as access controls, user monitoring, vulnerability management, and secure virtualization techniques to prevent insider abuse, unauthorized access to shared resources, and misuse of CC capabilities for malicious purposes.

20.3.2 PRIVACY THREATS

The appropriate handling, processing, storage, and deletion of data necessitate the diligent consideration of individuals' rights concerning the utilization of their personal information [27]. This entails compliance with contractual obligations, organizational policies, and adherence to governing regulations or laws, such as the GDPR in Europe, which has been in effect since 2020. These threats are more concentrated on how they can breach and/or invade the privacy of the users within the relevant system because the issue of privacy primarily impacts the users of the system.

Various privacy threats exist within the realm of CoT systems, with implications for user privacy and data governance as mentioned in Ferdous et al. [9] and Ari et al. [4]. One such threat is the unnoticed capture and unaware identification of individuals, exemplified by inconspicuous IoT devices like a discreet camera at a building entrance that captures facial images of individuals without their knowledge or consent. The collected data can be subsequently utilized inappropriately to identify users and violate their privacy. Furthermore, users may face challenges in accessing their recorded IoT data, as it may be stored in cloud providers' repositories, leading to IoT data inaccessibility.

The lack of control and transparency is another significant concern, where users have limited control over their uploaded data once it reaches the cloud. Users may remain unaware of any potential misuse by data collectors, and expressing explicit consent for data collection, processing, analysis, presentation, and sharing becomes complex due to the continuous stream of data captured by ubiquitous sensing processes [28]. Consequently, establishing effective access control policies that preserve privacy within the system becomes a daunting task.

Loss of governance is a pertinent threat associated with cloud infrastructures, as users relinquish authority to cloud providers, which can impact their privacy. Profiling and tracking are possible through the aggregation of data collected from various sensors, enabling the creation of user profiles and monitoring their activities across multiple domains without their awareness. Additionally, the extensive computing power in CoT systems introduces the risk of unforeseen inferences, potentially invading user privacy or enabling the prediction of future events that were previously unattainable in conventional IoT systems.

Unauthorized disclosure of data is another notable concern, as gathered data may be made public or transferred to a CoT system for archiving and analysis. This data gathering, storage, and analysis can pose a risk of unauthorized disclosure, making it challenging to obtain user consent and raise their awareness about it. Compounding these threats is the fact that privacy hazards are not limited to cybercriminals, as trusted companies may also misuse private information for their own gain, highlighting the importance of prioritizing security and privacy in CoT systems.

CoT brings forth various privacy threats that affect user privacy and data governance. The unnoticed capture and unaware identification of individuals, IoT data inaccessibility, lack of control and transparency, loss of governance, profiling and tracking, unforeseen inferences, and unauthorized disclosure all contribute to the complex landscape of privacy concerns in CoT. Safeguarding security and privacy should be of paramount importance in the development and implementation of CoT systems [4].

20.4 ETHICAL CONSIDERATIONS AND DATA GOVERNANCE IN THE CLOUD OF THINGS

CoT has tendencies to revolutionize data analytics by utilizing cloud-based infrastructure for IoT devices. However, a proper CoT implementation requires resolving crucial ethical issues and creating strong data governance norms. The development of regulations pertaining to global Internet and data governance is undergoing a dynamic process influenced by diverse national interests and constituencies. This intricate landscape has resulted in a complex amalgamation of regulatory frameworks that reflect distinct perspectives on the optimal functioning of digital infrastructure, the management of big data, and the responsible parties involved in this process [29].

Cloud governance encompasses a range of critical issues that need to be addressed to ensure the secure and efficient operation of cloud services.

Levite and Kalwani [30] highlight several key concerns in this regard. Firstly, security and robustness encompass the preparedness of cloud service providers (CSPs) to counteract security threats originating from malicious actions, as well as risks stemming from natural events, technical failures, and human errors. The ability to actively defend against such threats is crucial for maintaining the integrity and reliability of cloud services.

Resilience is another significant aspect of cloud governance, involving the implementation of measures to mitigate the adverse impact of service interruptions, failures, and distortions in cloud-based services. Backup plans, insurance policies, and other strategies are employed to ensure the continuity and availability of services, minimizing disruptions and downtime.

Consumer protection is a pertinent concern, considering the inherent power asymmetry between CSPs and consumers, as well as the oligopolistic nature of the CSP market. Safeguarding the rights and interests of consumers becomes essential, necessitating the development of policies and regulations that promote fair practices, transparency, and accountability within the cloud service sector.

Prosperity and sustainability represent a broader set of concerns related to the economic implications of cloud services. This encompasses the examination of the cloud's impact on local and global economic systems, as well as the formulation of policies to address any adverse effects resulting from the market's growth and reliance on CSPs. Considerations include employment, economic growth, innovation, welfare, and environmental sustainability.

It raises yet concerns regarding human and civil rights. As the cloud evolves into a substantial repository of data and provider of crucial services, the potential for privacy infringements and rights violations becomes a pressing issue. Individuals and social groups may find their privacy compromised, data and services disrupted, manipulated, or exploited by various actors, including governments, commercial entities, and criminal elements. Ensuring the protection of human and civil rights within the cloud ecosystem is paramount to maintain trust and uphold fundamental values.

In addition to the aforementioned governance issues, ethical considerations also play a crucial role in the effective management of cloud services. Allison-Hope and Schuchard [31] identified three main ethical dimensions that are closely linked to cloud governance.

Firstly, the location dimension highlights the importance of having a well-defined siting strategy that takes into account legal and jurisdictional concerns related to the physical location of data centers and network architecture. This is essential for both customers and providers of cloud services to ensure compliance with applicable laws and regulations, as well as to address potential conflicts and privacy concerns associated with different geographical jurisdictions.

The function of law enforcement is another significant ethical dimension. CSPs must navigate the complex landscape of interactions with law enforcement agencies. This involves upholding the right to due process, resisting requests from law enforcement that may compromise human rights, and actively supporting ethical leadership and the rule of law. Balancing the needs of law enforcement with the protection of user rights and privacy is a critical ethical challenge in cloud governance.

Spreading awareness about the implications of cloud services is vital. Users who may have limited knowledge about the impact of the cloud on privacy and freedom of expression have a responsibility to provide guidance and direction to cloud service providers, who are regarded as experts in cloud-related issues. By fostering awareness and education, users can actively contribute to the ethical development and responsible use of cloud-based services.

To address these ethical dimensions, robust security measures and data protection practices must be implemented within the CoT. By adhering to best practices, conducting regular security audits, and promoting employee awareness, CoT can ensure the confidentiality, integrity, and availability of data. These security and data protection efforts are essential for building trust among users, mitigating risks, and safeguarding the overall security of the CoT ecosystem [31].

Some of the major considerations that need to be made in order to guarantee the ethical use of data and uphold data governance in the CoT are examined in this section.

20.4.1 Data Privacy and Security

It is crucial to protect data privacy and ensure security precautions are robust in the CoT. The CoT leverages cloud-based infrastructure for IoT devices; therefore, it is essential to have security measures in place to protect data while it is being transported and stored.

1. **Encryption and Secure Communication Protocols**: By encrypting data during transmission, sensitive information is transformed into an unreadable format, preventing unauthorized access. Establishing secure connections between IoT devices and the cloud infrastructure, such as through Transport Layer Security, further enhances data security.
2. **Access Controls and Authentication Procedures**: These are essential to prevent unauthorized access to data. Role-based access control (RBAC) allows specific privileges and permissions to be assigned based on individuals' or devices' roles and responsibilities. Implementing multi-factor authentication (MFA), which requires users to provide multiple forms of verification like passwords and unique codes sent to their mobile devices, adds an extra layer of protection.
3. **Data Backup and Disaster Recovery**: In CoT, it's crucial to have data backup and disaster recovery systems. Service providers should put in place reliable backup plans to guarantee data availability and resilience in the event of system failures, natural catastrophes, or cyberattacks. To ensure that data

can be recovered and business continuity is maintained, regular backups should be made and data restoration procedures tested.

4. **Data Encryption at Rest**: By encrypting data stored in databases or storage systems, unauthorized access to the data is prevented. Encryption at rest ensures that even if the data is stolen, it remains unreadable without the decryption keys. Effective key management procedures and encryption methods should be employed to maintain the confidentiality and integrity of data.

5. **Regular Security Audits and Updates:** These play a vital role in identifying and addressing potential vulnerabilities in CoT infrastructure. Conducting penetration testing and vulnerability assessments enables the proactive detection and resolution of system weaknesses. Timely implementation of security patches and upgrades is crucial to ensure the system remains secure against known vulnerabilities and exploits.

6. **Effective Incident Response Plan**: It is essential to manage data breaches and minimize their impact. Despite robust security measures, breaches can still occur. The incident response plan should include procedures for detecting, containing, investigating, and disclosing data breaches. Prompt communication with affected individuals and relevant authorities is a key component of the plan to ensure that appropriate actions are taken to mitigate harm.

7. **Secure Communication and Network Infrastructure:** In CoT, reliable network architecture and secure communication protocols are essential. For the protection of data transferred between IoT devices, cloud platforms, and other components, service providers should use secure communication routes, such as encrypted connections. To identify and counteract potential threats and assaults, network security measures such as firewalls, intrusion detection systems, and frequent vulnerability assessments should be put in place.

20.4.2 Consent and User Control

Gaining users' informed consent and giving them control over their data are essential in CoT. The collection, usage, and sharing of users' data should be transparent to them, and they should be able to choose how they want it to be used.

1. **Transparency in Data Collection and Use:** In order to give users clear information about the types of data that will be gathered, the goal of collection, and how it will be utilized, CoT systems should be open about their data collecting practices. By being transparent, users are informed and empowered to decide how their data will be used.

2. **Consent Mechanisms:** A key component of data governance in CoT is obtaining user consent. Users should be able to understand the data being collected and why it is being gathered with the help of transparent consent procedures that are simple to use. Users should have complete control over the consent process, including the freedom to grant or remove consent at any time.

3. **User-Centric Privacy Settings and Preferences:** The CoT should offer user-centric privacy settings and choices to improve user control. Users should be able to personalize their privacy settings, choosing which information they want to share and with whom. Options to set data sharing preferences, select the extent of data anonymization, or even choose to forego specific data collection practices might be offered here.

4. **Data Access and Deletion Rights:** The right of users to access their own data produced by the CoT should exist. CSPs ought to offer ways for users to quickly retrieve and access the data that their IoT devices have collected. Users should additionally be able to ask for the erasure of their data when it is no longer required or when they withdraw consent.

5. **Data Portability:** A key component of user control in CoT is data portability. To avoid being tied to a single provider, users should have the ability to move their data between several cloud services or platforms. A seamless data transfer can be made possible by standardized formats and protocols, allowing consumers to exercise their control and change service providers without losing access to their data.

6. **Data Lifespan and Retention Policies:** To specify the retention period for data gathered from IoT devices, data retention policies should be set within the CoT. These regulations specify the duration of data storage and the conditions under which data shall be erased or anonymized. The risk of data being held longer than necessary is reduced and principles of data protection are adhered to when there are clear criteria for data retention and erasure.

In order to empower users, give them control over their data, and build trust in the system, the CoT should prioritize openness, permission processes, user-centric privacy settings, data access rights, and data portability. These procedures ensure that individuals have a say in how their data is treated and are in line with ethical concerns around data governance in the CoT.

20.4.3 TRANSPARENCY AND ACCOUNTABILITY

To enable appropriate data processing and governance in CoT, transparency and accountability are fundamental principles. Service providers should be open and honest about how they acquire, share, and use customer data. To address any violations, malpractices, or ethical issues relating to data management, it is also necessary to develop clear accountability systems. The significance of openness and responsibility in the context of the IoT is examined in this subsection.

1. **Transparency in Data Collection and Use:** In order to give users a clear knowledge of the data being gathered, how it will be used, and who it will be shared with, CoT systems should be transparent about their data gathering practices. Service providers should make their privacy policies and terms of service available and simple to understand, including the objectives and

parameters of data collection and use. Users gain confidence from this transparency, which also enables them to make thoughtful decisions regarding their involvement in the CoT ecosystem.

2. **Communication of Data Handling Practices:** CSPs should clearly explain to users how they handle user data. This involves educating users on the different types of data gathered, the ways in which it is stored and processed, and the security precautions used with regard to data. In order to give users a thorough awareness of how their data is managed throughout its lifecycle, providers should also clearly define any data sharing or third-party engagement.

3. **Accountability Mechanisms:** In order to ensure that service providers are accountable for their activities regarding data processing, accountability is crucial in the CoT. In order to resolve any violations, malpractices, or ethical issues, providers should set up clear procedures. This entails having procedures in place for dealing with data breaches, reporting events to the appropriate authorities, and taking the necessary steps to put things right. Accountability procedures should also have provisions for openness and communication with those who are impacted, informing them of the actions taken to address any problems.

4. **Ethical Considerations and Impact Assessments:** Service providers in the CoT should do ethical impact analysis to examine any potential ethical ramifications of their data handling procedures. This entails evaluating aspects including data security, privacy, fairness, and societal impact. Providers can identify and manage potential hazards, lessen prejudices, and ensure that their practices comply with ethical standards by proactively taking ethical issues into account.

5. **Audits and Certification:** The CoT can offer assurance of compliance with data governance and privacy requirements through routine audits and certifications. Independent audits can confirm that service providers are using the right security precautions, abiding by data protection laws, and upholding moral standards. A dedication to openness, accountability, and appropriate data handling can be shown via certifications from reputable organizations or industry standards.

6. **User Feedback and Redress Mechanisms:** Service providers should set up channels for user feedback so that people can voice their complaints or make recommendations about how data is handled. Mechanisms for user feedback can assist in identifying areas for development and make sure that user viewpoints are taken into account during decision-making. Additionally, in order to address user complaints or grievances and enable a just settlement of problems with data treatment, suppliers should have redress mechanisms in place.

The CoT can provide responsible data handling and governance by embracing transparency, establishing accountability measures, conducting ethical effect assessments, and putting in place user feedback and redress procedures. These measures encourage user trust, data practices openness, and ethical issues surrounding the usage of data in the CoT ecosystem.

20.4.4 OTHER ETHICAL CONSIDERATIONS

In the CoT, ethical issues are crucial for ensuring that the use of data and technology is consistent with moral standards and cultural norms.

1. Fairness and Avoidance of Bias: Fairness is a fundamental ethical value in the IoT. Utilized algorithms and data analytics methods should be developed to reduce biases and prevent discriminating results. To guarantee that machine learning models are representative and unbiased, this necessitates careful assessment of the data utilized for training and evaluation. To keep an eye on and correct any potential biases that can develop, routine audits and evaluations should be carried out.

2. Social Impact and Benefit: The CoT should attempt to improve society and take into account the greater social impact. Service providers need to think about how their technology could affect people on an individual, community, and societal level. They ought to try to advance social good, address societal issues, and ensure that their deeds have a positive impact on the advancement of society.

3. Inclusion and Accessibility: Important ethical factors in the CoT include accessibility and inclusive design. Service providers ought to make an effort to make their technologies available to everyone, regardless of their financial status, age, or level of ability. By ensuring that everyone may profit from the developments in the CoT, inclusive design practices encourage equality and stop the emergence of digital divides.

4. Ethical Governance and Oversight: In the CoT, establishing moral governance and control mechanisms is essential. In order to offer direction, establish ethical standards, and ensure compliance, this requires the engagement of ethics boards, regulatory bodies, and industry standards organizations. To evaluate and address any ethical concerns that may surface throughout the development, implementation, and operation of CoT systems, routine ethical evaluations and audits should be carried out.

5. Employee Training and Awareness: To encourage a security-conscious culture, service providers should make investments in employee training and awareness programs. Employees should receive training on the CoT's possible hazards, data protection laws, and security best practices. Security issues brought on by human error or ignorance can be avoided with regular training sessions and awareness efforts.

The CoT can be in line with ethical values and encourage the responsible and advantageous use of data and technology by addressing privacy, fairness, transparency, accountability, social effect, inclusivity, and ethical governance. These moral considerations guarantee that the CoT ecosystem upholds moral principles, upholds the rights of individuals, and satisfies social requirements (see Box 20.1).

20.1 DATA BREACH CASE OF UBIQUITI

Ubiquiti, a global IoT device provider, experienced a data breach that compromised customer personally identifiable information (PII). The breach lasted for two months, starting in December 2020. The attackers gained access to an IT employee's LastPass account, obtaining root administrator access to all AWS accounts. This allowed them to manipulate resources such as S3 data buckets, application logs, and user database credentials. The breach was likely a result of basic misconfiguration errors rather than sophisticated tactics. The attackers obtained administrative read/write access to Ubiquiti servers on the AWS cloud and exfiltrated signing keys, cryptographic secrets, and control over the source code. These compromised privileges enabled the attacker to potentially add backdoors, manipulate firmware, and access encrypted data. Preventing such breaches requires adherence to security best practices. Recommendations include avoiding the use of root administrator accounts, implementing multi-factor authentication (MFA) everywhere, enforcing separation of duties, setting up alerts for root account access, and enabling proper access logging. Unfortunately, Ubiquiti lacked database access logging, making it difficult to determine the extent of the breach and which data might have been accessed. The incident raises ethical concerns regarding the protection of customer data and the potential compromise of devices deployed globally. Ubiquiti's lack of access logging and uncertain knowledge about the breach's scope highlight the importance of robust security measures and thorough monitoring.

The Ubiquiti data breach raises several ethical considerations:

Privacy: The compromise of customer personally identifiable information (PII), including names, email addresses, passwords, and contact details, infringes upon individuals' privacy rights. Ubiquiti had a responsibility to protect this sensitive information and safeguard the privacy of its customers.

Data Protection: The breach potentially exposed customer data, including information stored on S3 data buckets and user databases. This raises ethical concerns about data protection and the duty of organizations to implement robust security measures to prevent unauthorized access and safeguard sensitive data.

Transparency and Accountability: Ubiquiti's lack of database access logging and uncertainty about the extent of the breach raise questions about transparency and accountability. Ethically, organizations have a responsibility to promptly disclose breaches, accurately assess the impact, and inform affected individuals to allow them to take necessary actions to protect themselves.

Trust and Customer Confidence: The breach jeopardizes the trust and confidence customers place in Ubiquiti. Ethical considerations include the obligation to maintain trust by proactively implementing robust security measures, promptly addressing breaches, and providing transparent communication to customers about the incident and mitigation efforts.

Security Practices: The breach highlights the ethical responsibility of organizations to adhere to industry best practices for security, including implementing proper access controls, employing strong authentication mechanisms like MFA, and regularly monitoring and updating security measures to address emerging threats.

Duty of Care: Ubiquiti has a duty of care toward its customers and their devices deployed globally. The compromise of cryptographic secrets, source code control, and potential manipulation of firmware raises ethical concerns about fulfilling this duty by prioritizing the security and integrity of their products and services.

Ethical Responsibility of Third-Party Service Providers: The breach resulted from the compromise of an employee's LastPass account, emphasizing the importance of third-party service providers' ethical responsibility in implementing stringent security measures to protect sensitive credentials and prevent unauthorized access.

Source: Authors' based on Speiser [32].

20.5 CONCLUSION

CoT presents a transformative paradigm that combines the power of CC and IoT. This amalgamation has shown a growth trajectory of tremendous applications in all arenas. However, this also brings forth significant security, privacy, and ethical concerns that must be carefully addressed. This chapter has explored the various dimensions of these concerns, ranging from the threats faced in the CoT to the ethical considerations surrounding data governance. By implementing robust security measures, ensuring data privacy and user control, promoting transparency and accountability, and addressing other ethical considerations, it is possible to navigate the challenges and build a secure and ethically responsible CoT ecosystem. Continuous research, cooperation, and policy creation will be necessary to address new security, privacy, and ethical problems as technological development unfolds further. The CoT has enormous potential for allowing advanced applications while protecting the rights and well-being of individuals and benefits the society as a whole, provided that a comprehensive strategy that strikes a balance between innovation, protection, and ethical practices is adopted.

REFERENCES

1. X. Xu et al., "An edge computing-enabled computation offloading method with privacy preservation for internet of connected vehicles," Future Generation Computer Systems, vol. 96, pp. 89–100, Jul. 2019, doi: 10.1016/j.future.2019.01.012. Available: https://doi.org/10.1016/j.future.2019.01.012

2. S. Guerbouj, H. Gharsellaoui, and S. Bouamama, "A Comprehensive Survey on Privacy and Security Issues in Cloud Computing, Internet of Things and Cloud of Things,"

International Journal of Service Science, Management, Engineering, and Technology, vol. 10, no. 3, pp. 32–44, Jul. 2019, doi: 10.4018/ijssmet.2019070103. Available: https://doi.org/10.4018/ijssmet.2019070103

3. S. Sahmim and H. Gharsellaoui, "Privacy and security in Internet-based computing: cloud computing, Internet of Things, Cloud of Things: a review," Procedia Computer Science, vol. 112, pp. 1516–1522, Jan. 2017, doi: 10.1016/j.procs.2017.08.050. Available: https://doi.org/10.1016/j.procs.2017.08.050

4. A. A. A. Ari et al., "Enabling privacy and security in Cloud of Things: architecture, applications, security & privacy challenges," Applied Computing and Informatics, Jul. 2020, doi: 10.1016/j.aci.2019.11.005. Available: https://doi.org/10.1016/j.aci.2019.11.005

5. V. S. Narwane, R. D. Raut, B. B. Gardas, M. S. Kavre, and B. E. Narkhede, "Factors affecting the adoption of cloud of things," Journal of Systems and Information Technology, vol. 21, no. 4, pp. 397–418, Nov. 2019, doi: 10.1108/jsit-10-2018-0137. Available: https://doi.org/10.1108/jsit-10-2018-0137

6. S. Mishra, "Cloud of Things (CoT): Security, privacy & adoption," International Journal of Security and Its Applications, vol. 14, no. 3, pp. 1–14, 2019, doi: 10.33832/ijsia.2020.14.3.01

7. F. Alhaidari, A.-U. Rahman, and R. Zagrouba, "Cloud of Things: architecture, applications and challenges," Journal of Ambient Intelligence and Humanized Computing, vol. 14, no. 5, pp. 5957–5975, Aug. 2020, doi: 10.1007/s12652-020-02448-3. Available: https://doi.org/10.1007/s12652-020-02448-3

8. A. Botta, W. De Donato, V. Persico, and A. Pescape, "Integration of cloud computing and Internet of Things: a survey," Future Generation Computer Systems, vol. 56, pp. 684–700, Mar. 2016, doi: 10.1016/j.future.2015.09.021. Available: https://doi.org/10.1016/j.future.2015.09.021

9. Md. S. Ferdous, R. Hussein, M. O. Alassafi, and G. Wills, "Threat Taxonomy for Cloud of Things," in Internet of Things and Big Data Analysis: Recent Trends and Challenges, United Scholars Publications, Ed., United Scholars Publications, USA, 2016. Available: www.researchgate.net/publication/310350401_Threat_Taxonomy_for_Cloud_of_Things

10. P. L. Williams, I. K. Dutta, H. Daoud, and M. Bayoumi, "A survey on security in internet of things with a focus on the impact of emerging technologies," Internet of Things, vol. 19, p. 100564, Aug. 2022, doi: 10.1016/j.iot.2022.100564. Available: https://doi.org/10.1016/j.iot.2022.100564

11. M. Jensen, J. Schwenk, N. Gruschka, and L. Lo Iacono, On Technical Security Issues in Cloud Computing. 2009. doi: 10.1109/cloud.2009.60. Available: https://doi.org/10.1109/cloud.2009.60

12. P. N. Mahalle, S. D. Babar, N. R. Prasad, and R. Prasad, "Identity Management Framework towards Internet of Things (IoT): Roadmap and key Challenges," in Springer eBooks, 2010, pp. 430–439. doi: 10.1007/978-3-642-14478-3_43. Available: https://doi.org/10.1007/978-3-642-14478-3_43

13. Y. Li and F. Teraoka, Privacy Protection for Low-cost RFID Tags in IoT Systems. 2012. doi: 10.1145/2377310.2377335. Available: https://doi.org/10.1145/2377310.2377335

14. I. Stojmenovic and S. Wen, The Fog Computing Paradigm: Scenarios and Security Issues. 2014. doi: 10.15439/2014f503. Available: https://doi.org/10.15439/2014f503

15. P. Gope and T. Hwang, "Untraceable sensor movement in distributed IoT infrastructure," IEEE Sensors Journal, vol. 15, no. 9, pp. 5340–5348, Sep. 2015, doi: 10.1109/jsen.2015.2441113. Available: https://doi.org/10.1109/jsen.2015.2441113

16. R. Roman, J. Zhou, and J. Lopez, "On the features and challenges of security and privacy in distributed internet of things," Computer Networks, vol. 57, no. 10, pp. 2266–2279, Jul. 2013, doi: 10.1016/j.comnet.2012.12.018. Available: https://doi.org/10.1016/j.comnet.2012.12.018

17. D. Das, S. Maity, S. B. Nasir, S. Ghosh, A. Raychowdhury, and S. Sen, "High efficiency power side-channel attack immunity using noise injection in attenuated signature domain." 2017 IEEE International Symposium on Hardware Oriented Security and Trust (HOST), Mclean, VA, USA, 2017, pp. 62–67. doi: 10.1109/hst.2017.7951799. Available: https://doi.org/10.1109/hst.2017.7951799

18. W. Zhou and F. Kong, "Electromagnetic side channel attack against embedded encryption chips." 2019 IEEE 19th International Conference on Communication Technology (ICCT). doi: 10.1109/icct46805.2019.8947185. Available: https://doi.org/10.1109/icct46805.2019.8947185

19. D. Genkin, A. Shamir, and E. Tromer, "Acoustic cryptanalysis," Journal of Cryptology, vol. 30, no. 2, pp. 392–443, Feb. 2016, doi: 10.1007/s00145-015-9224-2. Available: https://doi.org/10.1007/s00145-015-9224-2

20. S. D. Babar, P. N. Mahalle, A. Stango, N. R. Prasad, and R. Prasad, "Proposed Security Model and Threat Taxonomy for the Internet of Things (IoT)," in Communications in Computer and InSormation science, Springer Science+Business Media, 2010, pp. 420–429. doi: 10.1007/978-3-642-14478-3_42. Available: https://doi.org/10.1007/978-3-642-14478-3_42

21. N. Komninos, E. Philippou, and A. Pitsillides, "Survey in smart grid and smart home security: issues, challenges and countermeasures," IEEE Communications Surveys & Tutorials, vol. 16, no. 4, pp. 1933–1954, Jan. 2014, doi: 10.1109/comst.2014.2320093. Available: https://doi.org/10.1109/comst.2014.2320093

22. S. Subashini and V. Kavitha, "A survey on security issues in service delivery models of cloud computing," Journal of Network and Computer Applications, vol. 34, no. 1, pp. 1–11, Jan. 2011, doi: 10.1016/j.jnca.2010.07.006. Available: https://doi.org/10.1016/j.jnca.2010.07.006

23. C. Modi, D. Patel, B. Borisaniya, A. Patel, and M. Rajarajan, "A survey on security issues and solutions at different layers of Cloud computing," The Journal of Supercomputing, vol. 63, no. 2, pp. 561–592, Oct. 2012, doi: 10.1007/s11227-012-0831-5. Available: https://doi.org/10.1007/s11227-012-0831-5

24. S. Pearson and A. Benameur, Privacy, Security and Trust Issues Arising from Cloud Computing. 2010. doi: 10.1109/cloudcom.2010.66. Available: https://doi.org/10.1109/cloudcom.2010.66

25. S. S. Tirumala, H. Sathu, and V. Naidu, Analysis and Prevention of Account Hijacking Based INCIDENTS in Cloud Environment. 2015. doi: 10.1109/icit.2015.29. Available: https://doi.org/10.1109/icit.2015.29

26. F. Qazi, "Application Programming Interface (API) security in cloud applications," EAI Endorsed Transactions on Cloud Systems, vol. 7, no. 23, p. e1, Oct. 2023, doi: 10.4108/eetcs.v7i23.3011. Available: https://doi.org/10.4108/eetcs.v7i23.3011

27. E. Schiller, A. Aidoo, J. Fuhrer, J. L. Stahl, M. Ziörjen, and B. Stiller, "Landscape of IoT security," Computer Science Review, vol. 44, p. 100467, May 2022, doi: 10.1016/j.cosrev.2022.100467. Available: https://doi.org/10.1016/j.cosrev.2022.100467

28. Y. Zhang, D. Zheng, and R. H. Deng, "Security and privacy in smart health: efficient policy-hiding attribute-based access control," IEEE Internet of Things Journal, vol. 5, no. 3, pp. 2130–2145, Jun. 2018, doi: 10.1109/jiot.2018.2825289. Available: https://doi.org/10.1109/jiot.2018.2825289

29. FP Analytics, "Data governance - Part 1," Foreign Policy, Jun. 03, 2020. Available: https://foreignpolicy.com/2020/05/13/data-governance-privacy-internet-regulation-localization-global-technology-power-map/

30. A. Levite and G. Kalwani, "Cloud Governance Challenges: A Survey of Policy and Regulatory Issues," Carnegie Endowment for International Peace, Nov. 2020. Available: https://carnegieendowment.org/2020/11/09/cloud-governance-challenges-survey-of-policy-and-regulatory-issues-pub-83124

31. D. Allison-Hope and R. Schuchard, "Taking ethics to the cloud," Sustainable Business Network and Consultancy (BSR), Apr. 10, 2012. Available: https://www.bsr.org/en/blog/taking-ethics-to-the-cloud

32. K. Speiser, "Cloud IoT device provider, Ubiquiti, has data breached," Sonrai | Enterprise Cloud Security Platform, Apr. 26, 2022. Available: https://sonraisecurity.com/blog/major-provider-cloud-iot-devices-breached/

Index